D1345220

RTC Limerick
3 9002 00022461 9

VHDL for Designers

VHDL for Designers

Stefan Sjoholm
ABB Industrial Systems, Sweden

and

Lennart Lindh
Malardalens University, Sweden

An imprint of **Pearson Education**

Harlow, England • London • New York • Reading, Massachusetts • San Francisco • Toronto • Don Mills, Ontario • Sydney
Tokyo • Singapore • Hong Kong • Seoul • Taipei • Cape Town • Madrid • Mexico City • Amsterdam • Munich • Paris • Milan

Pearson Education Limited
Edinburgh Gate
Harlow
Essex CM20 2JE
England

and Associated Companies throughout the world

Visit us on the World Wide Web at:
http://www.pearsoneduc.com

First published 1997 by
Prentice Hall Europe

Printed and bound in Great Britain by
Ashford Colour Press Ltd, Gosport, Hants

Library of Congress Cataloging-in-Publication Data

Available from the publisher.

British Library Cataloging-in-Publication Data

A catalogue record for this book is available from the British Library

ISBN 0-13-473414-9

10 9 8 7 6 5 4 3
05 04 03 02 01 00

To our children
Louise, Annelie and Tobias Sjöholm
Therese and Jens Lindh

Contents

Preface

This is our third book on VHDL (the first in English). It is longer than usual in order to cover new parts of VHDL (including the new VHDL-93 standard) and goes into greater detail. We are delighted that the book is now being used in universities and colleges, and that it is being read by technical strategists and development engineers in industry.

The aims of the book are to teach VHDL and to show how VHDL is used in practice for designing electronic systems with present-day development tools. This book also covers how to build environmental models (testbenches).

The book contains plenty of examples, enabling it to be used both as a textbook at various levels and, at a later stage, as a reference book. We have made a conscious effort to be concise and to the point.

VHDL has been used for design for a number of years now, and the book reflects the possibilities currently on offer. Tools, development methodology, synthesis tools and, not least, testing methodology are developing quickly. Behavioural synthesis, for example, is an area for which great importance is predicted. The book therefore provides an introduction to behavioural synthesis and the possibilities that it offers. We also aim to upgrade the book in a few years' time when new technical progress has been made.

VHDL is a fairly new language. In some ways, VHDL syntax is like other well-known languages such as C and Pascal, but its behaviour is completely different. C and Pascal are adapted to a CPU, i.e. a serial machine which performs one instruction at a time, while VHDL is adapted to general structures in hardware. The structure of hardware is largely parallel. The result means that, practically speaking, performance is always several degrees better with VHDL that with "conventional" programming languages for CPUs. This means that VHDL will be used more and more as an implementation language in place of software languages. We believe that, in the future, the majority of small microcontrollers will be designed in VHDL instead of with a CPU and machine code.

For some years now there have been simulators on the market which can "execute" VHDL code as a means of verifying behaviour. VHDL code can be transformed automatically into an FPGA (Field Programmable Gate Array) or gate matrix. This transformation from code to a technology is called synthesis. One prerequisite is that the synthesis tool should generate hardware with the same behaviour (functionally) as the VHDL code which was originally simulated. It is difficult to write general VHDL code for synthesis, so different synthesis tools can transform different parts of the language. The book therefore contains examples of three different synthesis programs: Synopsys, Autologic (Mentor Graphics) and ViewLogic. ViewLogic is a PC- or workstation-based system, while Synopsys and Autologic are both workstation-based systems.

For many hardware designers, describing behaviour in VHDL instead of with gates requires adjustment to begin with. This adjustment is a must, however, if high design productivity and quality are to be achieved. By describing the design in a technology-independent high-level language, VHDL, the designer can concentrate on making the design functionally correct. VHDL provides much greater "design efficiency" and robustness than gate-level design. The use of VHDL in conjunction with a synthesis tool also means that the technology can be changed quickly and easily. This facility can, for example, be exploited for what is known as rapid prototyping of ASICs (Application Specific Integrated Circuits) using FPGAs.

Methodical, robust, structured, testable and technology-independent design demands a good knowledge of VHDL, which is what this book offers. What is more, if the advantages of describing a design in a high-level language like VHDL are to be enjoyed, a good design methodology and well thought-out test methodology are required. That is why the book also contains chapters on design and test methodology.

All VHDL reserved words are printed in **bold** in the VHDL code.

Some of the automatic generated synthesis figures show the principal of the synthesis result and our intention is not to make the labels readable.

If you have any comments regarding the contents of the book, we would be pleased to hear from you:

Stefan Sjöholm, ABB Industrial Systems, 721 67 Västerås, Sweden
(E-mail: stefan.j.sjoholm@seisy.mail.abb.com) or

Lennart Lindh, Mälardalens högskola, IDt, Box 883, 721 23 Västerås, Sweden
(E-mail: lennart.lindh@mdh.se).

There is a WWW site for material concerning the book and addresses of other sites of interest. The address is:

http://www.mdh.se/avdelningar/idt/forskning/cus/books/VHDL/

We would like to thank Synopsys, ViewLogic, Mentor Graphics, Motorola, Texas Instrument, Xilinx and others for their support.

VHDL is an acronym of VHSIC Hardware Description Language and VHSIC is an acronym of Very High Speed Integrated Circuit.

Stefan Sjöholm and Lennart Lindh

1

Introduction and overview

VHDL originated in the early 1980s and is now accepted as one of the most important standard languages for specifying, verifying and designing electronics. Today, a number of high-tech companies work exclusively with VHDL for digital systems. Universities run VHDL courses for their students and there are several VHDL groups to support new users. In the next few years, small and medium-sized companies will probably start to use VHDL. VHDL and ASIC (Application Specific Integrated Circuits) have also started to compete with single-chip controllers. VHDL frees the designer from having to use von Neumann structures and allows him/her to work with real concurrency instead of sequential machines. This opens up completely new possibilities for the designer.

Two important reasons for using VHDL instead of traditional schematic design are shorter development times for electronic design and simpler maintenance.

VHDL started out as a specification and modelling language, with the first simulators being developed in the late 1980s. A few tools complied with the full VHDL standard at the end of the decade, and design with VHDL came a few years later. All the major tool manufacturers now support the VHDL standard. VHDL is now a standardized language, with the advantage that it is easy to move VHDL code between different commercial platforms (tools). This means that VHDL code for a simulator of a certain tool manufacturer can be moved to those of other manufacturers without having to be changed. Unfortunately, VHDL is not standardized for design (synthesis), although it is standardized for designing digital models. In recent years it has become possible to synthesize more and more parts of the language constructions.

Fortunately, good VHDL tools, and VHDL simulators in particular, have also been developed for PCs in recent years. This means that prices have fallen dramatically, enabling smaller companies to use VHDL too. There are also PC synthesis tools, primarily for FPGAs and PLDs, but their functionality is slightly more limited than that of workstation tools.

A new possibility, which has changed the world of electronics designers, is the way in which several thousand gates and flip-flops can be programmed as an IC circuit in just a few minutes on a single PC without the need for expensive equipment. This means that it is possible to generate a file automatically from VHDL for programming circuits; this method is called rapid prototyping. At present there are circuits with up to 100,000 gates in a single circuit. Rapid prototyping can also be used for small production series.

Writing VHDL code instead of using schematic components (e.g. gates) is a new way of designing. Working with VHDL does not simply mean writing code, there are also facilities for building hierarchies and designing with a component library. VHDL is also good for writing code for standard circuits (e.g. Motorola 680020). There are now companies which sell standard components for simulation. This means that models of entire circuit boards with standard components can be verified by using a computer simulation. If an entire circuit board is simulated, access times and invalid addresses, for example, can be detected in a very powerful manner. This is usually done by each component having a section of code which checks whether the interface signals arrive on time and whether valid addresses are used, etc.

Designing with VHDL means that the designer writes code and then verifies the function in a simulator, following which the code is synthesized into a netlist. Synthesis can be compared to a compiler which translates the code into machine code. In hardware, the VHDL code is translated into a schematic with gates and flip-flops.

It was the American Defense Department which initiated the development of VHDL in the early 1980s because the US military needed a standardized method of describing electronic systems. VHDL was standardized in 1987 by the IEEE (Institute of Electrical and Electronics Engineers). The reference manual is called *IEEE VHDL Language Reference Manual Draft Standard version 1076/B* and was ratified in December 1987 as IEEE 1076-1987. Note that VHDL is standardized for system specification but not for design.

VHDL has major similarities with the ADA software language. This is partly because the company (Intermetrics) which the Pentagon commissioned to specify the new language had had a great deal of experience with ADA. The objective is that the VHDL standard, just like all IEEE standards, should be revised every five years. The latest standard, VHDL-93, was delayed slightly and dates from 1993; it does not contain any major changes from the old VHDL-87 standard, although some new VHDL commands and attributes have been included, primarily for

VHDL modelling. In this book the new additions to the standard are marked with "VHDL-93".

The great success of VHDL is due in part to the fact that it is the only hardware language to have been standardized. ADA, for example, has far more trouble when competing with the other programming languages of the C and C++ type.

1.1. Why use VHDL?

Designing in VHDL offers advantages over traditional schematic design techniques. This section examines the advantages and disadvantages of the language.

VHDL supports the development environment for digital development. VHDL also supports various development methods, top-down, bottom-up or any mix. Many of today's electronic products have a life of more than ten years, and consequently have to be redesigned several times in order to exploit new technology. The simplest way to do this is to use technology-independent VHDL design, which means that it is possible to change technology using automatic tools. When an electronic product has a life of ten years, it is common for the electronics to be modified and new functions to be added. VHDL supports modifiability, as the language is easy to read, hierarchical and structured.

The language supports hierarchies (block diagram), reusable components, error management and verification. Hierarchies are described by using structural VHDL, procedures and functions. Structural VHDL can be compared with a block diagram. Many systems support graphical input which can translate to structural VHDL automatically. The language also supports concurrent and sequential language constructions. The language also supports everything from specification to gate description.

The design of VHDL components can be technology-independent or more-or-less technology-independent for a technical family. The components can be stored in a library for reuse in several different designs. VHDL models of commercial IC standard components can now be bought, which is a great advantage when it comes to verifying entire circuit boards. A component can also be made in such a way that it can be changed automatically. A FIFO component (First In First Out), for example, can be designed in such a way that only one component is needed to cover every eventuality (sizes, rows). These components are called generic components.

The code for a VHDL component can be verified functionally in a simulator. The simulator simulates ("executes") the VHDL code with input signals and produces a signal diagram and error messages on the basis of the components. The input signals are defined either in VHDL or in the simulator's language. When the

VHDL code is simulated, functional verification takes place. At a later stage, time verification of the design is also possible.

In the case of traditional schematic design, the designer has to keep a manual check on technology-specific factors such as timing, area, driving strength, component choice and fan-out. One of the great advantages of designing in VHDL is that the designer can concentrate on function, i.e. implement the requirement specification, and does not need to devote time and energy to technology-specific factors which do not affect function.

VHDL is standardized, and this makes it possible to move code between different development systems for modelling (simulation). For design it is harder to move the code because there is no standard. This book will therefore illustrate how synthesizable VHDL codes should be written for ViewLogic, Mentor Graphics and Synopsys synthesis tools. These three tools represent a large market share of the synthesis tools currently used in industry, universities and colleges. ViewLogic is a PC-based system (also available for workstations) and does not have the same functionality as the other two when it comes to synthesis. Synopsys and Mentor Graphics Autologic 2 represent workstation-based systems which are more expensive but also more powerful than ViewLogic. Synopsys has been the leading company in VHDL synthesis throughout the 1990s but is now facing serious competition for the first time from Autologic 2 as far as synthesis is concerned. Fortunately, Synopsys and Autologic 2 support much the same subset of VHDL (99 percent), so that VHDL code can be moved between these two leading synthesis tools without having to be modified.

As this book shows how synthesizable VHDL code should be written for everything from simple PC synthesis tools to the most advanced synthesis tools for workstations, the reader will have no difficulty using the book's synthesis templates for system tools other than the three mentioned above.

VHDL has not yet been standardized for analog electronics. However, standardization work is in progress on VHDL with an analog extension (AHDL) to allow analog systems to be described as well. This new standard will be based wholly on the VHDL standard and will have a number of additions for describing analog functions.

1.2. Development flow

This section describes the overall development methodology for a development process. It can be used to advantage for small-scale student laboratory experiments and projects.

The development phase for the product is one of the first in the product's lifecycle. During this phase the product is specified, designed and verified. One common development model is called the waterfall model. It starts with specification and leads in defined stages to a functioning prototype. The waterfall model is ideal for well-written specifications such as laboratory experiments.

The development flow from specification to a prototype can be divided up as follows (see Figure 1.1):

Development phase	Result	Documentation
Analysis	Specification	What has to be done?
Design	VHDL code	How?
Technology mapping	Netlist	
Prototyping	Prototype	What is the result?

Figure 1.1: Overview of the design flow.

- The analysis phase consists of writing a specification. The specification can be written in VHDL or ordinary language. The purpose of the specification is to find out WHAT HAS TO BE DONE. The specification can be described in VHDL and then verified in a VHDL simulator.

- The design phase means transforming the specification into an architecture and VHDL code. It is not yet possible to do this automatically. The phase starts with defining an architecture (block diagram). When the architecture is ready, the VHDL code is written for the various components (blocks) or ready-made components copied from a library. Then the function of the design is verified in a simulator. When the result agrees with the specification, the designer can go on to the next phase. In this phase the major question is HOW SHOULD THE ARCHITECTURE AND COMPONENTS BE DESIGNED?

- The next stage is technology mapping. It is parameters such as price, performance and supply, etc., that determine which technology will be selected. This phase is now largely automated. The time constraints are described in a format which can be read by the synthesis tool. If the synthesis tool cannot meet the time constraints, the design phase must be repeated in full or in part. An approved synthesis produces a technology-dependent netlist (schematic), which is an input file for other tools. It takes just five minutes to program an FPGA circuit. If, for example, a gate array is involved, it will take several weeks before a finished circuit can be supplied. If a gate array is used, test vectors also have to be included in the design. FPGAs are already tested at the factory. The work during this phase is largely the production of a manufacturing basis.

- A prototype is then built and compared with the specification. If the RESULT is the same as the SPECIFICATION, the circuit is ready. This comparison is called validation.

The documentation for a student project can be structured as follows:

1. Summary

2. Introduction

2.1 Project documentation
Timetable
History

3. Specification
Function and timing constraints

4. Description of function
Architecture definition
Component I/O
Component description (VHDL)

5. Synthesis result
Number of gates

6. Prototype verification
Pin layout of the chip

7. Conclusions

References

Appendices:
Specification in VHDL with simulation results
Testbench
Description of the component in VHDL
Simulation results
Timing constraints to synthesis
Test protocol

1.3. History

The history of electronics design is a very interesting area. The major revolution witnessed to date has been driven by integrated circuits becoming larger and cheaper, but new tools have also started to be used and developed at an ever-increasing pace.

If we go back a decade, a design was verified by building a physical prototype (usually wire-wrapped on a circuit board). A signal generator was used to generate input signals and an oscilloscope to monitor the signals from the prototype (see Figure 1.2). The prototype was like a black box with input/output signals that could be monitored on the oscilloscope screen.

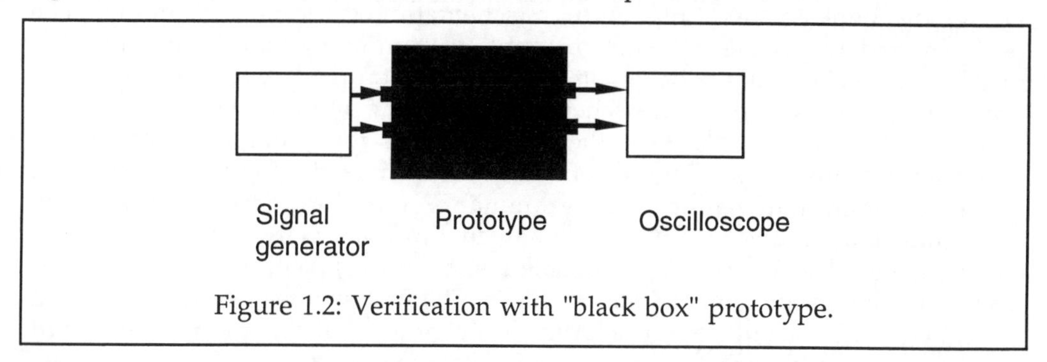

Figure 1.2: Verification with "black box" prototype.

One problem was that it was not possible to see ten non-periodic pulse diagrams on an oscilloscope. A change came about when logic analysers became available, making it possible to save pulse diagrams from a number of output signals.

Then development progressed to more complex prototypes (black boxes). This led to the problem of verifying internal nodes (interconnections) on the chip. The solution came with the next computer tools. These had an integrated prototype model, signal generator and logic analyser in the software. In the computer, the signal generator and oscilloscope are called the simulator (see Figure 1.3). The prototype was modelled with text or symbols. The computer could then simulate (execute) the model with delays, which are defined for each gate.

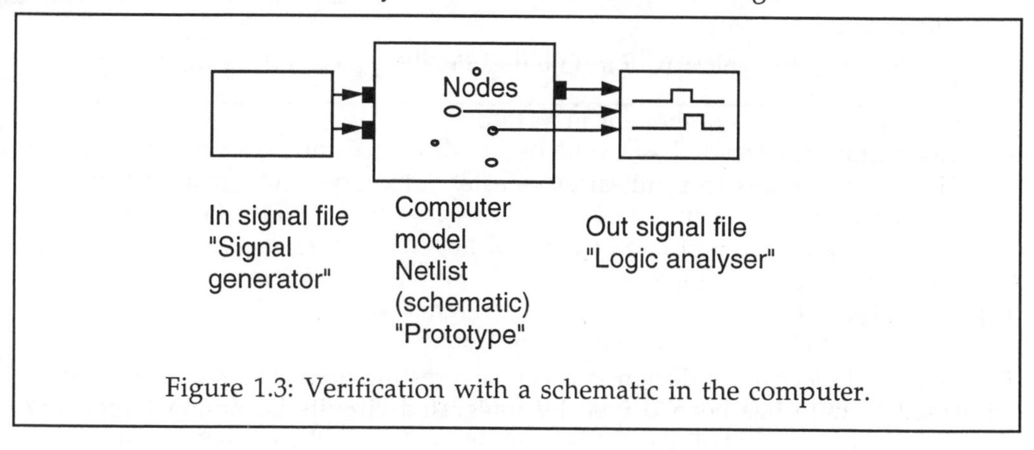

Figure 1.3: Verification with a schematic in the computer.

Now it was possible to save all the information from each node in the design for a whole simulation. The information is stored in a time base from the start time 0 to the finish time. In the beginning the signal information was just 1 or 0. Later X (undefined) and Z (high impedance) were added. It was also possible to simulate at different temperatures and production spreads to check whether the design worked throughout the temperature range and for different production consignments.

Designing Boolean functions with a schematic was time-consuming. The designer had to do the gate network and gate minimization manually. This problem was solved by the development of tools to translate a text file describing Boolean equations into a gate schematic (including minimization logarithms) automatically. These tools were the first synthesis tools and the process was called logic synthesis (see Figure 1.4). The first commercial logic synthesis tools came together with simple programmable circuits (gates only). Then programmable circuits which also contained flip-flops were developed. These circuits were called PLDs (Programmable Logic Devices). With these PLD circuits it was possible to implement state machines. The first synthesis programs, which translated state descriptions (including state minimization) to specific circuit families now had a market (see Chapter 9, "State machines").

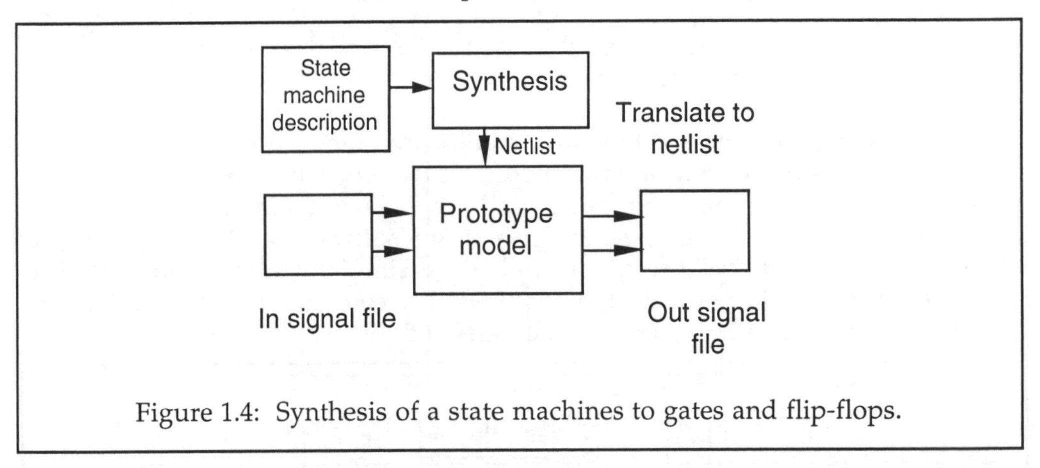

Figure 1.4: Synthesis of a state machines to gates and flip-flops.

The designers now faced new problems, with completely new circuit families appearing and disappearing at an ever-faster rate. One problem was that, if a circuit disappeared from the market, the designer was usually forced to do the entire code description again, as each tool had its own code syntax. Then new tools started to appear which were suitable for several different circuit families. Each tool used its own language. The designer was tool-dependent but had several different circuit families to choose from.

As designs became more complex, it became increasingly difficult to verify function simply by looking at pulse diagrams. Checking all the set-up times, protocols, invalid states, etc., for thousands of components is a time-consuming task which was frequently unsuccessful. Therefore, to obtain a flawless circuit, at least one redesign was usually necessary. Today the opposite is true, with the majority of designs working perfectly first time. Redesigning simply to change tools also caused many problems.

VHDL solved these problems because it is possible to synthesize certain basic elements of the language. It is still not possible, however, to move code between all the systems for synthesis, but the problem has been reduced. VHDL has no standard for synthesis. If the designer decides to use a subset of VHDL, the code may be movable for synthesis, but there are no guarantees. Verification with VHDL also provides an opportunity to build an error manager into each component (see Figure 1.5). RAM components, for example, can check access times and make sure that READ and WRITE are not active at the same time. If a signal breaks these requirements, an error message will appear on the simulator. The components have started to "talk" and can say what is not working during the verification phase.

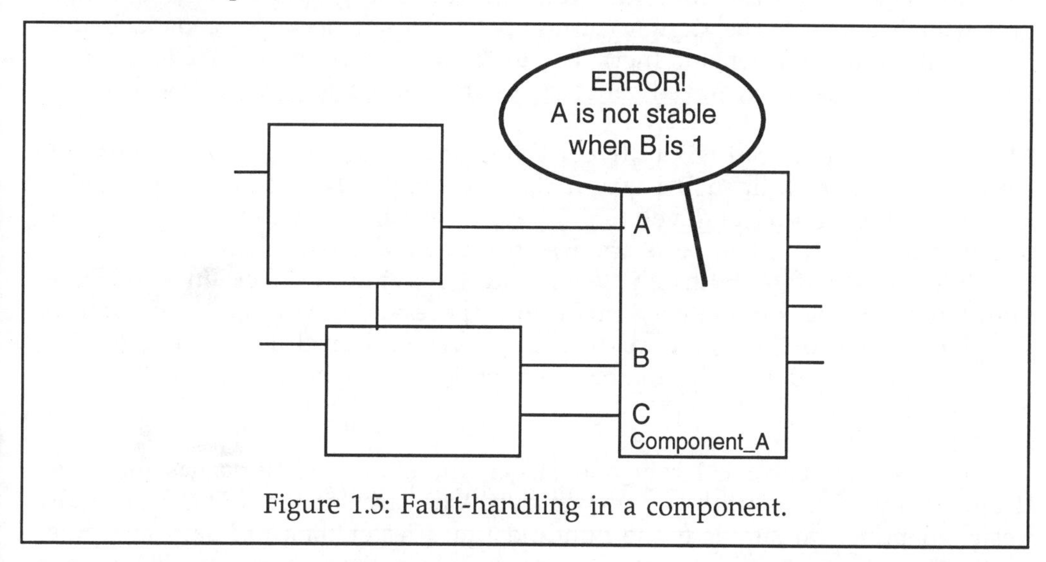

Figure 1.5: Fault-handling in a component.

When a standard language was created, it also became possible to start new companies for marketing components that could be simulated and that could be used by all the major tool suppliers of the day. Such components are usually of a high quality with an advanced error manager. These commercial VHDL components are only for simulation, not for synthesis.

As we mentioned, there were several languages like VHDL. The difference is that VHDL is standard, and the language has become a leading language both commercially and in the academic world. VHDL was created for writing specifications (modelling) and verifying (simulating) models, but it became apparent that certain parts of the VHDL language were also very suitable for design.

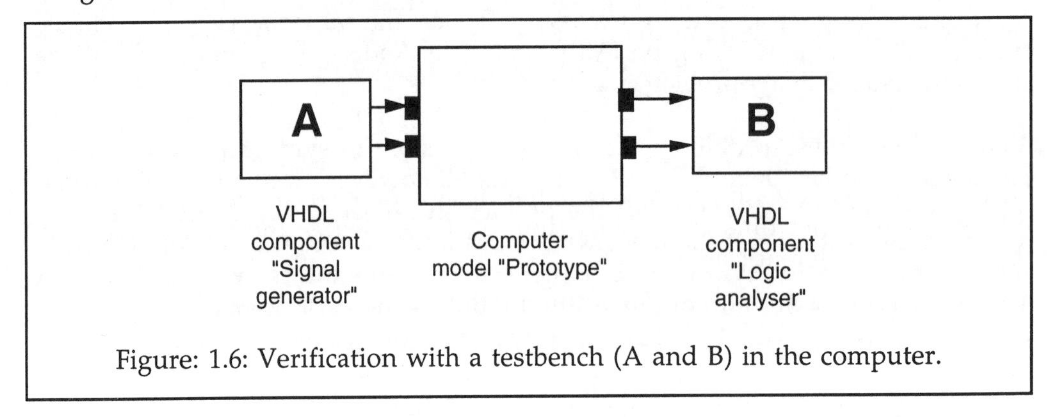

Figure: 1.6: Verification with a testbench (A and B) in the computer.

In order to verify subsystems there now exist methods that use what are known as testbenches. The structure of the testbench is shown graphically as the parts A and B in Figure 1.6 The design (prototype) does not need to be described in VHDL, it can be a schematic (net). The testbench is written in VHDL, with the advantage that it can be moved between systems (see Chapter 8, "Testbenches").

Mix level simulators can mix different levels of descriptions and languages. The simulators can work in such a way that, for example, the design is modelled at the highest behavioural level (e.g. the entire VHDL syntax) to verify the specification. Then certain parts are designed at RT (Register Transfer) level. It is possible to simulate both RT level and behavioural level on all VHDL simulators. Some simulators, including those of ViewLogic and Mentor Graphics, can also simulate a mixture of behavioural level, RT level, gate level with interconnect delays (on the basis of the layout), standard components and analog functions.

As designs have become larger and larger, the simulation time has increased dramatically. The solution to this is hardware accelerators. These specially designed units can simulate one hundred times faster than ordinary computer tools. Unfortunately, they are also one hundred time more expensive than ordinary computers.

Tools for formal verification have been marketed for verifying electronic systems. Formal verification can be used to verify whether a system is in what is known as deadlock. This expression comes from software and is now just as valid for hardware. Formal methods can also be used to verify whether two descriptions are identical (see Chapter 11, "Design methodology").

There are now tools for generating testable designs automatically. This means, for example, that the tool puts a shift register into the design which can be loaded and read off from pins on the package. The tool also produces a test file which can later be run on a tester at the IC factory (see Chapter 12, "Test methodology"). Unfortunately, there is no standard for the test file for the testers.

1.4. Synthesis

Using synthesis software means that the designer avoids having to translate, minimize and meet time constraints from VHDL code. Synthesis is defined for the following different classes:

- **Logic synthesis**, translates and minimizes Boolean functions only into gates.
- **RTL synthesis**, same as logic synthesis but also translates sequential language constructions into gates and flip-flops (state machines).
- **Behavioural synthesis**, can reuse one hardware component for more than one parallel sequential language construction.

We have already described logic synthesis, so we will start with RTL synthesis. A simple example is used below to explain the basics of how synthesis tools are working.

The following VHDL code is an example of what can be synthesized.

```
process(sel,a,b)
begin
  if sel= '1' then
    c <= b;
  else
    c<= a;
  end if;
end process;
```

The VHDL code is technology-independent. The VHDL code is easy to read: if sel is equal to 1, then signal *c* is assigned the value of signal *b*, otherwise *c* is assigned the value of *a*. The VHDL code is used for input data for the synthesis tool (see Figure 1.7).

Time constraints are specified in a file which is also input data for the synthesis tool. The technology library describes the technology into which the VHDL code is to be translated. The following example shows a simple command for translating a VHDL file to a schematic with gates and flip-flops (ViewLogic's synthesis tool):

```
VHDLDES VHDL_FIL -TEC=XC4000,
```

XC4000 is a pointer which gives the synthesis tool the information that it needs, e.g. how much driving strength the circuits have, speed, surface size in millimetres. XC4000 in this case is the Xilinx 4000 family. Other libraries which can be used include Actel, XC3000 (Xilinx 3000 family).

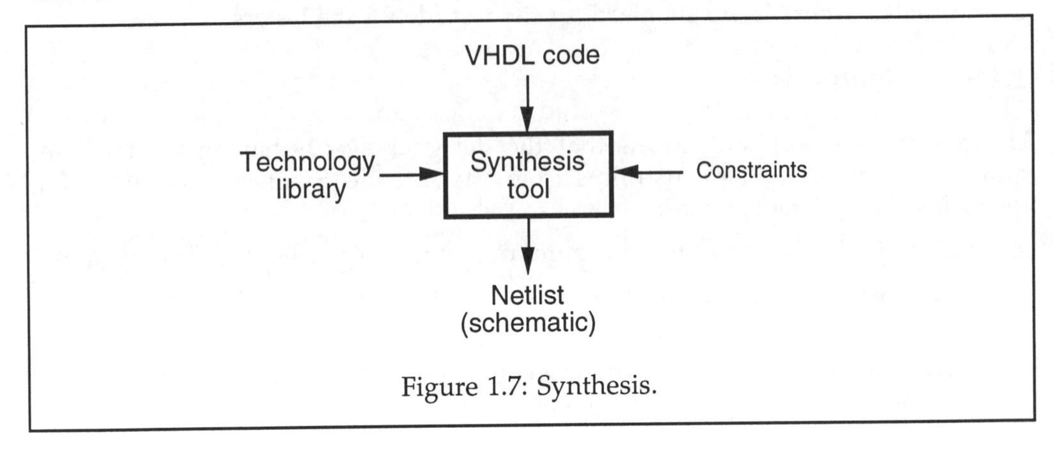

Figure 1.7: Synthesis.

The result of the VHDL code after synthesis will be a two-input multiplexor (see Figure 1.8). The VHDL code has been translated and minimized. Now the design is also technology-dependent. If another technology had been chosen, the result might have been a few gates instead.

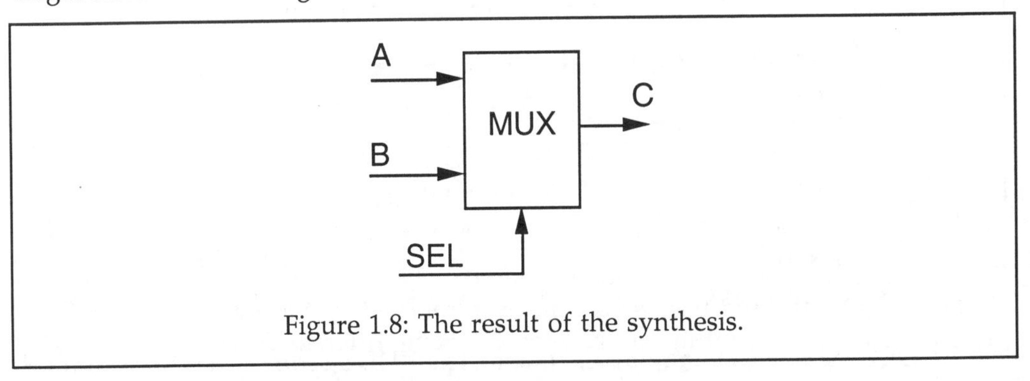

Figure 1.8: The result of the synthesis.

Another example which can be synthesized in a minute is as follows:

```
a:= b + c;
if a > 32 then
   count <= count + 1;
end if;
```

The above example shows some of the advantages of VHDL: it can be translated into an appropriate technology in less than a minute. This book will also describe other aspects of the language that will make VHDL even more interesting.

When it comes to large designs, there are redundant functions in several places, e.g. adders, multipliers, sorters and algorithms. In software it is usual not to write the code for a function more than once (procedure or function), and arithmetic functions in the machine code use one ALU unit.

In hardware it is slightly harder to share the same function among several users, but tools are now starting to appear which will resolve this dilemma. Users today are often an algorithm and the shared functions are add and multiply, but they could be a subsystem and the shared functions could be procedures and RAMs.

This process to share instead of having redundant hardware units (functions) is called behavioural synthesis. Behavioural synthesis works a great deal with timetables which determine when the different users are to utilize the shared component.

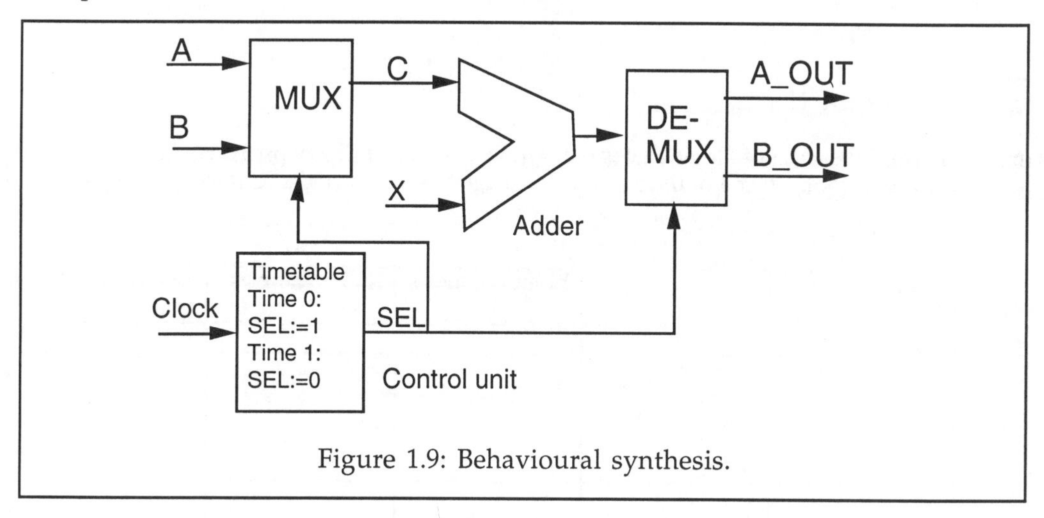

Figure 1.9: Behavioural synthesis.

Figure 1.9 shows how an adder is shared between A and B. The control unit ensures that only one user at a time is connected to the adder. Instead of an adder, a complex algorithm can be shared among different users.

A future development may be for large function blocks to be used by several users, just as present-day compilers for software make it possible. When synthesis tools become just as powerful as today's compilers, the designer will be able to benefit from the natural concurrency in electronics and will then have access to a very powerful tool (see Chapter 17, "Behavioural synthesis").

Perhaps it might be possible in the future to build a hardware operating system to support hardware units (program) - quite a challenge! One of the questions would then be: what is the difference between hardware and software?

1.5. Exercises

1. What are the advantages of VHDL compared with traditional schematic design?

2. Who was the driving force behind standardization of VHDL?

3. Why was a language like VHDL standardized?

4. Describe what simulating a design on computer means?

5. Synthesize the following VHDL code manually. Use only NAND gates.

(a)
```
a_out <= not (a_ir and b_in and c_in);
```

(b)
```
if a_in = '1' then
    a_out <= '0';
else
    a_out <= '1';
end if;
```

6. If you have access to logic (gates only), RTL (gates and flip-flops) and behavioural synthesis tools, which of the tools will manage to synthesize the following examples?

Mark the suggested tools with a cross.

Exercise		Logic synthesis	RTL synthesis	Behavioural synthesis
(a)	a:= b and c;			

(b)	if b='1' then
	c:='1';
	else
	c:='0';
	end if;			

(c)	wait until clk='1'; d<= '1';			
(d)	a<=(b*d) + (c*e);			
(e)	wait for 1 ms;			

Guidance to exercise 6:

(exercise c) *d* is assigned the value 1 on *clk*'s positive flank

(exercise d) * = multiplication

(exercise e) The language statement waits for a millisecond, then the program continues. "1 ms" is a time specification.

2

Introduction to VHDL

This chapter describes the following basic terms:

- The language's abstraction levels
- Hierarchies - structural description
- Components

VHDL is a language with many constructions. It can describe entire systems right down to individual gates, which sometimes makes it difficult to grasp as a whole. Learning to use VHDL for small designs is easy, but much greater knowledge is needed to use the language for complex designs.

In order to understand the language, terms like abstraction, concurrent and sequential statements, hierarchy, library, etc., are required. VHDL has inherited a great deal from the programming language ADA, but has also acquired elements such as concurrency and time which are specific to electronics.

2.1. VHDL language abstractions

VHDL is rich in language constructions, in addition to which the language can be used to describe different abstraction levels, from functions right down to a gate description. Abstraction levels are a means of concealing details. If the designer wants to multiply two numbers (A = B * C), there are several different description methods:

- Using the operator "*" in VHDL, i.e. a<= b * c;
- Designing the multiplier at gate level
- Designing the multiplier at layout level

The above three examples show how the same function can be implemented at three different abstraction levels: RTL, logic (gate level) and layout level.

Figure 2.1 defines the names of different abstraction levels for hardware description languages.

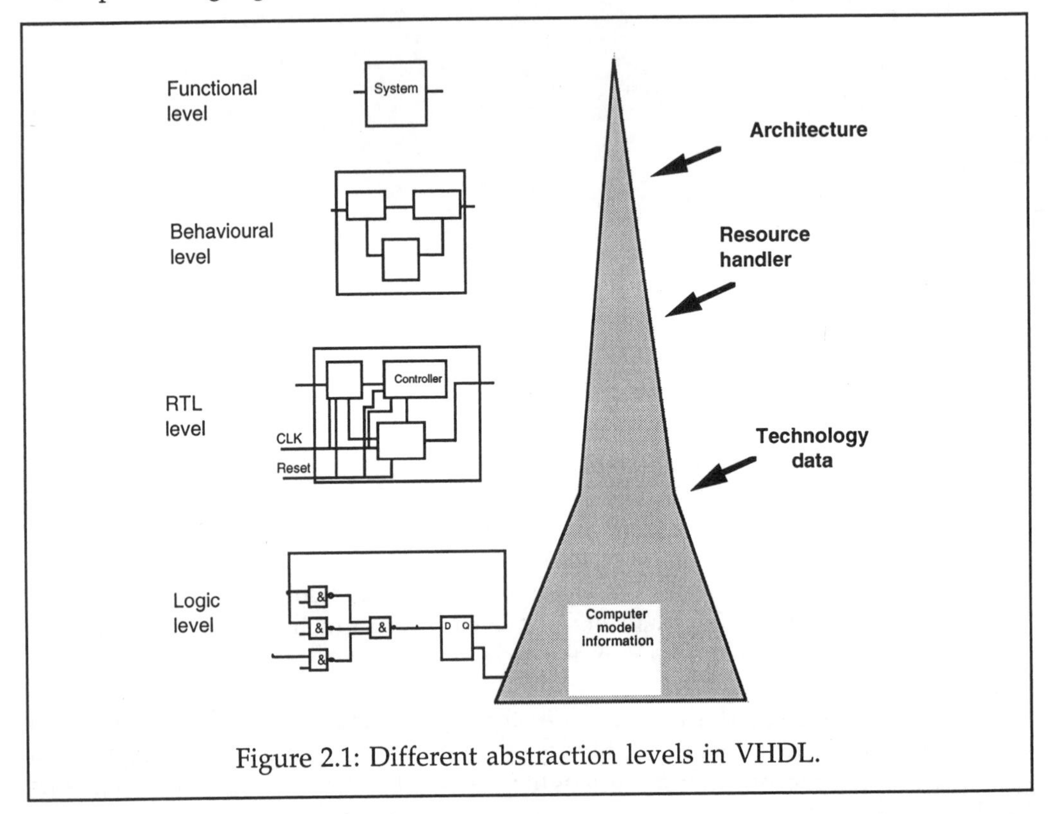

Figure 2.1: Different abstraction levels in VHDL.

Algorithms can be described at **functional level**, e.g. a controller algorithm can be described and simulated on the computer. An algorithm does not need to contain any time information. Specifications written in VHDL will be able to be simulated.

Behaviour and time are described at **behavioural level**. No architecture is required at behavioural level, e.g. implementation of registers, operators, interface with RAM, etc., are not defined. The advantage of models at this level is that models for simulation (verification) can be built quickly. A behavioural model can be described as functional modules and an interface between them; the modules contain one or more functions and time relations. In certain cases an architecture can be defined. Problems can arise if the functions have to share certain resources (e.g. multipliers), as the architecture will be changed at RT level. One solution is to pack functions which use the same resources, e.g. multipliers, together, as the architecture will not have to be changed. Several research projects are using behavioural synthesis tools, and there are a few such tools on the market.

RTL (Register Transfer Level) consists of a language which describes behaviour in asynchronous and synchronous state machines, data paths, operators (+,*,<,>..), registers, etc. In VHDL there are a number of different language constructions which can be synthesized at this level, e.g. generic components, instantiation, structural VHDL and overloading. At RT level, all registers are defined in the VHDL code.

Logic or **gate level** is a description with Boolean algebra or a gate network.

There are also lower levels such as transistor (also called electrical) and layout level. These levels are not supported by VHDL. At transistor level there are models of transistors, capacitances and resistances. At layout level, models are made of the physical process.

Synthesis is done between each level, e.g. between RT and logic levels the VHDL code is translated into gates and flip-flops, following the design which is minimized. These activities generate more information, e.g. technology information is required to synthesize from RT to gate level.

The volume of information increases between the various abstraction levels. Each transition (synthesis) generates more information. In other words, the complexity increases. Ideally, complexity would not be increased, but unfortunately that is impossible. In order to implement a function in an ASIC, technology information, wiring, gate information and set-up times, etc., are required. A 16-bit adder at RT level can be described with "+", but a page may be needed to describe the same adder at logic level.

Why are different abstraction levels used? If we draw a comparison with programming languages for CPUs, then we can see that they have different abstraction levels too, e.g. microcode, machine code, assembler, C and C++. If a program is being designed with a requirement for very short execution times, assembler is used. If, on the other hand, a complex program has to be designed, it is written in C or ADA. It is usually the requirements that determine the abstraction level at which the information is to be described. The same is true of VHDL. If a short development time is required, a high abstraction level should be chosen as the description language. In practice RT level (and parts of behavioural) can be synthesized automatically to gate level.

Describing an adder with "+" is much faster than describing the adder at gate level. If the design is at the idea stage, or if it is to be used as a model, behavioural or functional level should be chosen. If the design has tough performance requirements or has to be very small, RT level is often the best description method, provided that an advanced synthesis tool is used. It is difficult to achieve better results at gate level, though it is possible. Layout level can sometimes be effective, particularly for full custom ASICs, for achieving maximum power and minimum area. There is usually a trade-off between the various requirements which determine the level at which the design is to be described.

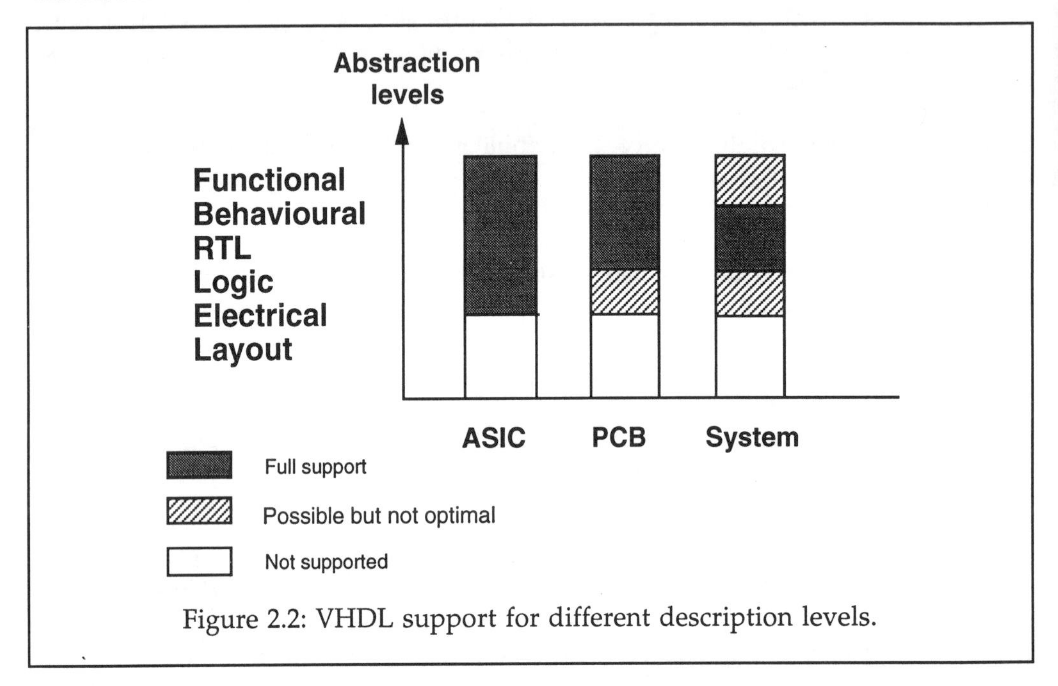

Figure 2.2: VHDL support for different description levels.

Referring to Figure 2.2:

- **ASIC** (Application Specific Integrated Circuit) usually includes gate array, standard cell and full custom designs. In Figure 2.2, **FPGAs** (Field Programmable Gate Arrays) are included in the ASIC column.

- **PCB** means Printed Circuit Board design. On a circuit board there are usually several ASICs together with a microprocessor and its infrastructure.

- **System** can mean different things depending on the person asked. In the figure it refers to the linking of a number of PCBs.

Figure 2.2 shows that VHDL can be used to build models from the highest abstraction level down to and including logic level. VHDL is ideal for building models and designing ASICs, i.e. from functional to gate level. When it comes to PCBs, the simulation times are so long at logic level that it is usually impossible to verify the design at gate level. Therefore, the components at PCB level should be described at RT level at least in order to achieve faster simulation times. As regards large systems, it is difficult to describe complex functionality with VHDL.

Abstraction levels in the VHDL language should not be confused with the abstraction hierarchies of the design. Both the abstractions support the designer working with complexity.

2.1.1. Simulation

Simulating models is an effective way of verifying the design. The model in the computer is only a time-discrete model, however, while reality is continuous.

The computer model is more or less like reality. It is least like reality at a high abstraction level (behavioural) and most like it at the lowest level (layout).

Abstraction levels	Description	Time granularity
Behavioural level	Behavioural level	microsecond
RT level	RTL language	clock cycles
Logic level	Boolean equations	nanosecond

Table 2.1: Abstraction levels.

A computer model has several refinements which are better than a physical prototype. Setting the worst case for process parameters and temperature range provides better verification than building a prototype. With a prototype, only the prototype is verified.

In a static timing analser or simulator it is possible to verify/simulate different timing cases:

- **Worst case:** Lowest voltage (e.g. 4.5 V), highest temperature (e.g. 125°C) and slowest process characteristic.
- **Typical case:** Normal voltage (5.0 V), normal temperature (e.g. 25°C) and normal process characteristic.
- **Best case:** Highest voltage (e.g. 5.5 V), lowest temperature (e.g. -55°C) and fastest process characteristic.

Simulation of behavioural models is for verifying functionality. Functionality and sometimes verification are possible on RTL models. Time constraints for the next level can be set for RTL synthesis, but only the physical model provides a good analysis of actual time behaviour. It is also possible to calculate/simulate power consumption, etc., at different levels.

The problem of simulations taking more and more time as the accuracy of models increases can mean that large designs (models) may be impossible to simulate (because simulation takes too long) at layout level. The simulation time also increases if various checks are described, e.g. set-up, hold and spike tests. If the verification has long input stimuli strings for verifying the design, there will also be long simulation times. The design should therefore be verified functionally with as high a description level as possible, i.e. normally RT level. The timing (including set-up and hold requirements) is analysed most effectively and quickly when using a timing analyser at gate level (see Chapter 11, "Design methodology").

There are now simulators (mix level simulators) which can handle models at different abstraction levels and in different hardware description languages. The advantage of such simulators is that the design can be modelled at a low level (gate delay, load delay and interconnect delay) and the nearby circuits at a high level (VHDL behavioural level, testbenches) to reduce the simulation time.

2.1.2. Other languages for describing electronics

There are several languages which are used to describe electronic designs. One popular language is called VERILOG (see References). VERILOG is used from RT level down.

Structuring the levels is not done without difficulties, because in some languages, for example, there are no hierarchies, which causes major problems when working on complex assignments.

Examples of languages developed by universities or research centres are:

SLIDE Structured Language for Interface Description and Evaluation (Parker and Wallace, 1981)
CONLAN CONsensus LANguage (Piloty *et al.*, 1983)
ISPS Instruction Set Processor Specification (Barbacci *et al.*, 1979
ADLIB A Design Language for Indicating Behaviour [Hill *et al.*, 1979]
HILL HIerarchical Layout Language (Lengauer and Melhorn, 1984)
OODE Object Oriented Description Environment for computer hardware (Takeuchi, 1981)
BORIS Block-ORiented Interacting Simulation system (Decker and Maierhofer, 1984)

2.2. Design hierarchies - reducing complexity

Complex designs need a mechanism to reduce their complexity for the designer. It is difficult to understand a design with hundreds of components (RAM, state machines, control logic, etc.); it is easier to understand the same design with only a few components, e.g. a FIFO and a state machine called FIFO_Controller (First In First Out).

Using **hierarchies** to handle complexities is an old method. It does not mean that the design becomes less complex (sometimes it becomes more complex instead), but it becomes easier to understand for the designer.

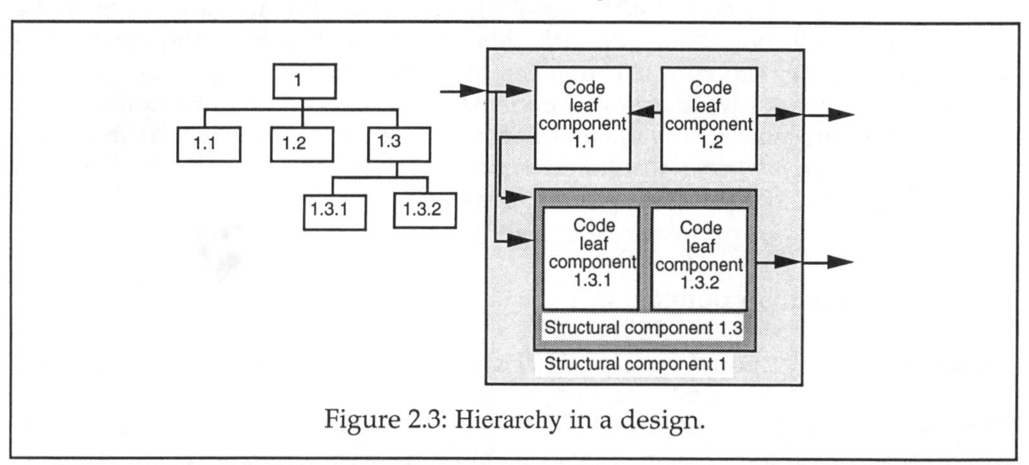

Figure 2.3: Hierarchy in a design.

There are several mechanisms by which to reduce complexity:

- Language abstractions (see section 2.1) use the language to describe complex matters without having to describe small details. Functions and procedures are important parts of the language in order to handle complexity.

- Design hierarchy (see Chapter 6, "Structural VHDL") uses components in order to conceal details - the black box principle. The term black box means that only the inputs/outputs of a component are visible at a certain level. It is the designer who decides how many different hierarchies there are to be in the design.

Both language abstractions and design hierarchy have the mechanism for reducing the degree of detail the higher up in the hierarchy (abstraction) you go.

Design hierarchies consist of components which contain other components, VHDL code alone, or a mixture of both. A hierarchy can be compared to an upside-down tree. Right out on the branches there are leaf components written in sequential and concurrent statements of VHDL code. Further in there are structural components consisting of other components written in structural VHDL code.

Figure 2.3 shows the design hierarchy of a whole design: which components the design contains and how they are structured. Information on the interface and what components there are at the next hierarchical level are contained in the structural description for the components.

Design work usually starts with defining the interface (entity) for the root (or top) component). Then the component is divided into new subcomponents with interconnections. The new structure can also be called architecture. There is no direct commercial tool which supports this process of dividing components into subcomponents (partitioning). The general rule is that it is done in such a way that the interface between the components is as small and uncomplicated as possible. The subcomponents must not be made too small, however, as this often results in a worse synthesis result and unnecessary VHDL code for describing the hierarchy (see Chapter 11, "Design methodology").

2.3. VHDL component

Components are a central concept in VHDL. Components are used, among other things, to build up component libraries, e.g. of microprocessors, special user circuits and other standard circuits. If a "good" component has been designed, it can be saved in a component library, enabling it to be copied as many times as required, i.e. components are reusable (see Figure 2.4). In the language of computer science, this copying is called creating instances of the component, i.e. creating the component in a schematic (or in the text file).

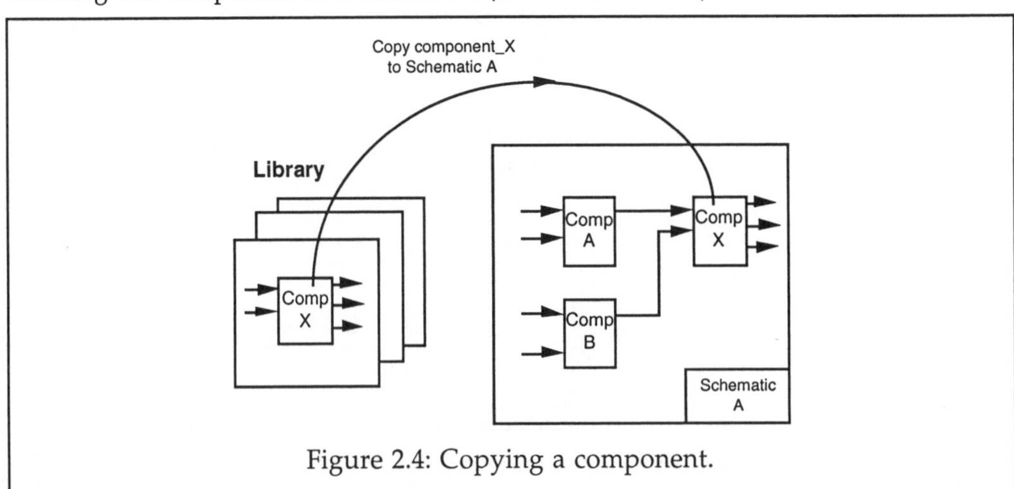

Figure 2.4: Copying a component.

Staying with computer science a while longer, VHDL is an object-based language, i.e. what separates VHDL from object-oriented languages is that the language does not have inheritance. Generic components and instantiation are typical for object-based languages. Generic components are components which can be modified before **instantiation**, e.g. a generic component which copes with different widths for the input and output signals.

The internal structure of a component can be concealed from the designer - the black box principle. In some cases there is absolutely no need to know how the component is structured. The designer is usually only interested in inputs and outputs, a specification of function and access times. The majority of hardware designers are used to working with black boxes such as the 74LSXX circuit family, for example.

A component in the library can consist of other components. It is possible for the library to contain large, complex components: type regulators, processors and communication circuits.

2.3.1. Entity and architecture

Components are a central concept in VHDL, and can be a complete design or a small part of a system This section examines the parts of a component.

A component is made up of two main parts (Figure 2.4):

- **Entity**: Port declaration for inputs and outputs
- **Architecture**: Structural or behavioural description

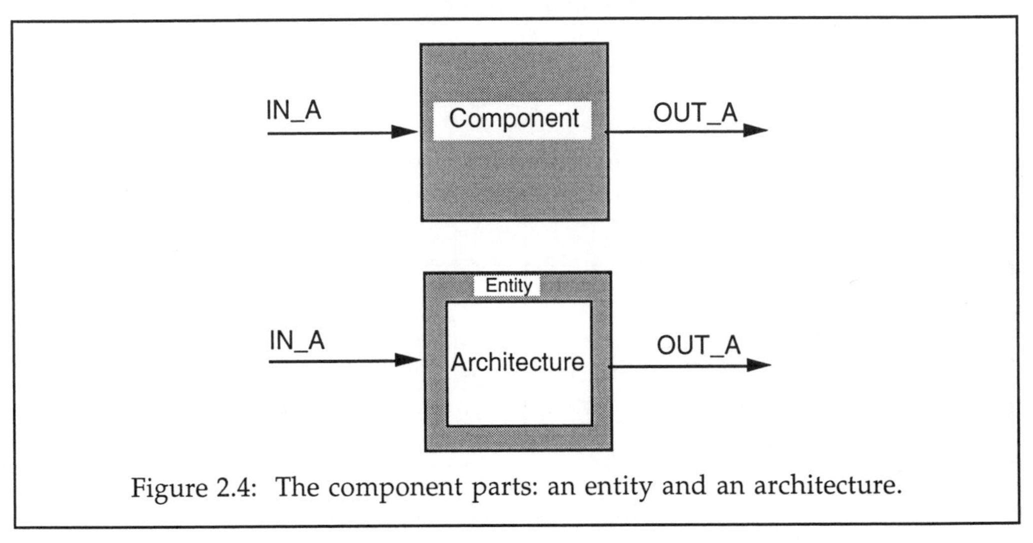

Figure 2.4: The component parts: an entity and an architecture.

Behaviour is defined as a collective name for functions, operations, behaviour and relations. Behaviour can also be a structural description, i.e. the component consists of other components.

The entity can be regarded as a black box with inputs and outputs. A microprocessor, for example, has an entity consisting of data, address, control and bus signals. The behaviour is to be found in the architecture, which may be structural VHDL code, i.e. it consists of other components, e.g. ALU and status registers.

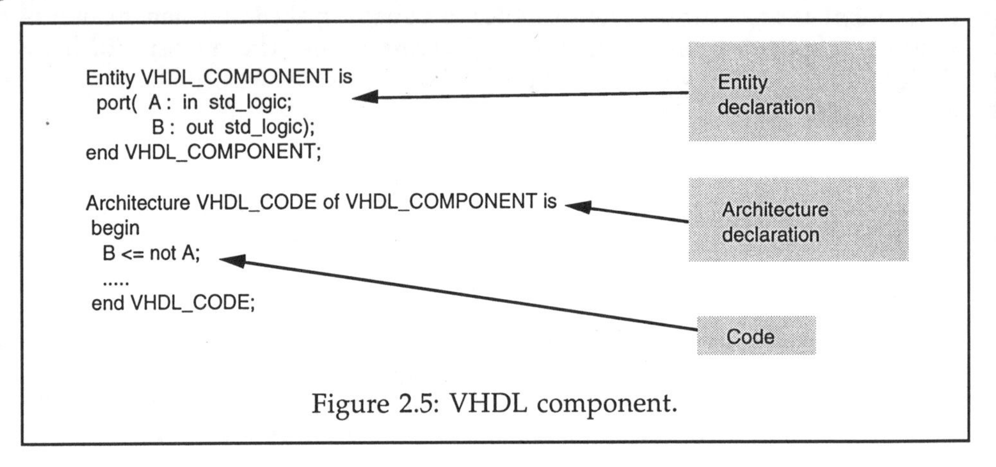

Figure 2.5: VHDL component.

Figure 2.5 shows a **component** with a simple Boolean function. The **entity** is called *vhdl_component* and the **architecture** *vhdl_code*.

Two names are specified in the architecture declaration: vhdl_component, which describes which entity the architecture belongs to, and vhdl_code, which is the name of the architecture.

Entity

An **entity** declaration defines the interface between an entity and the environment in which it is used. The **entity** name is the same as the **component** name.

```
Syntax:

entity <identifier_name> is
port (  [signal] <identifier> : [<mode>] <type_indication> ;
        [signal] <identifier> : [<mode>] <type_indication>) ;
end [ <identifier_name> ] ;
```

<mode>	=	**in, out, inout, buffer, linkage**
in	=	Component only read the signal
out	=	Component only write to the signal
inout	=	Component read or write to the signal (bidirectional signals)
buffer	=	Component write and read back the signal (no bidirectional signals, the signal is going out from the component)
linkage	=	Used only in the documentation

In the following example there is an input signal *a_in* and an output signal *b_out*.

```
Entity vhdl_component is
 port(signal a_in:  in  std_logic ;    -- input
      signal b_out: out std_logic);    -- output
end vhdl_component;
```

Mode **inout** should only be used in the case of bidirectional signals. If the signal has to be reread, either mode **buffer** or an internal dummy signal should be used (see Chapter 14, "Common design errors in VHDL and how to avoid them"). The word **signal** is normally left out of the port declaration, as it does not add any information. Mode **in** and the name of the **entity** after **end** can also be left out. The following two examples are therefore identical.

```
Entity ex1 is
 port( signal a,b:  in   std_logic;
       signal c:    out  std_logic);
end ex1;

Entity ex1 is
 port( a,b:     std_logic;
       c: out   std_logic);
end;
```

Note that VHDL does not differentiate between upper-case and lower-case letters.

An example of an 8080A (Intel processor) entity is as follows:

```
Entity uP8080A is
 port( clk1,clk2:  in   std_logic;  -- Clock inputs
       RESET:      in   std_logic;  -- Initializes processor
       HOLD:       in   std_logic;  -- Suspends processor
       INT:        in   std_logic;  -- Interrupts processor
```

```
            READY:   in     std_logic;
            D:       inout  byte;          -- Data bus
            A:       out    word;          -- Address bus
            INTE:    inout  std_logic      -- Interrupt enable
            DBIN:    out    std_logic;     -- Data bus in
            WR_N:    out    std_logic;     -- Data bus out
            SYNC:    out    std_logic;     -- Start of processor
                                              cycle
            HLDA:    out    std_logic;     -- Hold acknowledge
            WAITs:   out    std_logic);    -- Wait output
    end;
```

In the VHDL-93 standard, **end entity** can be written instead of just **end** as in the VHDL-87 standard:

Syntax (VHDL-93):

entity <identifier_name> **is**
port ([signal] <identifier> : [<mode>] <type_indication> ;
[signal] <identifier> : [<mode>] <type_indication>) ;
end [entity] [<identifier_name>] ;

Example (VHDL-93):

```
Entity ex is
  port( a,b:  in   std_logic;
        c:    out  std_logic);
end entity ex;
```

Architecture
An **architecture** defines a body for a component entity. An architecture body specifies a behaviour between inputs and outputs. The architecture name is not the same as the component name: instead an **architecture** is tied to an **entity**.

Syntax:

architecture <architecture_name> **of** <entity_identifier> **is**
[<architecture_declarative_part>]
begin
<architecture_statement_part>
end [<architecture_name>] ;

The architecture declaration part must be defined before the first **begin** and can consist of, for example, types, subprograms, components and signal declarations. The syntax for the architecture part for an inverter is as follows:

```
Architecture vhdl_architecture of vhdl_component is
begin
  b_out<= not a_in ;
end vhdl_architecture;
```

An entity can be linked to several architectures such as behavioural and RTL descriptions, for example.

The same addition as that introduced for entity was also introduced for architecture in VHDL-93.

```
Syntax (VHDL-93):

architecture <architecture_name> of <entity_identifier> is
[<architecture_declarative_part>]
begin
      <architecture_statement_part>
end [architecture] [ <architecture_name> ] ;
```

A complete example for a **NAND** component is as follows:

```
--****************************************
--*
--*  Filename       : NAND.VHD
--*  File type      : VHDL for design (RTL)
--*  Date           : 14/11-96
--*  Description    : NAND gate
--*  Author         : Mr X
--*  State          : Verified
--*  Error          : None (I hope)
--*  History        : A) started with the design 12/10

Entity nand_comp is
  port( a_in, b_in :  in  std_logic;   -- inputs
        a_out :       out std_logic ); -- output
end;

Architecture nand_behv of nand_comp is
signal int: std_logic;          -- Internal signal declaration
begin
```

```
  int<= a_in and b_in;
  a_out <= not int;
end;
```

A double dash "--" indicates that the rest of the line is commentary. The above example shows how a file header and commentary should be done in code. The above example merely illustrates the principle, with realistic VHDL components normally having tens or hundreds of lines in the architecture.

Component instantiation is done as follows:

```
U1: nand_comp port map(a_in => givar_in_a,
                       b_in => givar_in_b
                       a_out => out_lamp);
```

The component *nand_comp* is instanced and given the name *U1*. Each name of a new instance of *nand_comp* must be an unique name (see chapter 6, "Structural VHDL").

Instead of declaring the internal signal int, the operator NAND could be used directly in the VHDL code. The following logical operators are defined in VHDL:

```
NOT
AND
NAND
OR
NOR
XOR
XNOR          -- VHDL-93
```

Example:

```
Architecture rtl of ex is
signal int: std_logic;
begin
  int <= not (((a nand b) nor (c or d)) xor e);
  A_OUT<=int and f;
end;
```

2.4. Exercises

1. Describe the various language hierarchies in VHDL.
2. Which language level can be used to build models for simulation only? Give reasons for your answer.

3. Which language level can be used for rapid design and simulation? Give reasons for your answer.

4. Write an entity (*Component_A*) and an architecture (**rtl**) for the following function:

```
d_out <= (a_in and b_in) and c_in;
```

"Type_indication" is std_logic (explained in Chapter 3).

5. Describe the difference between design hierarchy and language abstractions.

6. Explain the words package, instance, generic and structural VHDL.

7. What does "**--**" mean in the code?

8. If a product has to have a very long life, then commentary, among other things, is very important. What do you think a file header with commentary for an entity should look like? Write an example.

2.5. References

Barbacci, M.R., W.B. Dietz und L.J. Szewerenko (1979) 'Specifications, evaluation and validation of computer architecture using instruction set processor descriptions', *Proceedings of the 4th International Symposium on Computer Hardware Description Languages*, IEEE.

Decker, H. and J. Maierhofer (1984) 'Very high level model description and simulation', *Proceedings of the Soc. Comput. Simulation Conference, San Diego.*

Hill, D.D. (1979) 'ADLIB: A modular, strongly-typed computer design language, *Proceedings of the 4th International Symposium on Computer Hardware Description Language,* IEEE.

Lengauer, T. and K. Melhorn (1984) 'The HILL System: A design environment for the hierarchical specification, compaction and simulation of integrated circuit layout', *Proceedings MIT Conference on Advanced Research in VLSI*, Artech House Company.

Parker, A. and J. Wallace (1981) 'SLIDE, an I/O Hardware Description Language', *IEEE Transactions on Computers,* vol. C-30.

Piloty, R., M. Barbacci, D. Borrione, D. Dietmeyer, F. Hill and P. Skelly (1983) *CONLAN Report*, Springer-Verlag.

Takeuchi, A. (1981) 'Object-oriented description environment for computer hardware', *Proceedings of the IFIP International Conference and their Applications, Kaiserslauten.*

VERILOG, *VERILOG HDL Compiler Reference Manual*, Gateway Design Automation Corp.

3

Concurrent VHDL

There are a number of concurrent language constructions in VHDL and it is important to understand how such constructions work. This chapter illustrates various concurrent language constructions and provides basic information on concurrent behaviour. It also describes how error management can be built into the component (not synthesizable). Error management provides many opportunities for creating a component library whose components which check input signals and send an error message to the simulator if an error occurs. Finally, various data types are examined.

3.1. Signal assignment

Variables are sequential objects and signals are concurrent objects. In hardware it usually feels natural to work concurrently using **signal** assignment statements.

> Syntax:
>
> Signal assignment:
> <target_identifier> '<=' <selected_expression>';'

Example:

```
Entity NAND_comp is
  port( a,b: in   std_logic;
          c: out  std_logic);
end;
```

```
Architecture rtl of NAND_comp is
begin
   c<=a nand b;
end;
```

The above example has no component delay. This is the normal method of description for VHDL code written for synthesis. Alternatively, the code could have been written with a component delay, e.g. 5 ns, using the **after** command in VHDL:

```
Architecture NAND_beh of NAND_comp is
begin
   c<=a nand b after 5 ns;      -- Component delay = 5 ns
end;                            -- inertial delay.
```

It is possible to define several values in one signal assignment. The values are enumerated one after the other followed by a comma. For example:

```
c<= '1',
      '0' after 10 ns,
      b after 20 ns;
```

The output signal above will have the following values:

```
Time   Value
0      '1'
10     '0'
20      b              -- Value of signal b
```

The signal assignment above will also be executed again as soon as signal *b* changes value. In this way a waveform can be obtained for signal *c*. This description method is not supported by the synthesis tools. The delay after the **after** command must be in ascending order. The following example is therefore incorrect:

```
c<= '1',
      '1' after 10 ns,
      '0' after 5 ns;          -- 5 ns less than 10 ns-> Error
```

A correct example would be:

```
c<= '1',
      '0' after 5 ns,
      '1' after 10 ns;
```

3.2. Transport and inertial delay

There are two types of propagation delay: **inertial** and **transport**.

- **Inertial** delay:
  - Default in VHDL
  - Spikes are not propagated (if **after** is used)
  - Often used for component delays

- **Transport** delay:
  - Pulses are propagated, irrespective of width
  - Good for interconnect delays

If transport delay mode is to be used, the command has to be written out in the VHDL code. Inertial, on the other hand, is the default and so does not have to be specified.

Example:

```
q1<=a after 5 ns;                -- Inertial delay
q2<=a transport after 5 ns;-- Transport delay
```

If a spike of, say, 2 ns occurs in a CMOS (Complementary Metal Oxide Semiconductor) component which has a delay of 10 ns, this spike is not normally visible at the output. That is why inertial delay is usually used for modelling component delays. Problems arise, however, if you want to model a component which has a 10 ns delay and all the spikes at the input which are equal to or greater than, say, 5 ns are visible at the output. This problem has been solved by VHDL-93 introducing a new command: **reject**. **Reject** can be used to define the length of spike at which the component is to let the spike through or not.

Example (VHDL-93):

```
q3<=reject 4 ns inertial a after 10 ns;
```

The above delay will ignore all spikes at the input which are less than 4 ns. Pulses which are 4 ns or larger will be visible at the output (q3) after 10 ns. Note that **inertial** has to be specified if the **reject** command is used.

The synthesis tools do not support any of these delay models. They are best used when making VHDL models for simulation. However, inertial delay can be used in design. All that happens is that the synthesis tool ignores the delay. Transport delay will cause errors in the majority of synthesis tools. The timing constraints must be set up in the synthesis tool when designing. (See Chapter 11, "Design methodology"). An example is given in Figure 3.1.

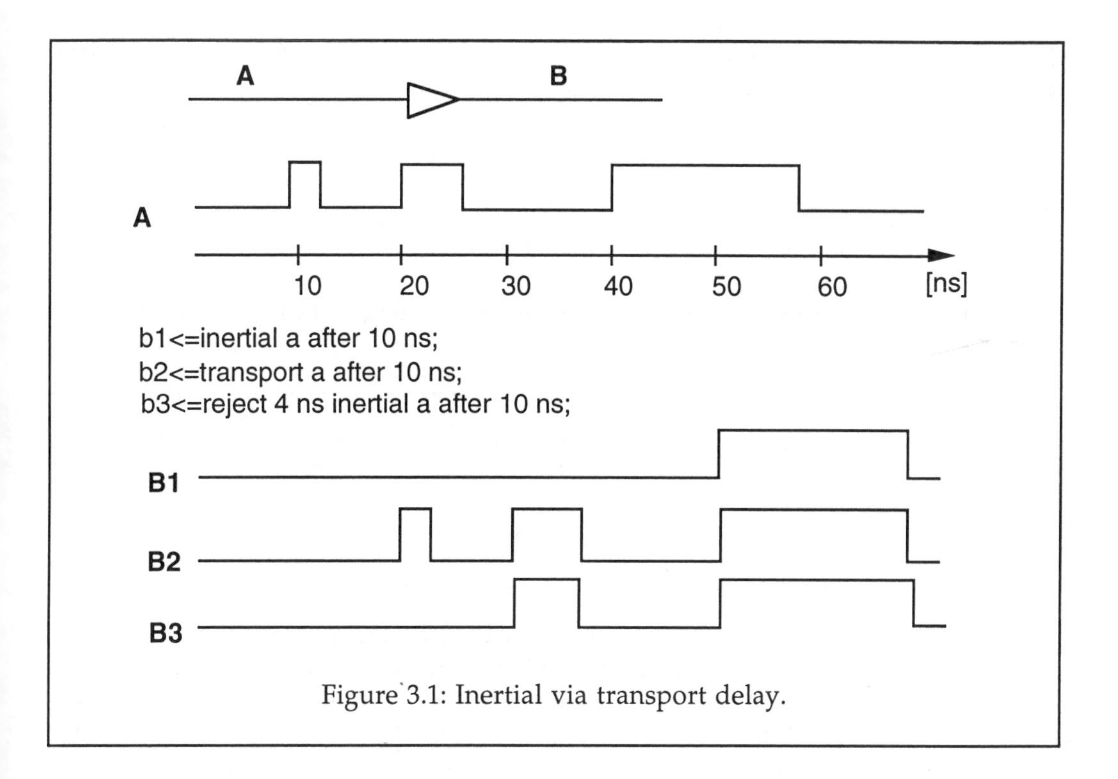
A
B
A
10
20
30
40
50
60
[ns]
b1<=inertial a after 10 ns;
b2<=transport a after 10 ns;
b3<=reject 4 ns inertial a after 10 ns;
B1
B2
B3

Figure 3.1: Inertial via transport delay.

3.3. Concurrency

Hardware is parallel by nature, and so is VHDL. This means that all concurrent language constructions in VHDL code can be executed concurrently, making the order in which the code is written irrelevant.

Example:

```
Architecture example of ex is
begin
  a<=b;
  b<=c;
end example;
```

is equivalent to the following:

```
Architecture example2 of ex is
begin
  b<=c;
  a<=b;
end example2;
```

The two examples above are identical from a functional point of view because the concurrent VHDL commands are event-controlled:

```
b<=c;            -- Is executed when c  changes  value
```

This means that when input signal *c* changes value, all lines in which *c* is to the right of the assignment symbol are executed, i.e. b<=c.

If synthesis is performed on the above example, the result shown in Figure 3.2 is obtained.

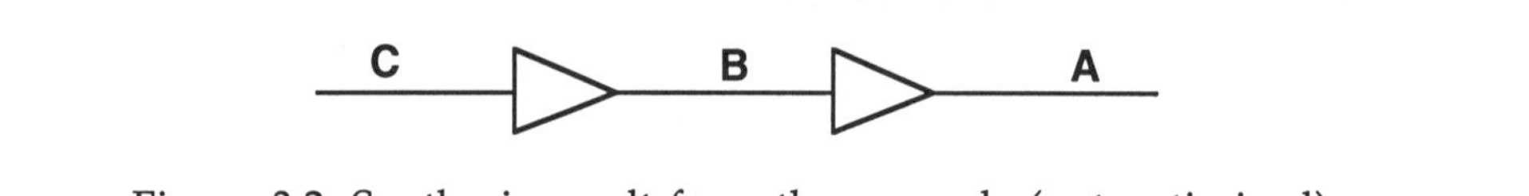

Figure 3.2: Synthesis result from the example (not optimized).

As the hardware shows, signal *a* does not change before signal *b* changes value. Similarly, signal *b* does not change value before signal *c* changes value. It is the same in the VHDL code, as it is event-controlled. This means that all the concurrent VHDL commands can be written in any order without the function of the design being changed.

All concurrent statements can be identified with a label:

```
Label_name:  b<=a;
```

This label is only used for documentation and has no functional significance. Unlike in some programming languages (e.g. Basic), it is not possible to jump to a label.

3.4. Delta time

Delta time is used in VHDL for "queuing up" sequential events. The time between two sequential events is called a delta delay. A delta delay has no equivalent in real time but is executed while the simulation clock is standing still. In a signal assignment, the value is not assigned to the signal directly, but after a delta delay at the earliest:

```
b<=a;    -- Signal b assign value of signal a after one
            delta  delay.
```

An example is provided in Figure 3.3.

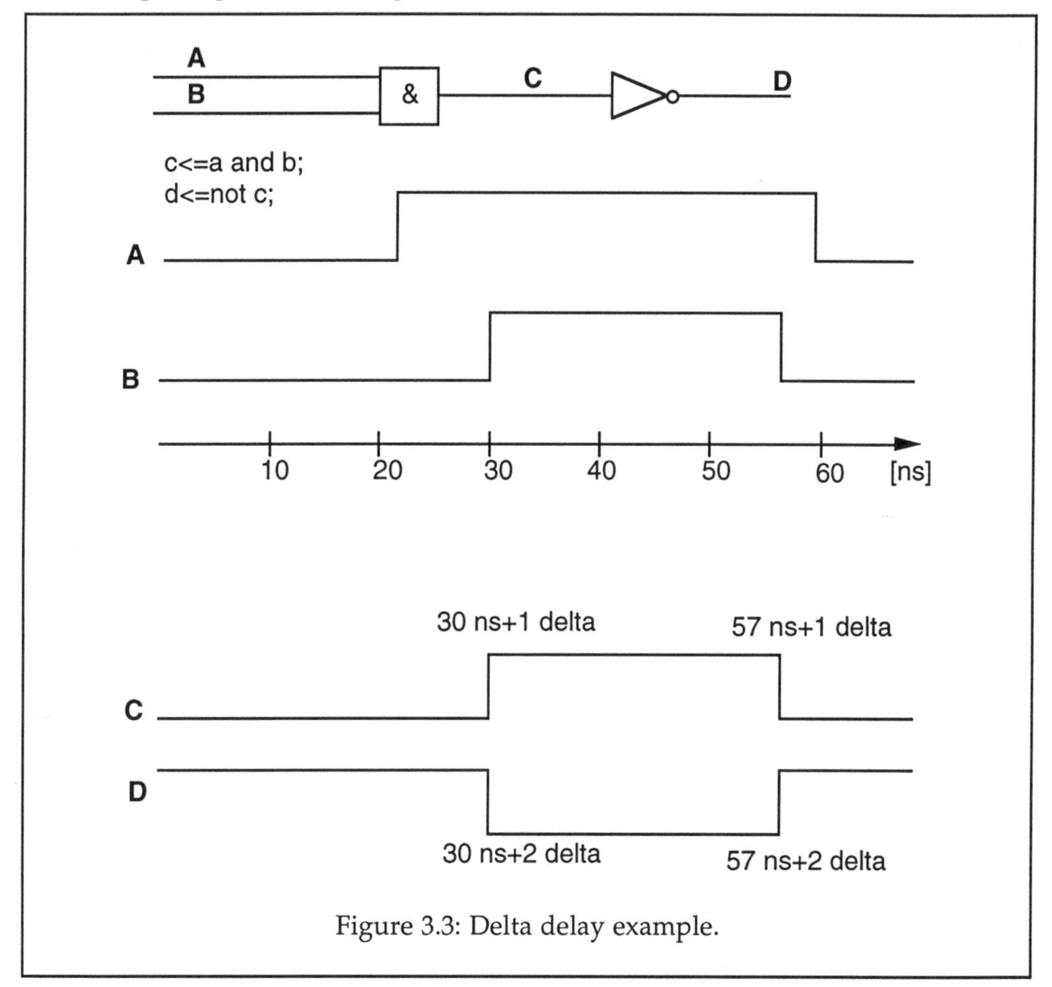

Figure 3.3: Delta delay example.

In a combinational logic block, where all elements have a 0 ns delay, all assignments will take place at 0 ns, but it may contain many delta delays. The VHDL simulator will count up the number of delta times until all signals are stable. If the VHDL code is written incorrectly, there is a risk that the design will oscillate *ad infinitum*. To prevent VHDL simulators jamming, most simulators stop after 1000 delta times, for example. The number can usually be set. The following example is correct VHDL, but generates a design which oscillates *ad infinitum*:

```
q<= not q;
```

Signal *q* will be updated to the inverse of itself after a delta time. This causes the line to be executed again, and *q* will be updated after another delta time, etc. This problem is easy to solve by inserting a delay:

```
q<= not q after 10 ns;
```

Note that, if the above code is synthesized, it will lead to asynchronous feedback, which is not usually desirable in design (see Chapter 12, "Test methodology").

3.5. When statement

Syntax:

<target> <= <expression> [**after** <expression>] **when** <condition>
<expression> [**after** <expression>] .... ;

It is permissible to have several **when else** lines. The <target> signal must always be assigned a value irrespective of the value of <expression>. This means that the command has to be ended with an **else** <expression>.

Example:

```
Entity ex is
  port( a,b,c:  in    std_logic;
        data:   in    std_logic_vector(1 downto 0);
        q:      out   std_logic);
end;
Architecture rtl of ex is
begin
 q<=  a when data="00" else
      b when data="11" else
      c;
end;
```

In VHDL-93, however, the final **else** <expression> can be left out (this is not supported by the majority of synthesis tools, though).

Example (VHDL-93):

```
Architecture rtl of ex is
begin
  q<=  a when data="00" else
       b when data="11"; -- No else <expression>, only
end;                     -- valid in VHDL-93
```

The **when** command is very useful, e.g. when designing a three-state buffer. An example is given in Figure 3.4.

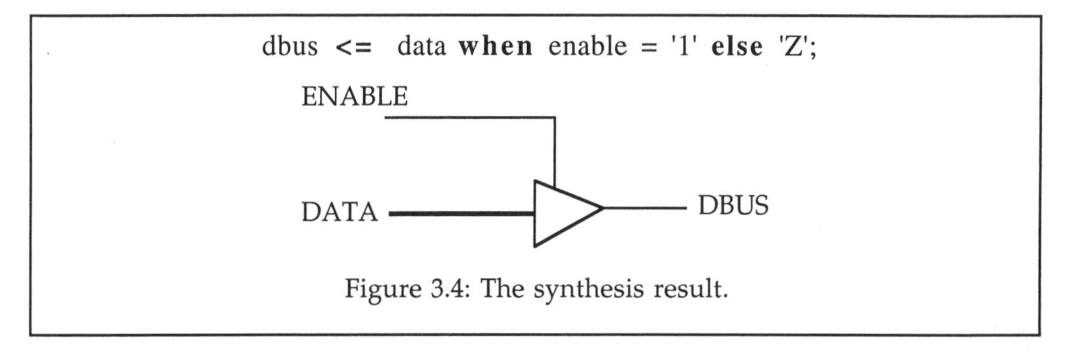
dbus <= data when enable = '1' else 'Z';
ENABLE
DATA
DBUS
Figure 3.4: The synthesis result.

If *dbus* and *data* are changed to vectors in their declaration, the entire vector (bus) will be given a three-state buffer per output. However, the code in the architecture has to be modified slightly:

```
dbus <= data when enable = '1' else (others=>'Z');
```

By using (**others**=>'Z') assignment, the entire vector, regardless of the vector length, will be assigned the value 'Z'. This manner of assigning the entire vector a value is very effective and makes the code easier to maintain. If the length of the vector is to be changed, only the declaration for the vectors has to be modified and not the code in the architecture.

It is permissible to use several different signals in the <condition> expression, making the command very flexible and useful for design.

Example:

```
q<= a when en='0' else
    b when data="11" else
    c when enable='1' else
    d;
```

Note that the conditions are checked in the order in which they are enumerated. The conditions are evaluated line by line until a condition is true. This means that if line two (data="11") and line three (enable='1'), for example, are transposed, the result, if both conditions are true simultaneously, will be different. Suppose that en='0', data="11" and enable='1'. This means that *q* will be given the value of signal *b* in the above example. If lines two and three are transposed, *q* will be given the value of signal *c*. It is only between concurrent commands that order is unimportant. Inside a concurrent command, e.g. in **when else**, order is important.

VHDL syntax does not differentiate between upper-case and lower-case letters. The only exception is inside single quotation marks (' ') or double quotation marks (" "), e.g. when the value 'Z' is assigned to a signal of type std_logic or vlbit, the 'Z' must be upper-case:

```
sig<='Z';        -- Good
sig<='z'         -- Bad
```

3.6. With statement

Syntax:

with <expression> **select**
<target> **<=** <expression> **when** <chose> ;

All possible <choices> must be enumerated. If you want to collect together all the remaining <choices> on a line, **when others** can be used. In this case, **others** must be the last <choice> alternative.

Example:

```
Entity ex is
  port( a,b,c: in    std_logic;
         data: in    std_logic_vector(1 downto 0);
         q:    out   std_logic);
end;
Architecture rtl of ex is
begin
   with data select
      q<= a when "00",
          b when "11",
          c when others;
end;
```

When compared with the **when else** command, the **with select** command is not as flexible, as only one <expression> is allowed. However, the command normally results in code that is relatively easy to read and structured.

3.7. Example of a behavioural model of a multiplexor

The two behavioural models of a multiplexor which follow are equivalent. The first architecture uses a **when** statement and the second a **with** statement. Both models can also be synthesized. The synthesis tool will ignore the 10 ns delay, however:

```
Entity mux2 is
  port( sel_0,a,b: in   std_logic;
        c:         out  std_logic);
end mux2;

Architecture mux2_beh of mux2 is
begin
  c<= a after 10 ns when sel_0='0' else
       b after 10 ns;
end mux2_beh;

Architecture mux2_beh2 of mux2 is
begin
   with sel_0 select
  c<= a after 10 ns when '0',
       b after 10 ns when others;
end mux2_beh2.
```

Thus VHDL provides several different construction options for describing the same behaviour. Chapter 15, "Design examples and design tips" looks at which alternative should be chosen.

3.8. Generics

Generics can be used to introduce information into a model, e.g. timing information. The **generic** must be declared in the **entity** before the **port** command. A **generic** can have any data type; there are, however, some restrictions with regard to what synthesis tools will accept.

Example:

```
Entity generic_ex is
  generic (delay:time:=10 ns);
  port( a,b: in   std_logic;
        c:   out  std_logic);
end generic_ex;
```

```
Architecture generic_beh of generic_ex is
begin
  c<=a and b after delay;
end generic_beh;
```

The component above has a **generic** delay of *delay* ns. The delay has been defined as 10 ns. This value can be changed if the component is instanced. Generics can also be used to design generic components. (See Chapter 6, "Structural VHDL".)

3.9. The assert command - error management in VHDL

Assert is an interesting language construction which makes it possible to test function and time constraints on a model inside a VHDL component.

If the condition for an **assert** is not met during simulation of a VHDL design, a message of a certain **severity** is sent to the user (the simulator). Using **assert** it is possible to test prohibited signal combinations or whether a time constraint is not being met, type set-up, or whether there are unconnected inputs to the component.

```
Syntax:
Assert <condition>
Report <message>
Severity <error_level>
```

If the condition is not met, the **assert** message is sent to the user plus the name of the (unit) process which was cancelled. There are four different **severity** levels for the message (error levels):

- Note
- Warning
- Error
- Failure

The message and error level are reported to the simulator and are usually displayed in plain text in the VHDL simulator's command window. It is normally possible to set the error level at which the VHDL simulator should abort the simulation. The default for most simulators is severity level Error.

Assert can be used to verify external signals to the component or the internal behaviour, both time and function verification. As **assert** is both a sequential and a concurrent command, it can be used virtually anywhere in the code (see Figure 3.5).

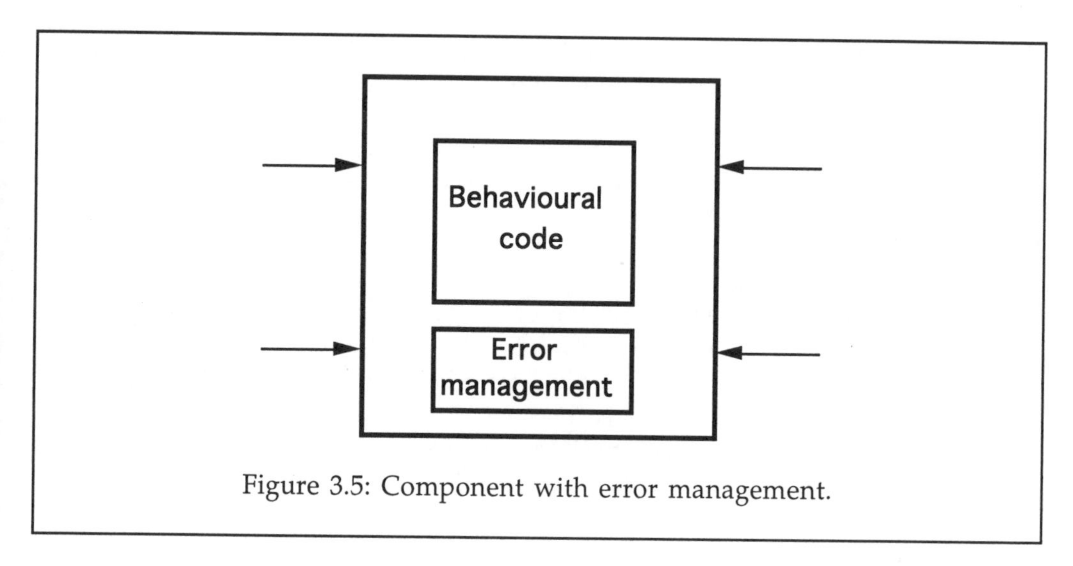

Figure 3.5: Component with error management.

The error management code is only present during the VHDL simulation. It is not included in the synthesis.

An example of error management code is as follows:

```
assert in_0 /= 'X' and in_1 /= 'X'
report "in is not connected"
severity Error;
```

The error management code checks that in_0 and in_1 are connected, or that they have an undefined value. If they are not connected, the simulator will stop and display "in is not connected".

Variable *now* is defined in the VHDL standard. It contains the simulator's internal absolute time. With the **assert** command it is possible to check that the simulator time does not exceed, say 900 ns:

```
process(clk)
begin
  assert now < 900 ns
  report "stopping simulator (max simulation time 900 ns)"
  severity Failure;
end process;
```

If the **assert** command is used in the concurrent part of the VHDL code, the command will only be executed if an event on its "sensitivity list" takes place. If the **assert** command only checks the time with variable *now*, no event will occur and the **assert** command will never be executed. Therefore, the **assert** command in the above example has been placed in a **process** which is executed at each clock edge. There is more on processes in Chapter 4, "Sequential VHDL".

Where the **assert** command is used in the concurrent part of the VHDL code, it can be placed in the same category as a passive process, and can be used in the entity in the following example:

```
Entity ex is
  port( a,b:  in   std_logic;
        q:    out  std_logic);
begin
  assert a/='1' or b/='1'
  report "a='1' and b='1' at the same time"
  severity warning;
end;

Architecture
  ...
```

If you build a test environment, assert is good for verifying the response from the circuit. More on this in Chapter 8, "Testbenches".

If the **report** command is used in the sequential part of VHDL code the **assert** command can be left out in VHDL-93 standard.

Example (VHDL-87):

```
process(a,b)
begin
  if a='1' and b='1' then
    assert false
    report "a='1' and b='1'";
  end if;
  ...
end process;
```

Example (VHDL-93):

```
process(a,b)
begin
  if a='1' and b='1' then
    report "a='1' and b='1'";
  end if;
  ...
end process;
```

3.10. Behaviour and dataflow

Two design approaches can be used when writing VHDL code: behavioural and dataflow design.

An example of dataflow is:

```
q <= c3 and c2 and c1 or c0;
c0 <= not c0;
```

A behavioural description might be:

```
if sign = '0' then
  q0 <= q0+1;
else
  q0 <= q0-1;
end if;
```

Both of the above examples can be synthesized. Often, however, the behavioural code is written in such a way that it can only be simulated and not synthesized (see Chapter 11, "Design methodology").

3.11. Objects, class and type

Some of the contents of this section have already been discussed, but terms are described more systematically below. Defining an object in VHDL means instancing a **constant, variable** or **signal**.

An object contains a value of a specific type.

Each object has a type and a class. Type indicates what type of data the object contains. Class indicates what can be done with the object. VHDL is strongly typed. This means that different types cannot be mixed without type conversion.

Class	*Object*	*Data type*
Signal	a:	std_logic;

Class
There are three different class objects:

- Constant
- Variable
- Signal "a wire"

Constant does not change value. It can be declared in all parts, and can be of any type whatsoever. For example:

```
constant a: a_type:="1001";
```

Variable changes value. Can be declared in process and subprogram, and can be of any type whatsoever. For example:

```
a:=John;
```

Signal can change value in relation to time. For example:

```
b<=input1;
```

> Hint:
> A good name conversion is to end all type names with "**_type**":
> **ex John_type**

3.11.1. Data types

Each signal must have a declared data **type** which determines the values that the signal can assume. The commonest data types are described below.

Boolean type

The **Boolean** data type can only be true or false. Constants, signals and variables can be declared as Boolean:

```
Architecture rtl of ex is
begin
 process(...)
 variable John: boolean;
 begin
   John:= a < b;
   if John then ...
   ...
 end process;
end;
```

Synthesis tools will translate the value false to "0" and true to "1". Assigning a signal of, for example, std_logic, to the value true or false without using a conversion function is not permitted.

Integer type

> The VHDL-87 standard does not define how long an integer should be. The length is implementation controlled, i.e. dependent on which tool is used. However, the majority of tools use 32 bits. This means that an integer can vary between -2147483648 and 2147483647.
>
> Example 1: **constant** loop_number: integer := 345;
>
> Example 2: **signal** my_int: integer **range** 0 **to** 255;

Character type (VHDL-87)

```
type character is (
    NUL, SOH, STX, ETX, EOT, ENQ, ACK, BEL,
    BS,  HT,  LF,  VT,  FF,  CR,  SO,  SI,
    DLE, DC1, DC2, DC3, DC4, NAK, SYN, ETB,
    CAN, EM,  SUB, ESC, FSP, GSP, RSP, USP,
    ' ', '!', '"', '#', '$', '%', '&', ''',
    '(', ')', '*', '+', ',', '-', '.', '/',
    '0', '1', '2', '3', '4', '5', '6', '7',
    '8', '9', ':', ';', '<', '=', '>', '?',
    '@', 'A', 'B', 'C', 'D', 'E', 'F', 'G',
    'H', 'I', 'J', 'K', 'L', 'M', 'N', 'O',
    'P', 'Q', 'R', 'S', 'T', 'U', 'V', 'W',
    'X', 'Y', 'Z', '[', '\', ']', '^', '_',
    '`', 'a', 'b', 'c', 'd', 'e', 'f', 'g',
    'h', 'i', 'j', 'k', 'l', 'm', 'n', 'o',
    'p', 'q', 'r', 's', 't', 'u', 'v', 'w',
    'x', 'y', 'z', '{', '|', '}', '~', DEL);
```

In VHDL-93, the number of permitted characters has been increased and several national characters have been added.

Character type (VHDL-93)

```
type character is (
        nul, soh, stx, etx, eot, enq, ack, bel,
        bs, ht, lf, vt, ff, cr, so, si,
        dle, dc1, dc2, dc3, dc4, nak, syn, etb,
        can, em, sub, esc, fsp, gsp, rsp, usp,

        ' ', '!', '"', '#', '$', '%', '&', ''',
        '(', ')', '*', '+', ',', '-', '.', '/',
        '0', '1', '2', '3', '4', '5', '6', '7',
        '8', '9', ':', ';', '<', '=', '>', '?',

        '@', 'A', 'B', 'C', 'D', 'E', 'F', 'G',
        'H', 'I', 'J', 'K', 'L', 'M', 'N', 'O',
        'P', 'Q', 'R', 'S', 'T', 'U', 'V', 'W',
        'X', 'Y', 'Z', '[', '\', ']', '^', '_',
        '`', 'a', 'b', 'c', 'd', 'e', 'f', 'g',
        'h', 'i', 'j', 'k', 'l', 'm', 'n', 'o',
        'p', 'q', 'r', 's', 't', 'u', 'v', 'w',
        'x', 'y', 'z', '{', '|', '}', '~', del,

        c128, c129, c130, c131, c132, c133, c134, c135,
        c136, c137, c138, c139, c140, c141, c142, c143,
        c144, c145, c146, c147, c148, c149, c150, c151,
        c152, c153, c154, c155, c156, c157, c158, c159,

        -- the character code for 160 is there (NBSP),
        -- but prints as no char

        The last 64 characters are special characters
            (example national characters)...
```

At present, the majority of synthesis tools support only the VHDL-87 character type and not the expanded version in VHDL-93.

Bit type

> Bit is a type which is defined in the standard in VHDL.
> Bit can only assume the values "0" or "1".

Vlbit type

> Vlbit is a type which ViewLogic uses and is an expansion of the BIT type in standard VHDL. Vlbit has the value 'X', 'Z', '0' or '1'. During simulation, 'X' means unknown. The synthesis translates 'X' to "don't care".
>
> Example: **if** John(6 **downto** 0) = "1XXXX0X" **then** ...
> is equal (synthesis) to: **if** (John(6)='1' **and** John(1)) = '0' **then** ...
>
> 'Z' means high impedance in both the simulation and the synthesis. You have to be careful about what you are synthesizing too, e.g. certain FPGAs (Field Programmable Gate Arrays), which do not have three-state. This problem can be solved by multiplexing the signal.

Only bit and bit_vector are defined in the VHDL standard. This leads to several restrictions in design if bit and bit_vector are used:

- Not possible to describe three-state
- Not possible to have several drivers for the same signal
- Not possible to assign unknown to a signal
- Not possible to assign don't care to a signal

As bit can assume only the values '0' and '1', three-state and unknown, for example, cannot be described using bit or bit_vector. In the case of a three-state signal, there are often several signals which can drive the signal. This means that a method has to be found for deciding which value a signal will be given if there are several drivers. This is not defined in the VHDL standard for the bit or bit_vector data type. Assigning the value *don't care* may also be effective in certain situations. This makes it possible for the synthesis tool to choose the value so that the set optimization goals can be achieved more easily.

In the design of simulation models in VHDL, the use of bit and bit_vector also leads to the following restrictions:

- Not possible to describe pull up (weak one)
- Not possible to describe pull down (weak zero)
- Not possible to describe that the signal is uninitialized

In order to solve these problems, VHDL suppliers have created their own data types. This in itself solved the above problems, but gave rise to new ones:

- The VHDL code became tool-dependent
- It was not possible to join two designs from two different tools

As these data types are only defined in their respective tools, problems arise if the VHDL code is moved to another simulator. Normally, of course, it is possible to define this data type in the new simulator. This, however, runs contrary to VHDL's philosophy of being platform- and simulator-independent. If these user-defined data types are used, it is not possible to join two designs or sub-blocks either. As previously mentioned, VHDL is a strongly typed language: it is not permissible to join two signals (e.g. in hierarchical design) if they do not have the same data type. In order to make VHDL an effective platform- and simulator-independent standard, IEEE-defined data types were introduced: std_logic and std_ulogic. These data types have become an industry standard for design. Std_logic and std_ulogic can assume exactly the same values, the difference being that std_logic is a resolved subtype of std_ulogic (see Chapter 5, "Library, package and subprograms").

Std_ulogic type

Std_ulogic is a type which is declared in the ieee package std_logic_1164. Std_ulogic can assume the following values:

```
1   'U'  -- Uninitialized
2   'X'  -- Forcing Unknown
3   '0'  -- Forcing 0
4   '1'  -- Forcing 1
5   'Z'  -- High Impedance
6   'W'  -- Weak Unknown
7   'L'  -- Weak 0
8   'H'  -- Weak 1
9   '-'  -- Don't care
```

Std_logic type

Std_ulogic is a type which is declared in the ieee package std_logic_1164. Std_ulogic can assume the same values as std_logic. The difference is that std_logic is defined as:

```
subtype std_logic is resolved std_ ulogic;
```

The fact that std_logic is resolved means that, if a signal is driven by several drivers, a predefined resolution function is called up which resolves the conflict and decides which value the signal will be given. This means that std_logic can, for example, be used for three-state buses in which several drivers can drive the same conductor on the bus (Figure 3.6), but not normally simultaneously. If a std_ulogic signal is driven by more than one driver, it will result in an error (Figure 3.7), as VHDL does not permit an unresolved signal to be driven by more than one driver.

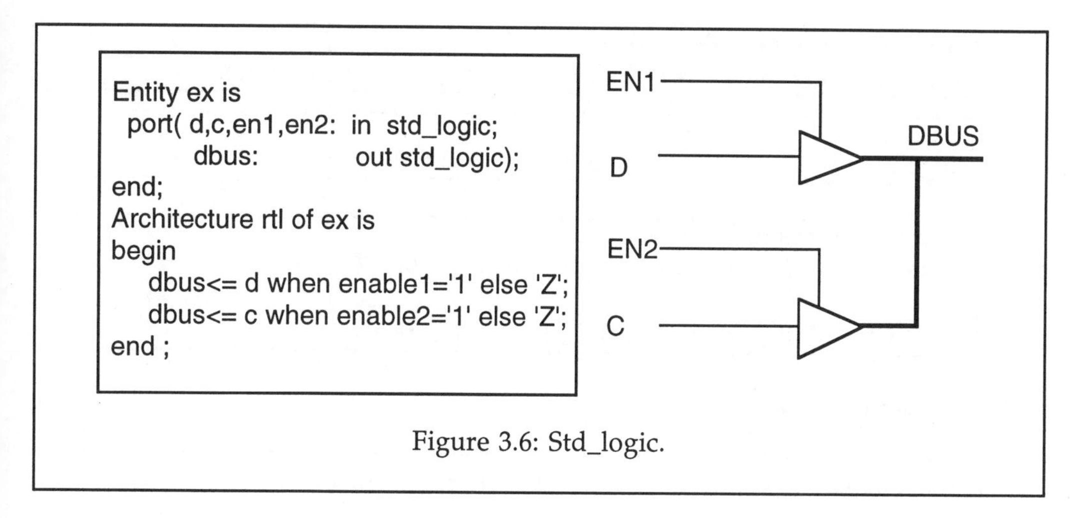
Entity ex is
port(d,c,en1,en2: in std_logic;
dbus: out std_logic);
end;
Architecture rtl of ex is
begin
dbus<= d when enable1='1' else 'Z';
dbus<= c when enable2='1' else 'Z';
end ;
EN1
D
EN2
C
DBUS

Figure 3.6: Std_logic.

```
Entity ex is
  port( d,c,en1,en2:  in  std_ulogic;
        dbus:             out std_ulogic);
end;
Architecture rtl of ex is
begin
   dbus<= d when enable='1' else 'Z';
   dbus<= c when enable2='1' else 'Z';
end ;
```

→ ERROR

Figure 3.7: Std_ulogic.

This restriction on std_ulogic has led to data type std_logic being preferred. It is generally easiest to use the same data type throughout the design, as this avoids type conversions. The disadvantage of std_logic is that no error occurs if you have two drivers for the same signal in the VHDL code by mistake. The signal will be given the value 'X' in the simulation, so the error should be discovered.

Array type

Std_ulogic_vector is defined as:
type std_ulogic_vector **is array** (natural **range** <>) **of** std_ulogic;

Std_logic_vector is defined as:

```
type std_logic_vector is array (natural range <>) of std_logic;
```

Bit_vector is defined as:

```
type bit_vector is array (natural range <>) of bit;
```

Vlbit_vector is defined as:

```
type vlbit_vector is array (natural range <>) of vlbit;
```

The types vlbit_vector and vlbit_1d are identical.

Type declaration

It is possible to create your own types in the architecture.

```
type <identifier> is type_definition> ;

type_definition = array <index_value> of element_type;
```

Example:

```
type A_type is array (3 downto 0) of  std_logic;
type my_int is integer range 0 to 15;
signal d: A_type;
signal q: my_int;
```

The above integer could also have been defined directly in the signal declaration:

```
signal q: integer range 0 to 15;
```

It is important to declare the range of an integer, as otherwise the synthesis tool will assume that the integer is 32 bit. If, for example, the integer is clocked into a register, 32 bits will be given if the range is not defined, as against just 4 for the integer in the example given above. The same also applies to vectors.

Vectors

Vectors can be decelerated in the signal declaration part:

```
Entity ex is
 port( a:  in     std_logic_vector(3 downto 0);
       b:  in     bit_vector(0 to 3);
       c:  out    std_ulogic_vector(4 downto 0));
end;
```

```
Architecture rtl of ex is
signal i1: std_logic_vector(3 downto 0);
signal i2: vlbit_vector(3 downto 0);
signal i3: std_ulogic_vector(3 downto 0);
begin
...
end;
```

It is permissible to define vectors with both **to** and **downto**. If **to** is used, the **vector** must be defined as 0 **to** 3. If downto is used, the **vector** must be defined as 3 **downto** 0. In designs it is usually easiest to use the same definition consistently. Moreover, **downto** better suits a hardware designer's normal way of thinking, i.e. the bit furthest to the left is the MSB (Most Significant Bit). Therefore, the majority of designs written for synthesis use **downto** in all **vector** definitions.

Time

```
Example:

constant delay : time := 0 ns;
a <= b after delay;
```

Enumerated type

It is possible to define your own data types in VHDL. In principle, these data types can assume any value whatsoever. State machines are a common area of use for enumerated data types.

```
Example:

type state_type is (start, idle, waiting, run);
signal state:state_type;
```

In VHDL simulation, the signal state will then be given one of its five defined values which are displayed in the waveform window in the simulator. These self-explanatory signal names make it easier to simulate the design. The majority of synthesis tools also support enumerated data types. In the above example, state will be converted to a 3-bit vector by the synthesis tool, as it could assume five different values.

3.11.2. Synthesizable data types

Table 3.1 shows which types can currently (1995) be synthesized in ViewLogic, Synopsys and Autologic 2. "_d" means the one-dimensional array, and "_2d" is a two-dimensional data array; V stands for ViewLogic, A for Autologic 2 and S for Synopsys.

Data types	Fully supported	Partly supported	Typically not supported	No meaning with support
Boolean	V, A, S			
boolean_1d	V, A, S			
boolean_2d	A, S		V	
String		V, A, S		
Integer	A, S	V		
integer_1d	A, S	V		
integer_2d	A, S	V		
Std_logic	A, S		V	
std_logic_1d	A, S		V	
std_logic_2d	A, S		V	
std_ulogic	A, S		V	
std_ulogic_1d	A, S		V	
std_ulogic_2d	A, S		V	
Vlbit	V	A, S		
vlbit_1d	V	A, S		
vlbit_2d		V, A, S		
Character		V, A, S		
character_1d		V, A, S		
character_2d		A, S	V	
Time				V, A, S
time_1d				V, A, S
time_2d				V, A, S
enumeration	V, A, S			
enum._array	V, A, S			
Record	A, S		V	
record_array	A, S		V	
Text				V, A, S

Table 3.1: Synthesizable data types with ViewLogic, Autologic 2 and Synopsys.

The other synthesis tools support roughly the same subset.

3.12. Vector assignment

When designing with VHDL, vectors are usually used as the data type. A **vector** can be assigned a value in many different ways.

If a binary value is to be assigned, it is done as follows:

```
a_vect<="10011";
```

Note that vectors are written inside double quotation marks and individual bits inside single quotation marks.

If logical operators, e.g. **and**, are used on vectors, the result will be a bitwise **and**.

Example:

```
Architecture rtl of ex is
signal a,b:std_logic_vector(3 downto 0);
signal c:std_logic_vector(3 downto 0);
begin
   a<="0110";
   b<="1101";
   c<= a and b;
end;
```

The vector *c* will be assigned the value "0100":

```
      "0110"
and   "1101"
      "0100"
```

The requirement with regard to logical operators is that both vectors should be of the same length and that the assigning vectors should also be of the correct length.

3.12.1. Bit string literals

For bit_vector there are predefined bit string literals which can be used to assign a bit_vector as follows:

	Example
Binary	B"11000"
Octal	O"456"
Hex	X"FFA5"
Decimal	239 only for constants
Real	4.6E-4 not supported for synthesis

The advantage of bit string literals is that the VHDL code often becomes easier to read if a vector is assigned a hexadecimal value rather than a binary one. From the synthesis point of view, however, there is no difference.

If a vector is to be assigned a binary value, the B does not have to be written out before the vector, precisely as shown earlier in the chapter. The two vector assignments below are therefore identical:

```
a_vect<="10011";
a_vect<=B"10011";
```

The number base example above only works on bit_vectors. ViewLogic has modified the standard slightly, however, which means that the above number bases also work on vlbit_vector in ViewLogic's tool.

Underline (_) can be used in a bit vector to improve readability. If underline is used, bit string literals must be written out. It is also permissible to use underline for bit string literals other than bit_vector (B).

Example:

```
a_vect<=B"1100_0011_0011_1100";--    Good
a_vect<="1100001100111100";     --   Identical to first
                                --   assignment
a_vect<=X"C33C";                --   Identical to first
                                --   assignment
a_vect<=X"C3_3C";               --   Identical to first
                                --   assignment
a_vect<="1100_0011_0011_1100";  --   Error, B missing
a_vect<="C33C";                 --   Error, X missing
```

VHDL is a strongly typed language. This means that it is not permissible to assign a bit_vector to a std_logic_vector, for example, without using a conversion function (see "IEEE package" in Appendix B).

Example of conversion:

```
Library ieee;
Use ieee.std_logic_1164.ALL;

Entity ex is
 port( a:   in    std_logic_vector(2 downto 0);
       b:   out   bit_vector(2 downto 0));
end;
```

```
Architecture rtl of ex is
begin
  b<=to_bitvector(a,0);
end;
```

As it is normal to use std_logic_vector or std_ulogic_vector when designing, a conversion function must be used in order to have bit string literals.

Example:

```
Library ieee;
Use ieee.std_logic_1164.ALL;
Use ieee.std_logic_unsigned.ALL;

Entity ex is
   port( a: out std_logic_vector(15 downto 0);
         b: out bit_vector(15 downto 0));
end;

Architecture rtl of ex is
begin
  a<=to_stdlogicvector(X"F6");   -- Conversion function must
                                 -- be used

  b<=X"E4";                      -- Bit_vector
end;
```

As the above example shows, X"F6" had to be converted to a std_logic_vector with the conversion function to_stdlogicvector. This is because X"F6" produces a bit_vector, and, as VHDL is a strongly typed language, a bit_vector cannot be assigned to a std_logic vector without first being converted to a std_logic_vector. The conversion function to be used depends on which package is declared at the top of the VHDL code (see Chapter 5, "Library, package and subprograms"). In this instance the conversion function to_stdlogicvector, which was declared in package ieee.std_logic_unsigned, was used. The conversion function used does not affect the result of simulation or synthesis (if they are functionally the same). The choice should be made on the basis of which function the synthesis tool supports.

If signal *a* in the above example had been assigned the value X"F6" directly, an error would have occurred in the compilation of the VHDL code. In the VHDL-93 standard, bit string literal has been enlarged. In VHDL-93 it is permissible to assign a std_logic_vector and a std_ulogic_vector the value directly without a conversion function. This makes the VHDL code easier to read and faster to write:

```
signal a: std_logic_vector(7 downto 0);
a<=to_stdlogicvector(X"F4");        -- VHDL-87
a<=X"F6";                           -- VHDL-93
```

The majority of synthesis tools do not support VHDL-93, however, which means that bit string literals for std_logic_vector can normally only be used for simulation, e.g. in testbenches.

With bit string literal, the assignment of a vector is just as flexible as when an integer is to be assigned a value. Both the data types (vectors and integers) can be assigned with any number base. An integer, for example, can be assigned by specifying the number base with a digit directly followed by the value inside two hash marks (#).

Example:

```
Architecture rtl of ex is
constant myint:integer:=16#FF#;         -- myint=255
signal int1,int2,int3: integer range 0 to 1023;
begin
  int1<=16#FE#;                         -- 16#FE# = 254
  int2<=2#100110#;                      -- 2#100110# = 38
  int3<=8#17#;                          -- 8#17# = 15
  ...
end;
```

3.12.2. Slice of array

If you only want to assign a value to a bit or part of the vector, it can be done as follows:

```
Architecture rtl of ex is
signal a_vect: std_logic_vector(4 downto 0);
signal b_vect: std_logic_vector(0 to 4);

begin
  a_vect(4)<='1';
  a_vect(3 downto 0)<="0110";
  b_vect(4)<='0';
  b_vect(0 to 3)<="1001";
end;
```

Note that when a slice of array is assigned a value, the slice direction must be the same as in the declaration, i.e. **downto** or **to**. It is also permissible to assign a vector a slice of another vector.

Example:

```
Architecture rtl of ex is
signal a_vect:   std_logic_vector(4  downto 0);
signal b_vect:   std_logic_vector(0  to 4);
signal c:        std_logic;
begin
  a_vect<="01101";
  b_vect(4)<=c;
  b_vect(0  to  3)<=a_vect(3  downto  0);
end;
```

The data type of the vectors must be the same on both sides of the assignment symbol, as previously mentioned, and the vector length also has to be equal. The following example is therefore wrong.

Example (bad):

```
Architecture bad of ex is
signal a: std_logic_vector(2  downto  0);
signal b: std_logic_vector(3  downto  0);
begin
   a<=b;  -- Left side = 3 bits, right side = 4 bits -> error
end;
```

The above assignment has to be rewritten so that both sides of the assignment symbol are of the same length, e.g.:

```
a<=b(2  downto  0);
```

Alternatively (dependent on the desired function):

```
a<=b(3  downto  1);
```

If the index order for the vectors has been defined differently for two vectors, it is still possible to assign one vector the value of the other. However, you have to keep a check on which bit receives which value.

Example:

```
Architecture rtl of ex is
signal a,b,c: std_logic_vector(2 downto 0);
signal d:     std_logic_vector(0 to 2);
  a<=d;
  b<=c;
end;
```

In the above vector assignments, vectors *a* and *b* are given the following values:

```
a(2)<=d(0);        b(2)<=c(2);
a(1)<=d(1);        b(1)<=c(1);
a(0)<=d(2);        b(0)<=c(0);
```

3.12.3. Concatenation

Ampersand (&) means concatenation in VHDL. Concatenation can be very useful in vector assignment.

Example:

```
Architecture rtl of ex is
signal a:     std_logic_vector( 5 downto 0);
signal b,c,d: std_logic_vector(2 downto 0);
begin
  b<='0' & c(1) & d(2);
  a<=c & d;
end;
```

If, for example, vector *c* has the value "011" and vector *d* the value "101", vector *b* is given the value '0' & '1' & '1', i.e. "011" and vector *a* the value "011" & "101", i.e. "011101".

Concatenation can also be used for **if** statements, for example:

```
if c & d = "001100" then
...
```

As previously mentioned, both sides of the vector assignment symbol must be of equal length. It is not permissible to use concatenation on the left of the assignment symbol in order to achieve this.

Example (bad):

```
Architecture bad of ex is
signal a:std_logic_vector(2 downto 0);
signal b:std_logic_vector(3 downto 0);
begin
  '0' & a<=b;          -- Error
end;
```

3.12.4. Aggregate

If you are designing with slightly larger vectors and want to assign the same value to the entire vector, this can be done as follows:

```
Architecture rtl of ex is
signal a: std_logic_vector(4 downto 0);
begin
  a<=(others=>'0');
end;
```

This command is identical to a<="00000";. The advantages of the a<=(**others**=>'0') method are that there is less to write in the case of large vector assignments, and that the assignment is completely independent of the vector length. Assignment to the vector with **others** is what is known as aggregate assignment. It is also permissible to assign some bits in the vector and then use **others** to assign the remaining bits:

```
a<=(1=>'1', 3=>'1', others=>'0'));
```

The above assignment means that bits 1 and 3 in vector *a* are assigned the value '1' and other bits in the vector are assigned the value '0'. It is also possible with aggregate to assign the vector the value of other signals:

```
a<=(1=>c(2), 3=>c(1), others=>d(0));
```

As an alternative to the above assignment to vector *a*, concatenation can be used (suppose that *a* has a length of 5 bits):

```
a<=d(0) & d(0) & c(1) & d(0) & c(2) & d(0);
```

The disadvantage of this description method is that the assignment is now length-dependent and must be converted if *a*'s length is changed. From the synthesis point of view, there will be no difference in the result. Both description methods are supported by the majority of synthesis tools.

3.12.5. Qualifier

Sometimes it is not clear which data type an expression has. If the compiler cannot decide unambiguously which data type an expression or value has, an error will be generated. To make the data type clear to the compiler, a qualifier can be used. A qualifier means that the data type is stated explicitly followed by a tick mark (') and the expression:

```
datatype'expression or value
```

Example:

```
ROM_type'("01","10","00");
```

3.13. Advanced data types

For advanced design, access to more advanced data types is needed. There follows a look at multidimensional data types, subtypes and records.

3.13.1. Subtypes

If a subset of a data type which has already been defined is declared, a **subtype** declaration should be used.

```
subtype identifier is basetype limit;
```

Specification is needed of what values the new **subtype** can assume, i.e. a subset of the base type's values. Alternatively, a limited length, e.g. vector length, of the base type is specified. For example:

```
subtype my_int is integer range 0 to 3215;              -- Good
subtype byte is std_logic_vector(7 downto 0);          -- Good
type byte2 is array (7 downto 0) of std_logic;         -- OK
type byte3 is std_logic_vector(7 downto 0);            -- Error
subtype byte4 is array (7 downto 0) of std_logic;      -- Error
```

As we see from the above example, **array** can only be used to define new types and not subtypes. Std_logic_vector(7 **downto** 0) cannot be used in new type declarations. *Byte* can also be declared as a new data type (byte2). The disadvantage of using type instead of **subtype** is that the declared type becomes a completely new type, i.e. it is not possible to assign a slice of a signal with the original data type to a signal with the new data type without converting the signal.

Example (bad):

```
Architecture bad of ex is
type byte is array (7 downto 0) of  std_logic; -- So far so good
                                               -- (but not perfect)
signal a:  byte;
signal c:  std_logic_vector(7  downto 0);
begin
  a<=c;         -- Error, a and c do not have the same data type
end;
```

If *byte* had been declared as a subtype, the above example would have passed the VHDL compiler without an error.

There are two predefined subtypes in VHDL:

```
subtype natural is integer range 0 to impl. defined     -- typical 2147483647
subtype positive is integer range 1 to impl. defined    -- typical 2147483647
```

From the synthesis point of view it does not matter whether a data type has been declared as a subtype or a separate type.

3.13.2. Multidimensional array

In principle it is possible to define any number of dimensions for a data type. Normally no more than two- or three-dimensional data types are used. When defining multidimensional data types, array is used in the type declaration.

Example:

```
type data4x8 is array (0 to 3) of std_logic_vector(7 downto 0);
type data3x4x8 is array (0 to 2) of data4x8;
```

Data type data4x8 is a two-dimensional data type (4 x 8 bits), while data3x4x8 is a three-dimensional data type (3 x 4 x 8 bits). The type declaration of data4x8, for example, could also have been written as follows:

```
type byte is array (7 downto 0) of std_logic;
type data4x8 is array ( integer range 0 to 3) of byte;
```

When a two-dimensional vector is to be assigned, it can be done in several ways. One array at a time can be assigned using index, or entire two-dimensional vectors can be assigned in one go using aggregate.

Example:

```
Architecture rtl of ex is
type data4x8 is array (0 to 3) of  std_logic_vector(7 downto 0);

signal d,e,f,g,h,i:    data4x8;
signal b1,b2,b3,b4: std_logic_vector(7 downto 0);
begin
  d(0)<="01010110";
  d(1)<="10101000";
  d(2)<="01110110";
  d(3)<="10111011";

  e<=(others=> (others=>'0')); -- Clear the whole 2-dim. signal

  f(0) (0)<='1';
  f(0) (1)<='0';
  ....

  g<=(b1, b2, b3, b4);

  h<=(others=>b1);

  process(h)
  begin
    l1: for n in 0 to 3 loop
      i(n)<=h(n);
    end loop;
  end process;

  ...
end;
```

The assignment method to be chosen is dependent on the application and the readability required. The majority of synthesis tools which support two-dimensional data types support all the above methods. The choice does not usually matter from a synthesis point of view.

3.13.3. Records

A **record** type contains elements of different data types:

```
type <identifier> is record
     record definition
end record;
```

Example:

```
Architecture beh of ex is
type data_date is record
   year:      integer range 1996 to 2099;
   month:     integer range 1 to 12;
   date:      integer range 1 to 31;
   hour:      integer range 0 to 23;
   minute:    integer range 0 to 59;
   second:    integer range 0 to 59;
   data:      std_logic_vector(31 downto 0);
end record;

signal d:data_date;
begin
   d.year<=1997;
   d.month<=4;
   d.date<=8;
   d.hour<=11;
   d.minute<=57;
   d.second<=22;
   d.data<=data_in;
end;
```

Alternatively, the above **record** could have been assigned with aggregate assignment:

```
d<=(1997, 4, 8, 11, 57, 22, data_in);
```

Record can be very powerful in certain types of design, both with regard to ASIC design and VHDL models for simulation. The advanced synthesis tools support records. There are some restrictions with regard to data types, of course.

3.14. Alias

Alias can be used to give alternative names to objects. **Alias** is often used to make the code easier to read.

Example:

```
alias data: std_logic_vector(7 downto 0) is data_in(15 downto 8);
```

In VHDL-87, **alias** is only permissible for objects. This restriction has been relaxed in VHDL-93, and it is now permissible to define aliases for functions, types, etc. The advanced synthesis tools support aliases of the VHDL-87 type.

3.15. Relational operators

In VHDL there are several defined relational operators:

Symbol	Operand
=	equal
/=	not equal
<	less than
>	greater than
<=	less than or equal
>=	greater than or equal

These operators produce a Boolean, i.e. true or false. The above operators can be used directly on integers, bit_vectors and std_logic_vectors, for example. The operators = and /= can also be used on all defined data types.

Example:

```
Architecture rtl of ex is
type my_type is (on,off,idle,start);
signal state1,state2: my_type;
signal a,b,c:          bit_vector(2 downto 0);
signal d,e,f,g:        std_logic_vector(2 downto 0);
signal my_int:         integer range 0 to 15;
begin
  a<=b when c="010" else b;
  d<=e when d/="111" else "000";
  e<=f when my_int<=12 else "001";
  f<="101" when e>=d else "110";
  g<=f when state1=state2 else "000";
  ...
end;
```

All relational operators are supported by the majority of synthesis tools. The simple synthesis tools may have certain restrictions with regard to which data type can be used, while the advanced synthesis tools support relational operators for all data types in principle.

3.16. Arithmetic operators

There are several predefined arithmetic operators in VHDL:

Symbol	Operator
+	addition
-	subtraction
*	multiplication
/	division
abs	absolute value
rem	remainder
mod	modulus
**	exponentiation

These arithmetic operators are predefined for data types integer, real (not mod and rem) and time. On the other hand, they are not defined for std_logic_vector and std_ulogic_vector, for example. If std_logic_vectors have to use the above operators, they must be defined for std_logic_vector, e.g. in a package. This has already been done in package ieee.std_logic_unsigned, for example.

```
Library ieee;
Use ieee.std_logic_1164.ALL;
Use ieee.std_logic_unsigned.ALL;    -- Needed for "+" with
                                    -- std_logic_vectors

Entity ex is
 port( a,b:  in   std_logic_vector(2 downto 0);
       c,d:  in   integer range 0 to 15;
       q1:   out  std_logic_vector(2 downto 0);
       q2:   out  integer range 0 to 30);
end;

Architecture rtl of ex is
begin
 q1<=a + b;
 q2<=c + d;
end;
```

The majority of synthesis tools support "+", "-" and "*" for integers. However, they are not supported for data types real and time. As far as operators for std_logic_vector are concerned, for example, the advanced synthesis tools support "+", "-", "*" and "**". Something which may vary is the package which the synthesis tool supports, i.e. where its operators for std_logic_vectors are defined. Something else which may need to be altered if the synthesis tool is changed is the package which has to be declared at the top of the VHDL code with regard to arithmetic operators (see Chapter 5, "Library, package and subprograms").

3.17. Init value

At time zero in the simulation all signals are given their initial value. This value varies depending on which data type they are declared as. Unless specified otherwise, all signals are given what is known as their datatype'left value. Bit and std_logic are defined as follows:

```
type bit is ('0', '1');
type std_logic is ('U', 'X', '0', '1', 'Z', 'W', 'L', 'H', '-');
```

The attribute 'left means the value of the data type's first (left-hand) value in the type declaration. This leads to:

```
bit'left='0'
std_logic'left='U'
```

A bit is therefore initiated to the value '0' and a std_logic to 'U'. If you want to know what a data type is initiated to, you must know how that data type is declared.

It is possible to change the default initial value in the VHDL code by defining a particular value, if desired. This should be done when the signal is declared, i.e. in the entity or in the signal declaration in the architecture.

Example:

```
Entity ex is
 port( a:   in   std_logic:='0';
       b,c: in   std_logic;
       q:   out  bit);
end;
Architecture behv of ex is
signal i1:     std_logic:='1';
signal i2:     std_logic:='H';
```

```
signal i_vect: std_logic_vector(3 downto 0):="00LL";
begin
 i1<=a or b;
 ...
end;
```

In the above example the input signals are initiated to the following values:

Signal	*Value*
a	'0'
b	'U'
c	'U'
q	'0'
i1	'1'
2	'H'
i_vect	"00LL"

If, for example, the internal signal *i1* is assigned the value '0' in the architecture at a time of 0 ns + 0 delta (i.e. directly), i1 will have the value '1' during a delta and then be given the value '0' at a time of 0 ns + 1 delta.

It is also permissible to define initial values for signals in the entity of mode out or inout. If several signals are to be initiated to the same value, this can be done in one line:

```
signal a,b: std_logic_vector(2 downto 0):="001";
```

Synthesis tools do not support initial values. However, the majority of tools accept their existence in the VHDL code, but ignore them with a warning. Therefore, it is not possible to initiate an input signal to 'H', for example, and suppose that the synthesis tool will connect a pull-up to the signal. Any pull-up in the design normally has to be instanced in the VHDL code.

There is a risk with initial values that the VHDL simulation and synthesis result will not behave the same, as a signal has been initiated to a value. In the hardware this input signal may have a different value at start-up, which can lead to a "mismatch" between VHDL simulation and gate-level simulation. If std_logic is used for design and no initiation value specified, it will be initiated to 'U'. This is a good initial value. If there is no logic driving the std_logic signal, other signals which are depending on the std_logic signal will be given the value 'X'. This error will then be easy to detect in the VHDL simulation. If the signal is initiated to '0', for example, the error is unambiguous but difficult to detect.

3.18. Exercises

1. What does the signal assignment symbol look like?

2. Which two delay modes are defined in the VHDL standard?

3. Draw output signals C1 and C2.

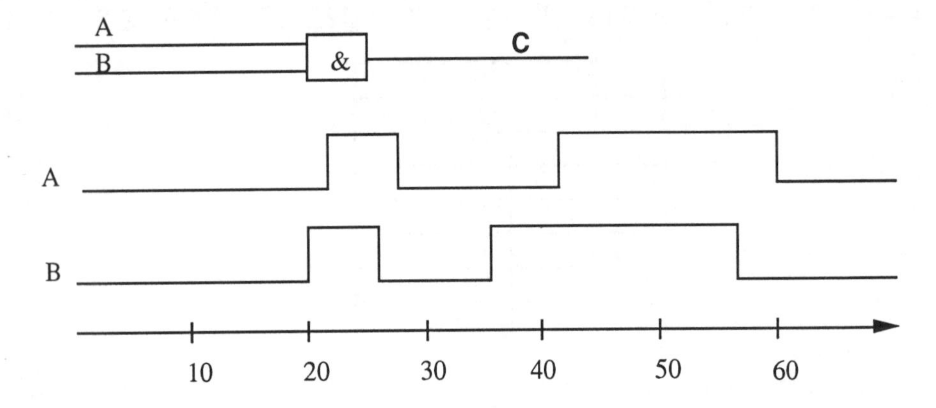


```
c1<=a and b after 10 ns;
c2<=transport a  and b after 10 ns;
```

C1

C2

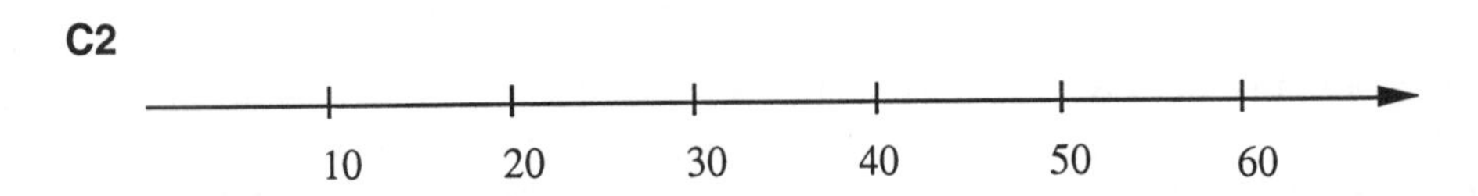

4. Complete the following tables. The waveform for signal *a* is given in both the examples.

```
Architecture rtl of ex is
begin
  b<=a;
  c<=b;
end;
```

	0	10	10+1delta	10+2delta	20	20+1delta	20+2delta
A	0	1	1	1	0	0	0
B	0						
C	0						

```
Architecture rtl of ex is
begin
  c<=b;
  b<=a;
end;
```

	0	10	10+1delta	10+2delta	20	20+1delta	20+2delta
A	0	1	1	1	0	0	0
B	0						
C	0						

5. (a) Why is the following line of VHDL incorrect?

```
q<=   a when  sel='0' else
      b when  sel='1';
```

(b) Rewrite the VHDL code, without changing the behaviour, so that it is correct.

(c) What change has been made in the VHDL-93 standard with regard to the **when else** command?

(d) Is example 5(a) valid in the VHDL-93 standard?

6. Which error levels exist in the **assert** command and which is default?

7. Rewrite the VHDL code below to include an **assert** command. The **assert** command should print out "too many ones" if vector *a* is equal to "111".

```
Architecture rtl of ex is
signal a: std_logic_vector(2 downto 0);
begin
    a<=c and b;
end;
```

8. (a) What are the three object classes in VHDL?

(b) Why are data types other than bit and bit_vector needed?

(c) Which two data types are IEEE-defined and what is the difference between them?

9. (a) What is bit string literal used for?

(b) For which data type is bit string literal defined in the VHDL-87 standard?

(c) What change has been made in the VHDL-93 standard?

10. Can an integer be assigned a value with any number base, and if so how?

11. What is a slice of array?

12. What is the advantage of using aggregate assignment for vector *a* in the following VHDL code?

```
Architecture rtl of ex is
signal a: std_logic_vector(31 downto 0);
begin
    a<=(others=>'1');
    ...
end;
```

13. What value will vector *a* be given after the following assignment?

```
Architecture rtl of ex is
signal a,b: std_logic_vector(4 downto 0);
signal c: std_logic_vector(0 to 1);
begin
    a<=(1=>'0', 3=>'1', others=>b(2));
    b<=(1=>'1', 3=>'0', others=>c(1);
    c<="10";
end;
```

14. (a) What value will signal *a* be initiated to at start-up in the following simulation?

```
type mytype is ('T', 'R', '0', '1');
signal a: mytype;
```

(b) What is a signal of data type bit initiated to?

(c) What is a signal of data type std_logic initiated to?

(d) Is it possible to affect a signal's initiation value, and if so how?

4

Sequential VHDL

We come now to sequential language constructions. Designing sequential processes presents a new design dimension to hardware designers who have worked at gate level. This chapter starts by looking at concurrent and sequential data processing. The difference between sequential and concurrent assignment is described, as it is important to understand the difference if you are to write code that synthesizes well. Then the sequential language constructions in VHDL are illustrated.

4.1. Concurrent and sequential data processing

Electronics are parallel by nature. In practice, all the gates in a circuit diagram can be seen as concurrently "executing" units. In VHDL there are language constructions which can handle both concurrent and sequential data processing. Many of the VHDL commands are only valid in the concurrent or sequential part of VHDL. The commonest VHDL commands in each part are given below:

Concurrent VHDL constructions

- Process statement
- When else statement
- With select statement
- Signal declaration
- Block statement

Sequential VHDL constructions

- If-then-else statement
- Case statement
- Variable declaration
- Variable assignment

- Loop statement
- Return statement
- Null statement
- Wait statement

There are also many VHDL commands which are valid in both the concurrent and the sequential parts of VHDL:

- Signal assignment
- Declaration of types and constants
- Function and procedure calls
- Assert statement
- After delay
- Signal attributes

As a VHDL designer, it is important to know where the various VHDL commands can be used and where the VHDL code is concurrent or sequential. Some examples of where the VHDL code is concurrent or sequential are given below. In brief, it can be said that the VHDL code is concurrent throughout the architecture except for in processes, functions and procedures.

```
Architecture rtl of ex is
concurrent declaration part
begin
 concurrent VHDL

 process(...)
 sequential declaration part
 begin
     sequential VHDL
 end process;

 concurrent VHDL
end;
```

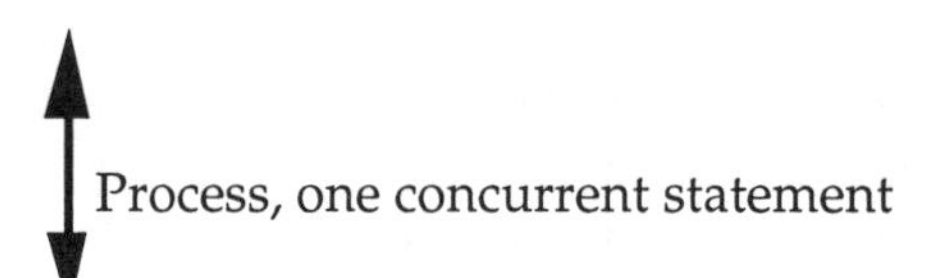

4.2. Signal and variable assignment

Signals and variables are used frequently in VHDL code, but they represent completely different implementations. Variables are used for sequential execution (like "normal" programs), while signals are used for concurrent execution.

```
<target_identifier> := <selected_expression>;

<selected_expression> =
< identifier >[(and ! or ..! xor) <identifier > ... ( and ! nor ..)
<identifier > ] ';'
```

Variable and **signal** assignment are differentiated using the symbols ":=" and "<=". A variable can only be declared and used in the sequential part of VHDL. Signals can only be declared in the concurrent part of VHDL but can be used in both the sequential and concurrent parts. A variable can be declared for exactly the same data types as a signal. It is also permissible to assign a variable the value of a signal and vice versa, provided that they have the same data type.

In a program for a CPU, the statements are executed sequentially. The following program is executed sequentially:

```
variable temp_a, diff: integer_type;

temp_a := in_1;
diff := temp_a - 2;
```

Suppose that the program is executed every delta t. Table 4.1 shows how three executions affect the variables. Delta t can be arbitrary.

Variable	T - delta t	T	T + delta t
in_1	2	3	3
temp_a	2	3	3
diff	0	1	1

Table 4.1: Sequential assignment statement.

At time T-delta t, *in_1* has the value 2; *temp_a* is given the value 2. The next statement gives *diff* the value (2-2) 0, etc. This is traditional programming.

If the same program as before is used and the statements are executed in the concurrent part of VHDL, signals will have to be used instead of variables:

```
signal temp_a, diff : integer_type;

temp_a <= in_1;
diff <= temp_a - 2;
```

If the time aspect is introduced, the program will look like this:

```
temp_a(t) <= in_1(t - ΔT);
diff(t) <= temp_a(t - ΔT) - 2;
```

in_1 is updated $T - \Delta T = 2$.

The two lines in the example are executed simultaneously. That means that *temp_a* on line 1 is updated after a delta time (ΔT). Then *temp_a* is given a value on line 2, but it takes another delta time until *diff* is updated (Table 4.2).

Variable	T - ΔT	T	T + ΔT
in_1	2	3	3
temp_a	old value	2	3
diff	old value	old value	0

Table 4.2: Parallel assignment statement.

The big difference between a signal and a variable is that the signal is only assigned its value after a delta delay, while the variable is given its value immediately.

Note also that no time elapses in the sequential part of VHDL. It is only when a **wait** statement or **end** (if a sensitivity list is used) is reached in the sequential part that delta delays, if any, take place and the simulator clock can run on. This is true even if there are thousands of lines in the sequential code:

Example:

```
process(a,b ...)
variable c: std_logic_vector(1 downto 0);
begin
  if a=12 then ...  -- Line 1
    e<="01";        -- Line 2
  ...
  if b<=10 then     -- Line 876
  ...
  c:="10";          -- Line 2876
  ...
end process;
```

If the above process is activated at time 22 ns + 4 delta, all lines in the process will be executed during that delta. Signal *e* will be assigned the value "01" at time 22 ns + 4 delta, just as the variable *c* will be assigned the value "10" at 22 ns + 4 delta. Signal *e* then updates its value to "01" at time 22 ns + 5 delta, while variable *c* is updated immediately, i.e. at 22 ns + 4 delta, to the value "10". The lines in the sequential VHDL code are executed line by line. That is why it is called sequential VHDL. In concurrent VHDL code, the lines are only executed when an event on the sensitivity list occurs.

Example:

Sum1 and sum2 are signals:

```
p0:process
begin
   wait for 10 ns;
   sum1<=sum1+1;
   sum2<=sum1+1;
end process;
```

Sum1 and sum2 are variables:

```
p1:process
variable sum1, sum2: integer;
begin
   wait for 10 ns;
   sum1:=sum1+1;
   sum2:=sum1+1;
end process;
```

Sum1, Sum2 = Signals

Time	Sum 1	Sum 2
0	0	0
10	0	0
10 + 1 delta	1	1
20	1	1
20 + 1 delta	2	2
30	2	2
30 + 1 delta	3	3

Sum1, Sum2 = Variables

Time	Sum 1	Sum 2
0	0	0
10	1	2
10 + 1 delta	1	2
20	2	3
20 + 1 delta	2	3
30	3	4
30 + 1 delta	3	4

Table 4.3: Signals versus variables.

As the above example shows, the behaviour is completely different depending on whether we use signals or variables. The above example cannot be synthesized, as **wait for** 10 ns is used. If it had been synthesized, the difference in behaviour between the two would have persisted after synthesis. It is therefore very important to know when to use concurrent commands and when to use sequential ones. It is not possible to say in general whether signals or variables should be used, the decision is totally dependent on what function is required. Variables, however, cannot transfer information outside the sequential part of VHDL in which it is declared, in this case process *p1*. If access is needed to the value of *sum1* or *sum2* as, for example, an output signal or in other parts of the architecture, they must be declared as signals or the value of the variable assigned to a signal.

Example:

```
Entity ex is
 port(sum1_sig, sum2_sig: out integer);
end;

Architecture behv of rtl is
begin
  p1:process
  variable sum1, sum2: integer;
```

```
    begin
      wait for 10 ns;
      sum1:=sum1+1;
      sum2:=sum1+1;
      sum1_sig<=sum1;
      sum2_sig<=sum2;
    end process;
  end;
```

Variables can only store temporary values inside a **process, function** or **procedure.**

The above reasoning regarding variables has been changed slightly in the VHDL-93 standard. Global variables (**shared variables**) have been introduced in VHDL-93 which can transfer information outside the process.

Example (VHDL-93):

```
Architecture behv of ex is
shared variable v: std_logic_vector(3 downto 0);
begin
  p0:process(a,b)
  begin
    v:=a & b;
  end process;

  p1:process(d)
  begin
    if v="0110" then
      c<=d;
    else
      c<=(others=>'0');
    end if;
  end process;

  ...
end;
```

Global variables are not accessible in the concurrent part of VHDL code, but only inside other processes. Nor can a global variable be included in a sensitivity list of a process. Global variables can give rise to non-determinism. An IEEE committee for VHDL is trying to solve this problem. When global variables were introduced in the VHDL-93 standard, the problem had still not been solved. Synthesis tools do not support global variables either. Great care must therefore be taken when using these variables.

4.3. Process statement

The **process** concept comes from software and can be compared to a sequential program. If there are several processes in an architecture, they are executed concurrently.

A process can be in Waiting or Executing state (Figure 4.1).

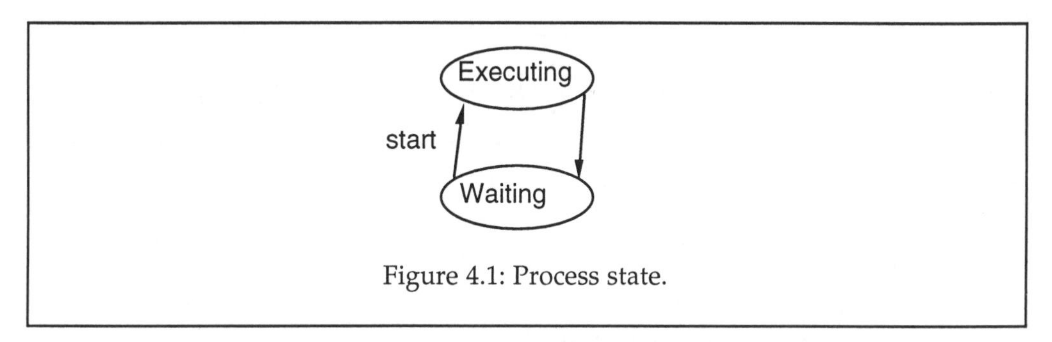

Figure 4.1: Process state.

If the state is Waiting, a condition must be satisfied, e.g. **wait until** clk='1';. This means that the process will start when clk has a rising edge. Then the process will be Executing. Once it has executed the code, it will wait for the next rising edge.

The syntax for the process is:

[<process_name> :] **process** [(<sensitivity_list >)]
    [<process_declarative_part>]
begin
    <process_statement_part>
end process [<process_name>];

Syntax (VHDL-93):

[<process_name> :] **process** [(<sensitivity_list >)] [is]
    ...
begin
    ...
end process [<process_name>];

Once the process has started, it takes the time delta t (the simulator's minimum resolution) for it to be moved back to Waiting state. This means that no time is taken to execute the process. A process should also be seen as an infinite loop between **begin** and **end process**;.

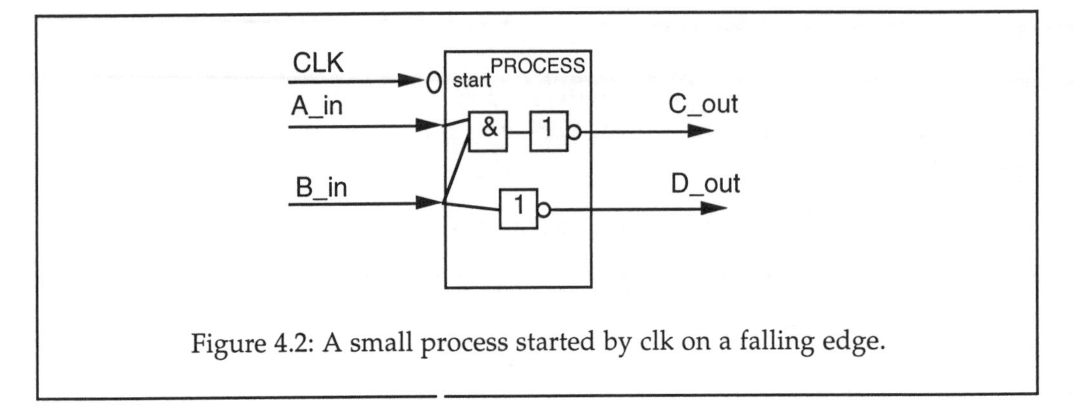

Figure 4.2: A small process started by clk on a falling edge.

The process in Figure 4.2 can be described as follows:

```
sync_process:process
begin
  wait until clk='0';
  c_out <= not (a_in and b_in);
  d_out <= not b_in;
end process sync_process;
```

The process is started when the signal clk goes low. It passes through two statements and waits for the next edge. The programmer does not have to add loop in the process: it will restart from the beginning in any case. In the model these two statements will be executed in a delta time which is equal to the resolution of the simulator. In practice the statements will take a different length of time to implement. This delay in the process can be modelled as follows:

```
c_out <= not(a_in and b_in) after 20 ns;
d_out <= not a_in after 10 ns;
```

This means that *c_out* will be affected 20 ns and *d_out* 10 ns after the start of the process. The simulator will place the result in a time event queue for *c_out* and *d_out* (Figure 4.3).

A_in = 1
B_in = 0

C_out =1 time=x +20ns

D_out =1 time=x +10ns

Figure 4.3: Time event queue in the simulator at time x.

If *a_in* or *b_in* is changed and *clk* gives a falling edge, another event will be linked into the queue relative to the time at which the change took place. If this is compared with the program area, there is at least one more state, namely ready. Ready is the state that the process is in when it needs a resource. In electronics, this part is omitted as the process already has access to a resource. The next part which differs is that the process is in EXECUTING state during a delta time. Delta time is used to enable the simulator to handle concurrency. In reality, delta time is equal to zero.

There are two types of process in VHDL:

Combinational processes
Clocked processes

Combinational processes are used to design combinational logic in hardware. Clocked processes result in flip-flops and possibly combinational logic.

4.3.1. Combinational processes

In a combinational process all the input signals must be contained in a sensitivity list (all to the right of <= or in **if/case** expression). If a signal is left out, the process will not behave like combinational logic in hardware. In real hardware, the output signals can change if one or more input signals to the combinational block are changed. If one of these signals is omitted from the sensitivity list, the process will not be activated when the value of the omitted signal is changed, with the result that a new value will not be assigned to the output signals from the process. The VHDL standard permits signals to be omitted from the sensitivity list. It is only in the design of simulatable models, and not for synthesis, that there is any point in omitting a signal from the sensitivity list. If a signal is omitted from the sensitivity list in a VHDL design for synthesis, the VHDL simulation and the synthesized hardware will behave differently. This is a serious error, as the VHDL code is meant to describe the hardware functionally.

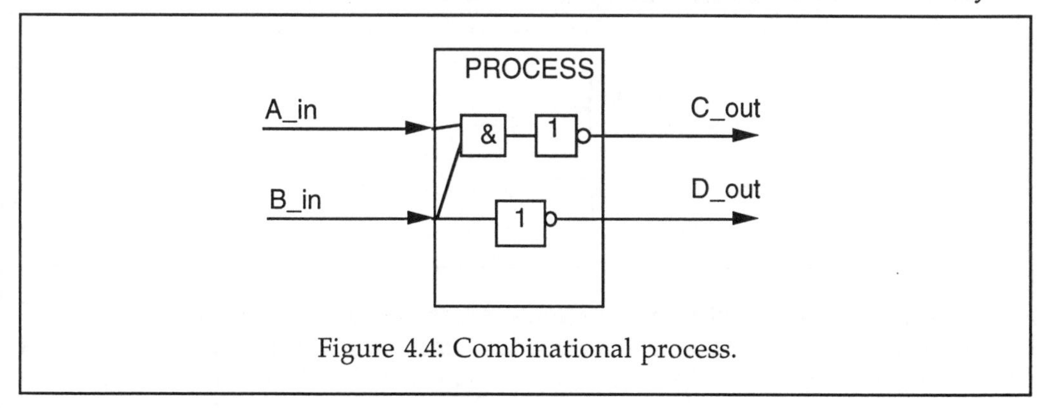

Figure 4.4: Combinational process.

The process in Figure 4.4 can be described as follows:

```
comb_process:process(a_in,b_in)
begin
  c_out <= not(a_in and b_in) after 20 ns;
  d_out <= not b_in after 10 ns;
end process comb_process;
```

In brackets after the process statement is a list of signals which start the process if there is a change.

If we omit *a_in* from the sensitivity list, *c_out* will retain its value and not update the value until *b_in* changes value. This will lead to the mismatch between VHDL simulation and hardware described above.

Example (bad):

```
comb_process:process(b_in)
begin
  c_out <= not(a_in and b_in) after 20 ns;
  d_out <= not b_in after 10 ns;
end process comb_process;
```

Combinational processes are good for dataflow design, e.g. data paths in a CPU.

Example (combinational process):

```
process (a,b,c)
begin
  d <= (a and b) or c;
end process;
```

The synthesis result is shown in Figure 4.5.

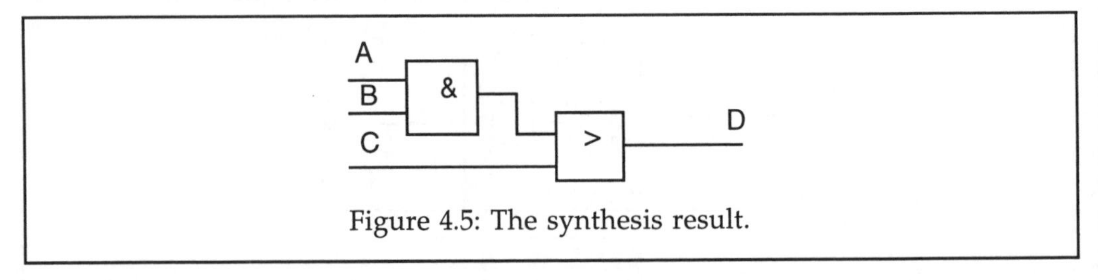

Figure 4.5: The synthesis result.

In the case of design with combinational processes, all the output signals from the process must be assigned a value each time the process is executed. If this condition is not satisfied, the signal will retain its value. The synthesis tool will perceive and resolve this requirement by inferring a latch for the output which is not assigned a value throughout the process. The latch will be closed when the old value for the signal has to be retained. Functionally, the VHDL code and the hardware will be identical. The aim of a combinational process is to generate combinational logic. If latches are inferred, the timing will deteriorate and the number of gates will increase. What is more, the latch will normally break the test rules for generating automatic test vectors (see Chapter 12, "Test methodology"). Processes which give rise to latches by mistake are called incomplete combinational processes (see Chapter 14, "Common design errors in VHDL and how to avoid them"). The solution is simple: include all the signals which are "read" inside the process in the sensitivity list for combinational processes.

4.3.2. Clocked processes

Clocked processes are synchronous and several such processes can be joined with the same clock. It is known that no process can start unless a falling edge occurs at the clock (*clk*). This means that data are stable when the clock starts the processes, and the next value is laid out for the next start of the processes.

Process A's output signal *d_out* is connected to the input of process B. The requirement is that *d_out* must be stable before the clock starts the processes. The longest time for process B to give the signal *d_out* a stable value once the clock has started determines the shortest period for the clock. In VHDL simulators this time is 10 ns (in this example).

The VHDL code for Figure 4.6 is:

```
A_process:process
begin
  wait until clk='0';
  c_out <= not(a_in and b_in);
  d_out <= not b_in after 10 ns;
end process A_process;

B_process:process
begin
  wait until clk='0';
  e_out <= not(d_out and c_in);
  f_out <= not c_in;
end process B_process;
```

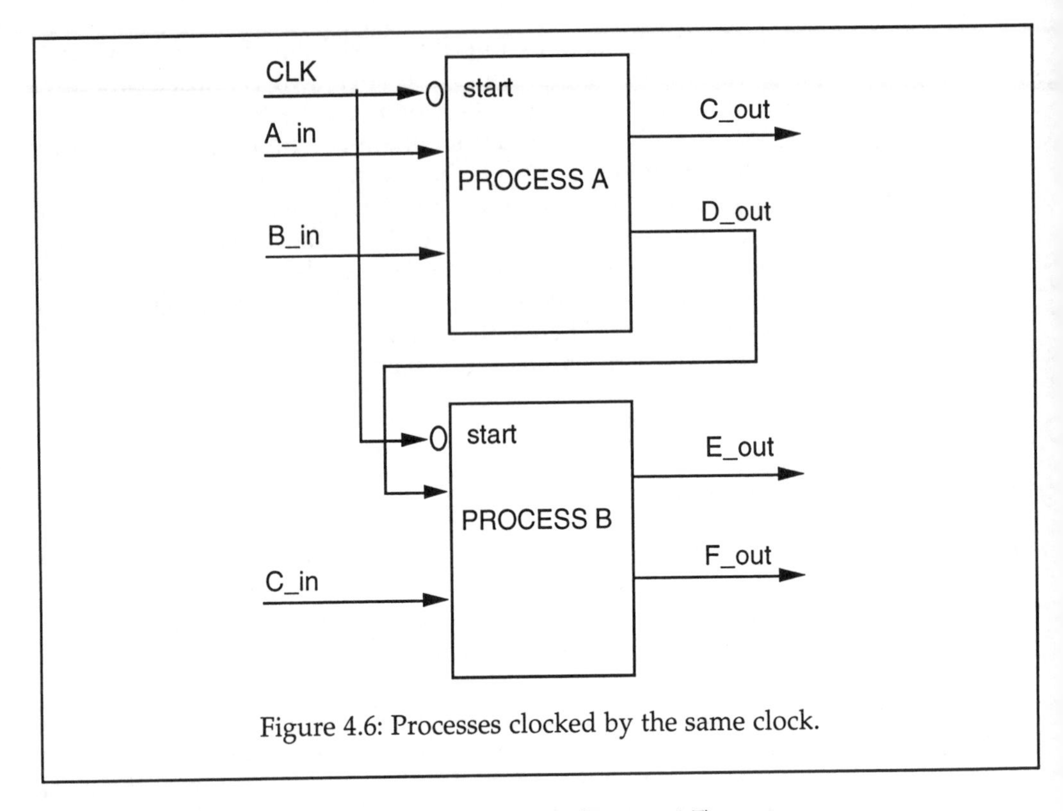

Figure 4.6: Processes clocked by the same clock.

These two processes can form a **component** (Figure 4.7).

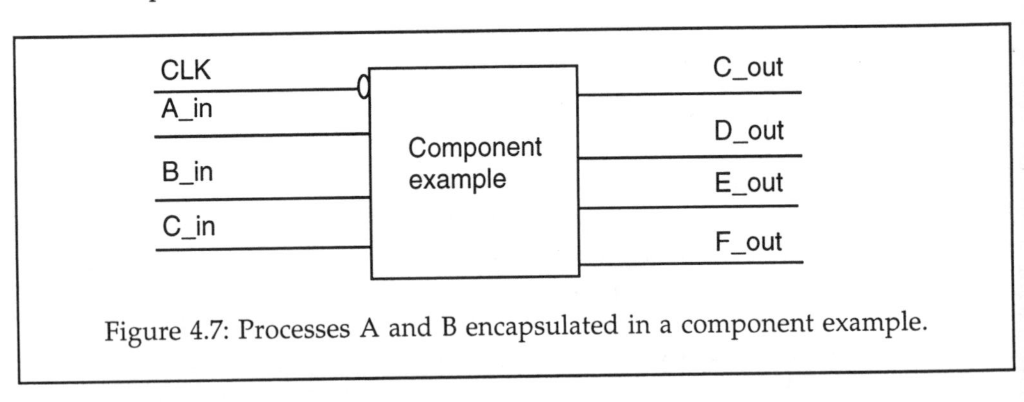

Figure 4.7: Processes A and B encapsulated in a component example.

```
Entity comp_ex is
   port( clk, a_in, b_in, c_in :  in    std_logic;
         d_out :                  out   std_logic;
         c_out, e_out, f_out :    out   std_logic);
end;
```

```
Architecture rtl of comp_ex is
--* internal signal declaration --*
begin
 A_process:process
 begin
 ...
 end process A_process;
 B_process:process
 begin
 ...
 end process B_process;
end;
```

Clocked processes lead to all the signals assigned inside the process resulting in a flip-flop.

Example of a clocked process:

```
example:process
begin
  wait until clk='1';
  dout <= din;
end process example;
```

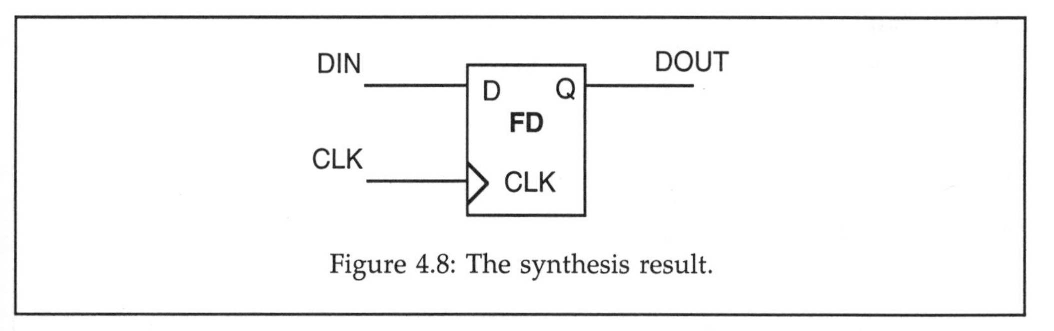

Figure 4.8: The synthesis result.

The preceding example shows how a clocked process is translated into a flip-flop. As Figure 4.8 shows, *din* will be translated to *dout* immediately. When it has been synthesized, the transfer will be equal to the flip-flop's specification. A more faithful model would have dout<=din **after** 1 ns.

Variables can also give rise to flip-flops in a clocked process. If a variable is read before it is assigned a value, it will result in a flip-flop for the variable.

Example:

```
process
variable count: std_logic_vector(1 downto 0);
begin
  wait until clk='1';
  count:=count + 1;
  if count="11" then
    q<='1';
  else
    q<='0';
  end if;
end process;
```

In the above example with variable count the synthesis result will show three flip-flops. One for signal *q* and two for variable *count*. This is because we read variable count before we assign it a value (count:=count + 1;).

If a signal is not assigned a value in a clocked process, the signal will retain the old value. The synthesis result from such a description will result in feedback of the signal from the output signal from the flip-flop via a multiplexor to itself (Figure 4.8). This method leads to the design remaining synchronous and testable (see Chapter 12, "Test methodology"). Alternatively, the clock could have been gated with signal *en* (Figure 4.9). However, this alternative is much worse than the first from the point of view of testability and design methodology.

Example (good):

```
process
begin
   wait until clk='1';
   if en='1' then
      q<=d;
   end if;
end process;
```

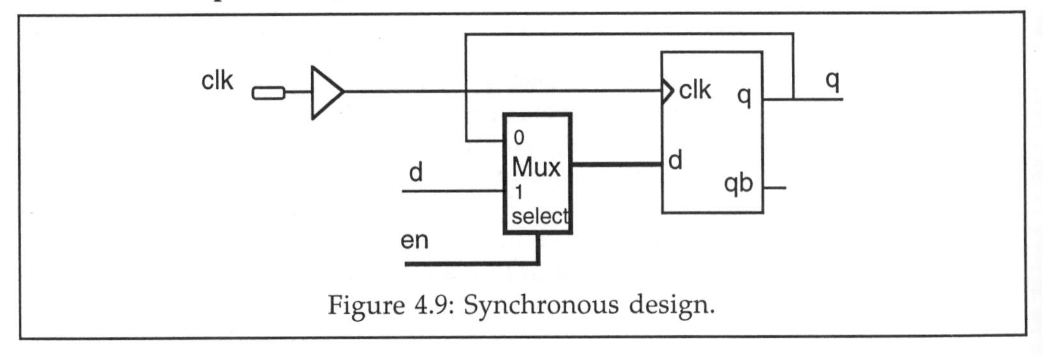

Figure 4.9: Synchronous design.

An example that is not so good, but which is still valid VHDL code, would be:

```
clk2<=clk and en;
process
begin
  wait until clk2='1';
  q<=d;
end process;
```

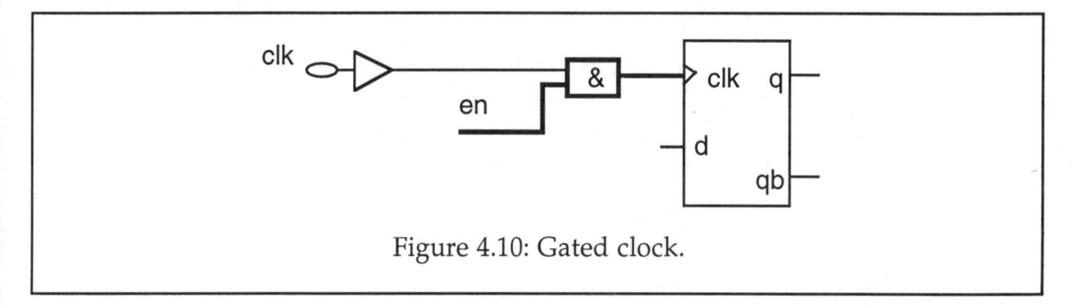

Figure 4.10: Gated clock.

A clocked process can result in combinational logic as well as flip-flops. It is also permissible and advisable for the assignment of the flip-flop's output to contain tests and complex expressions. This leads to less code and more readable code, and sometimes to a better synthesis result.

Example:

```
process(clk,resetn)
begin
 if reset='1' then
   q<=(others=>'0');
 elsif clk'event and clk='1' then
   if en='1' then
     q<=a + b;
   end if;
 end if;
end process;
```

All logic caused by a signal assignment in a clocked process will end up on the "left" of the flip-flops (before the flip-flop's input). The above example will therefore have an adder and a multiplexor at the flip-flop's input.

This leads to the logic structure shown in Figure 4.11, i.e. logic block B1 is included in the clocked process which describes flip-flop FD1. The VHDL code which describes logic block B2, on the other hand, should be included in a combinational process. It is possible, however, to include logic block B2 in the clocked process as well, but this is not advisable.

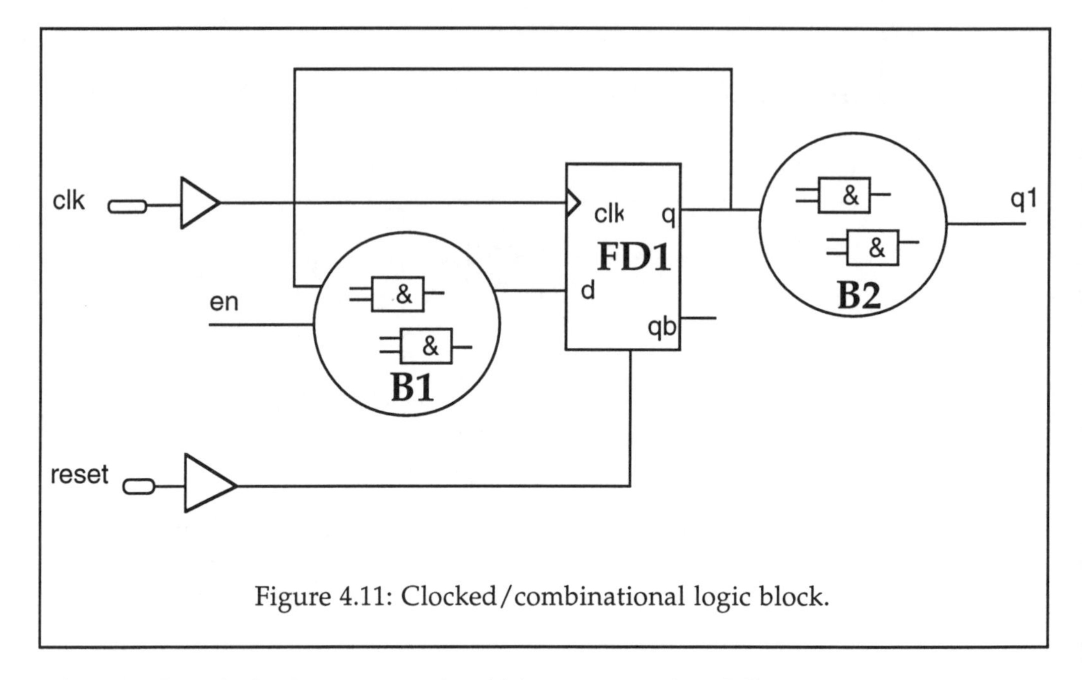

Figure 4.11: Clocked/combinational logic block.

The VHDL code for Figure 4.11 should be structured as follows:

```
Architecture rtl of ex is
signal q: std_logic;
begin
  FD1_B1:process(clk,reset)
  begin
    if reset='1' then
      q<='0';
    elsif clk'event and clk='1' then
      if en='1' then
        q<=....              -- some boolean expression (B1)
      end if;
    end if;
  end process;
  q1<=q and ...             -- some boolean expression (B2)
end;
```

Logic block B2 could also have been described in a combinational process.

4.4. If statement

An **if** statement chooses one or none of a sequence of statements. The choice is dependent on one or more conditions.

The **if** command corresponds to the **when else** command in the concurrent part of VHDL.

Syntax:

if <condition> **then** <sequence_of_statements>
[**elsif** <condition> **then** <sequence_of_statements>] ....
[**else** <sequence_of_statements>]
end if;

<condition>=
<expression> [<relational_operator> <expression>]..

<relational_operator>=
= /= < <= > >=

Example:

```
if sel = '1' then
  c <= b;
else
  c <= a;
end if;
```

The synthesis result is shown in Figure 4.12.

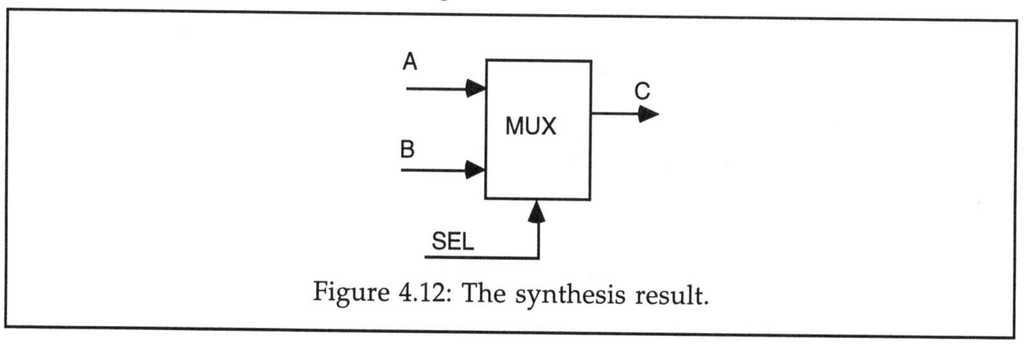

Figure 4.12: The synthesis result.

Example:

```
if sel = '1' then
  c<='1'; ...
```

```
elsif John = "1001" then
   c <= '0'; ...
elsif David > John then
   c <= '1'; ...
else
   c <= 'X'; ...
end if;
```

Several **else-ifs** are permitted, but just one **else**. Note **ELSIF** and **END IF**.

Example:

```
Entity ex_if is
   port( ssel:      in    std_logic;
         a,b,syn:   in    std_logic;
         sout:      out   std_logic);
end;

 Architecture rtl of ex_if is
 begin
  process(ssel,syn,a,b)
  begin
    if ssel='0' and syn='1' then
      sout<=a;
    else
      if ssel='1' and syn='0' then
        sout<=b;
      else
        sout<='0';
      end if;
    end if;
  end process;
end;
```

Example:

```
Entity ex_if2 is
 port (a,b:  in    integer;
          c:  out   boolean);
 end;

Architecture rtl of ex_if2 is
begin
   p1:process(a,b)
   begin
```

```
    if a>b then
        c<=true;
    else
        c<=false;
    end if;
  end process;
end;
```

4.5. Case statement

The **case** command corresponds to the **with select** command in the concurrent part of VHDL. Using a **case** statement often makes the VHDL code more readable.

```
Syntax:
case <expression> is
      when <choice> => <sequence_of_statements> ;
      when <choice> => <sequence_of_statements> ;
      ...
      [when others => [<sequence_of_statements>] ;]
end case ;
<choice>= <choice> [ | <choice> [ ...]]
| = or
```

Example:

```
Entity case_mux is
 port( a,b,sel:  in    bit;
       c:        out   bit);
end;

Architecture rtl of case_mux is
begin
  p1:process(sel,a,b)
  begin
    case sel is
      when '0' => c<=a;
      when '1'=>  c<=b;
    end case;
  end process;
end;
```

All the choices in **case** statements must be enumerated. It is not permissible for the same choice to be included in several **when** statements, i.e. to be enumerated several times. This also means that choices must not overlap. If the **case** expression contains many values, it may be laborious to enumerate all of them. In this case, **others** can be used.

Example:

```
Entity case_ex2 is
   port( a:  in   integer range 0 to 30;
         b:  out  integer range 0 to 6);
end;
Architecture rtl of case_ex2 is
begin
   p1:process(a)
   begin
      case a is
        when 0 =>          b<=3;
        when 1 | 2=>       b<=2;
        when others   =>   b<=0;
      end case;
   end process;
end;
```

Data type std_logic is usually used in design. As std_logic can assume nine different values, all the values must be covered in the **case** statement. This is also done using **others**.

Example:

```
Entity case_ex3 is
   port( a,b,sel: in    std_logic;
         c:       out   std_logic);
end;
Architecture rtl of case_ex3 is
begin
  p1:process(sel,a,b)
  begin
    case sel is
     when '0' =>       c<=a;
     when others=>  c<=b;
    end case;
  end process;
end;
```

If input signal *sel* in the above example assumes a value other than '0', signal *c* will be assigned the value of signal *b*. In VHDL simulation this value can be one of the other eight values which data type std_logic can assume. In hardware it can only be the value '1'. The synthesis tool understands that hardware can only assume the values '0' and '1', which leads to a correct synthesis result, i.e. a multiplexor.

It is also permissible to define ranges in the choice list. This can be done using **to** or **downto**.

Example:

```
Entity case_ex4 is
   port( a: in   integer range 0 to 30;
         q: out  integer range 0 to 6);
end;

Architecture rtl of case_ex4 is
begin
 p1:process(a)
 begin
   case a is
      when 0 =>                q<=3;
      when 1 to 17=>           q<=2;
      when 23 downto 18=>      q<=6;
      when others =>           q<=0;
   end case;
 end process;
end;
```

It is not permissible to define a **range** with a vector, as a vector does not have a range.

Example (bad):

```
Entity case_ex5 is
  port( a:  in   std_logic_vector(4 downto 0);
        q:  out  std_logic_vector(2 downto 0));
end;

Architecture rtl of case_ex5 is
begin
 p1:process(a)
 begin
```

```
    case a is
      when "00000" =>                q<="011";
      when "00001" to "11110"=> q<="010";        -- Error
      when others =>                 q<="000";
    end case;
  end process;
end;
```

If a range is wanted, std_logic_vector must first be converted to an integer. This can be done by declaring a variable of type integer in the process and then converting the vector using conversion function conv_integer. The same applies if you want to use an integer in the choice list for a vector.

Example (good):

```
Entity case_ex6 is
  port( a:  in   std_logic_vector(4 downto 0);
        q:  out  std_logic_vector(2 downto 0));
end;

Architecture rtl of case_ex6 is
begin
 p1:process(a)
 variable int: integer range 0 to 31;
 begin
    int:=conv_integer(a);

    case int is
      when 0 =>        q<="011";
      when 1 to 30=>  q<="010";
      when others => q<="000";
    end case;
 end process;
end;
```

The synthesis result is not affected by use of a conversion function. The decision as to whether a conversion function is to be used should be based on the documentation aspect and whether it is easier to write the VHDL code if an integer is used in the choice list.

If you want to include several vectors in a choice list in the **case** statement, it is not permissible to use concatenation to combine the vectors as one choice. This is because the **case** <expression> must be static.

Example (bad):

```
Entity case_ex7 is
  port( a,b: in   std_logic_vector(2 downto 0);
        q:   out  std_logic_vector(2 downto 0));
end;

Architecture rtl of case_ex7 is
begin
 p1:process(a,b)
 begin
   case a & b is                          -- Error
      when "000000" =>  q<="011";
      when  "001110"=>  q<="010";
      when others =>    q<="000";
   end case;
 end process;
end;
```

The solution is either to introduce a variable in the process to which the value *a & b* is assigned or to use what is known as a qualifier for the subtype.

Example (good):

```
Architecture rtl of case_ex8 is
begin
 p1:process(a,b)
 variable int: std_logic_vector(5 downto 0);
 begin
   int:=a & b;
   case int is
      when "000000" => q<="011";
      when "001110" => q<="010";
      when others =>   q<="000";
   end case;
 end process;
end;
```

Example (good):

```
Architecture rtl of case_ex9 is
begin
 p1:process(a,b)
 subtype mytype is  std_logic_vector(5 downto 0);
```

```
 begin
   case mytype'(a & b) is
      when "000000" => q<="011";
      when "001110" => q<="010";
      when others =>   q<="000";
   end case;
 end process;
end;
```

It is also permissible to give the **subtype** directly in the case expression:

4.6. Multiple assignment

In the concurrent part of VHDL you are always given a driver for each signal assignment, which is not normally desirable. In the sequential part of VHDL it is possible to assign the same signal several times in the same process without being given several drivers for the signal. This method of assigning a signal can be used to assign signals a default value in the process. This value can then be overwritten by another signal assignment. The following two examples are identical in terms of both VHDL simulation and the synthesis result.

Example 1:

```
Architecture rtl of ex1 is
begin
 p1:process(a)
 begin
   case a is
      when "00" =>      q1<='1';
                        q2<='0';
                        q3<='0';

      when "10" =>      q1<='0';
                        q2<='1';
                        q3<='1';

      when others =>    q1<='0';
                        q2<='0';
                        q3<='1';
   end case;
 end process;
end;
```

Example 2:

```
Architecture rtl of ex2 is
begin
 p1:process(a)
 begin
   q1<='0';
   q2<='0';
   q3<='0';
   case a is
      when "00" =>      q1<='1';
      when "10" =>      q2<='1';
                        q3<='1';
      when others  => q3<='1';
   end case;
 end process;
end;
```

If we compare the above examples, we see that there are fewer signal assignments in example 2. The same principle can be applied if, for example, **if-then-else** is used. The explanation for this is that no time passes inside a process, i.e. the signal assignments only overwrite each other in sequential VHDL code. If, on the other hand, the same signal is assigned in different processes, the signal will be given several drivers.

4.7. Null statement

In VHDL there is a statement which means "do nothing". This command can, for example, be used if default signal assignments have been used in a process and an alternative in the **case** statement must not change that value.

Example:

```
Architecture rtl of ex is
begin
 p1:process(a)
 begin
   q1<='0';
   q2<='0';
   q3<='0';
   case a is
      when "00" =>    q1<='1';
```

```
        when "01" =>  q2<='1';
                      q3<='1';

        when others => null;
      end case;
    end process;
  end;
```

In the above example, **null** could have been left out. Readability is increased if the **null** command is included. If **null** is omitted, there is a risk that, if someone else reads the VHDL code, they will be uncertain whether the VHDL designer has forgotten to make a signal assignment or whether the line should be empty.

4.8. Wait statement

There are four different ways of describing a **wait** statement in a process:

1. **process**(a,b)
2. **wait until** a=1;
3. **wait on** a,b;
4. **wait for** 10 ns;

The first and the third are identical if **wait on** a,b is placed at the end of the process. The following two processes are therefore identical.

Example 1:

```
Process(a,b)
begin
   if a>b then
      q<='1';
   else
      q<='0';
   end if;
end process;
```

Example 2:

```
Process
begin
   if a>b then
      q<='1';
   else
      q<='0';
   end if;
   wait on a,b;
end process;
```

In example 1 the process will be triggered each time that signal *a* or *b* changes value (a'event or b'event). **Wait on** a,b; has to be placed at the end of example 2 to be identical with example 1 because all processes are executed at start-up until they reach their first **wait** statement. If a sensitivity list had been used, it would, according to the VHDL standard, be the same as placing **wait on** at the end of the process. If **wait on** is placed anywhere else, the output signal´s value will be different when simulation is started (see the **wait** example below).

If a sensitivity list is used in a process, it is not permissible to use a **wait** command in the process. It is permissible, however, to have several **wait** commands in the same process.

Wait until a='1'; means that, for the **wait** condition to be satisfied and execution of the code to continue, it is necessary for signal a to have an event, i.e. change value, and the new value to be '1', i.e. a rising edge for signal *a*.

Wait on a,b; is satisfied when either signal *a* or *b* has an event (changes value).

Wait for 10 ns; means that the simulator will wait for 10 ns before continuing execution of the process. It is also permissible to use the **wait for** command as follows:

```
constant period:time:=10 ns;
wait for 2 * period;
```

Alternatives 2, 3 and 4 can also be combined into: **wait on** a **until** b='1' **for** 10 ns, but alternative 1 must never be combined with 2, 3 or 4. Suppose that the following **wait** command is used:

```
wait until a='1' for 10 ns;
```

The **wait** condition will be satisfied when *a* changes value or after a wait of max. 10 ns. The condition can therefore be regarded as an **or** condition, i.e. wait until signal *a* changes value or continue after 10 ns if signal *a* has still not changed value.

There follow below a number of examples of different **wait** commands and, in Figure 4.13, their simulation results. To ensure better comprehension of the example, the necessary VHDL knowledge will be repeated:

- At time 0 all signals are assigned the value type'left. This means that every signal is given the value declared at the far left of its data type declaration. Thus a signal of type bit is given the value '0', as bit is declared as type bit is ('0', '1');. All processes will also start to execute at time 0. All the processes will then execute line by line until a **wait** statement is reached.

- As we see in the wait example below, examples 1 and 2 are identical. In example 3 **wait on** a has not been placed at the end of the process, leading to a different value for output signal *c3* at start-up. In example 4 the **wait** condition is true only if signal *a* has a positive edge and is given the value '1'. The inverse of '1' is '0', so *c4* will be assigned the value '0' at times of 20 ns and 47 ns. In example 5, which is the most complex, the wait example is true after 10 ns or when *a* changes value.

Example (a,c1..c5=bit):

Example 1
```
process(a)
begin
  c1<=not a;
end process;
```

Example 2
```
process
begin
  c2<=not a;
  wait on a;
end process;
```

Example 3
```
process
begin
  wait on a;
  c3<=not a;
end process;
```

Example 4
```
process
begin
  wait until a='1';
  c4<=not a;
end process;
```

Example 5
```
process
begin
  c5<=not a;
  wait until a='1' for 10 ns;
end process;
```

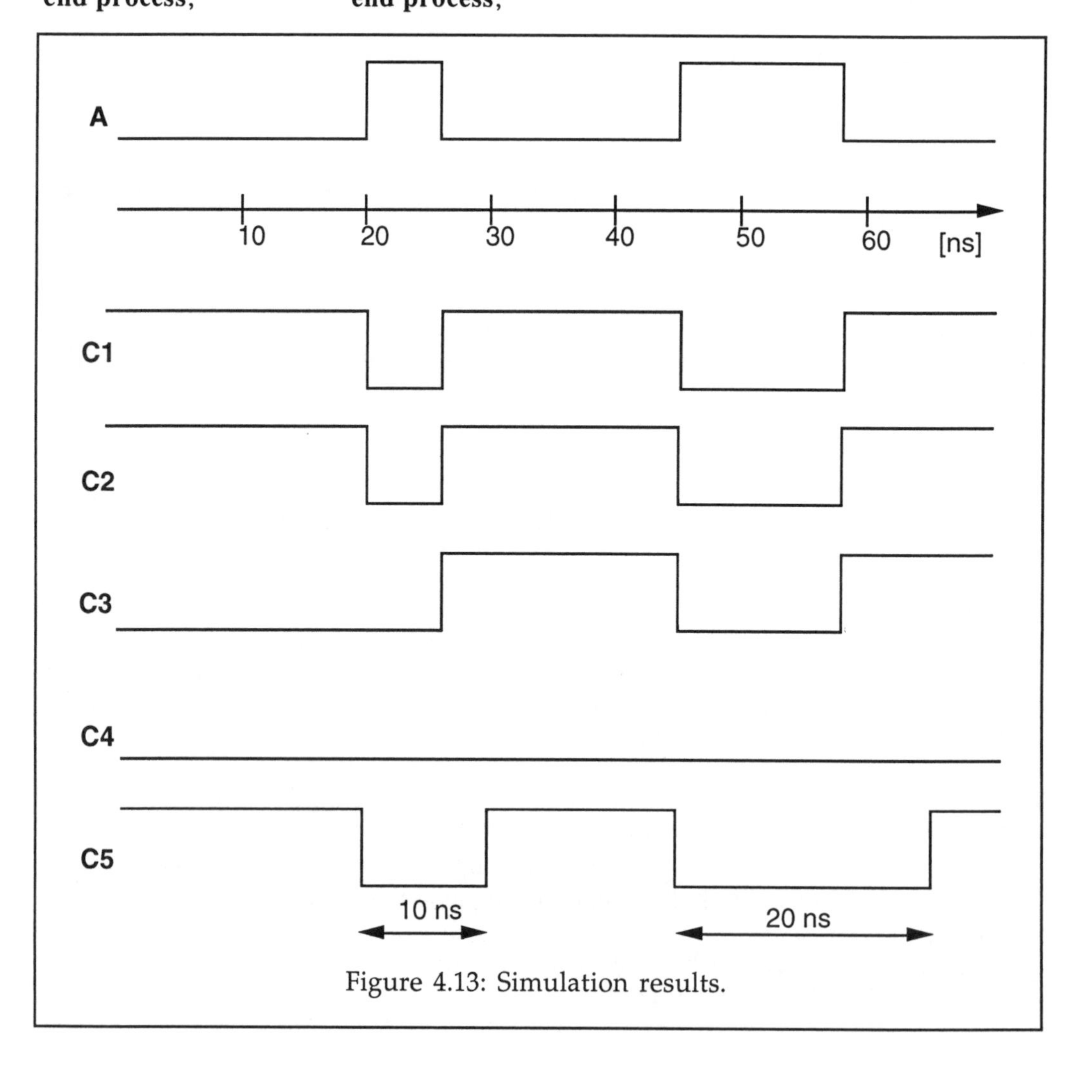

Figure 4.13: Simulation results.

Synthesis tools do not support use of **wait for** 10 ns;. This description method produces an error in the synthesis. What is more, the majority of synthesis tools only accept a sensitivity list being used for combinational processes, i.e. example 1, but not example 2, can be synthesized. Some advanced systems, however, will accept example 2, while example 3 is not permitted in synthesis. Example 4 is a clocked process and results in a D-type flip-flop after synthesis. Examples 3 and 5 can therefore only be used for simulation, and not for design.

As previously mentioned, it is permissible to have several waits inside the same process.

Example:

```
Process
begin
  wait until clk='1';
  ...
  wait until clk='1';
  ...
  wait until clk='1';
  ...
end process;
```

This description method is only accepted by a few synthesis tools, and those that do also require that only one clock should be used in the process (**wait until** clk='1';) and that the same clock edge is used in all the **wait** commands. The synthesis result will be reminiscent of a state machine, with the commands which come after the first **wait** statement being executed after the first clock edge and the commands which come after the second **wait** statement after the second clock edge, and so on. As the entire process command should be regarded as an infinite loop, the "program pointer" will jump back to the start of the process and continue execution when the end process; line is reached in the execution.

The wait command is a sequential command despite the fact that it is not permissible to use wait in functions. The **wait** command can, on the other hand, be used in procedures and processes.

4.9. Loop statement

There are two different **loop** statements in sequential VHDL:

- For loop
- While loop

4.9.1. For loop

Syntax:
[loop_label:] **for** <identifier> **in** <discrete_range> **loop**
  sequence_of_statement
end loop [loop_label];

The **loop** command is very useful when it comes to designing with one- or two-dimensional vectors.

Example:

```
Entity ex is
  port( a,b,c: in    std_logic_vector(4 downto 0);
             q:    out   std_logic_vector(4 downto 0));
end;

Architecture rtl of ex is
begin
  process(a,b,c)
  begin
    for i in 0 to 4  loop          -- Not allowed to declare the
                                   -- loop index 'i'.
      if a(i)='1' then
        q(i)<=b(i);
      else
        q(i)<=c(i);
      end if;
    end loop;
  end process;

end;
```

It is not permissible to change the value of the loop index variable in the process. The advanced system tools support the **for loop** command. They require the loop's range to be fixed.

The range in the **for loop** can be omitted, thereby making the loop an infinite loop. The loop can, however, be broken with an **exit** statement inside it. Each **loop** command can also be given a label.

Example:

```
Architecture rtl of ex is
begin
 process
 begin
  reset_loop: loop
    q<=(others=>'0');
    wait until  clk='1';
    exit reset_loop when  resetn='0';

    main_loop: loop
      wait until  clk='1';
      exit reset_loop  when  resetn='0';

      q<=a+b;
      wait until  clk='1';
      exit reset_loop when  resetn='0';

      while en='0' loop
        exit reset_loop when  resetn='0';
      end  loop:

      wait until  clk='1';
      exit reset_loop when  resetn='0';

      q<=b+c;

    end  loop;
  end  loop;
 end process;
end;
```

This description method is used a great deal in behavioural synthesis (see Chapter 17, "Behavioural synthesis").

4.9.2. While loop

Syntax:

[loop_label:] **while** <condition> **loop**

  sequence_of_statement

end loop [loop_label];

Example:

```
process(a,b,resetn)
variable i: integer range 0 to 31;
begin
  if resetn='0' then
    i:=0;
    q<=(others=>'0');
  else
    while q<12 loop
      if i=31 then
        exit;
      else
        i<=i+1;
        q<=a(i) + b(i) + q;
      end if;
    end loop;
  end if;
end process;
```

Only the advanced synthesis tools support **while loop**. The above example can only be simulated, however, not synthesized.

4.10. Postponed process

Postponed process is a new concept which was introduced in the VHDL-93 standard. Using postponed process it is possible to get a process to execute in the last delta cycle at a given time. As usual, a **postponed process** is activated by a sensitivity list or **wait** statement. The difference is that a **postponed process** does not start to execute at the delta event which activates the process, but waits until all signals are stable, i.e. there are no more delta enumerations at the time in question. This means that the process will be executed as the last event at that time. If several postponed processes are used, the VHDL-93 standard does not define which should be executed first. Use of more than one **postponed process** should therefore be avoided.

Postponed process is not supported for synthesis. It was primarily designers who make simulation models who wanted this addition to the new standard, the reason being that it is not permissible to write **wait for** 0 ns; in VHDL. With **postponed process** it is now possible to guarantee that signals will be assigned a value in the very last delta cycle, making it easier to describe certain simulation models.

Example:

```
Architecture behv of ex is
begin
 process
 begin
   ...
 end process;

 postponed process(a,b)
 begin
   if a>b then
     q<='1' after 5 ns;
   else
     q<='0' after 5 ns;
   end if;
 end process;
end;
```

A **postponed process** cannot give rise to any further delta delays. This means that all signal assignments in a **postponed process** must have an after delay.

4.11. Predefined signal attributes

Signal attributes are very useful when it comes to making VHDL models. In synthesizable VHDL code it is normal just to use the 'event attribute to check whether there is a clock edge. Some synthesis tools also use 'last_value to decide whether there is a clock edge (e.g. Mentor Graphics: Autologic 1).

An attribute is information which can be obtained on blocks, arrays, signals and types.

> Syntax:
>
> <name>'attribute_identifier

The commonest signal attributes are as follows:

<signal>'event
: returns value true or false if event occurred in present delta time period

<signal>'active
returns value true or false if activity occurred in present delta time period

<signal>'stable(t)
returns a signal value true or false if based on event in (t) time units

<signal>'quiet(t)
returns a signal value true or false if based on activity in (t) time units

<signal>'transaction
returns a event whenever there is an activity on the signal

<signal>'delay(t)
returns a signal delayed (t) time units

<signal>'last_event
returns amount of time since last event

<signal>'last_active
returns amount of time since last activity

<signal>'last_value
returns value equal to previous value

<signal>'range
returns the signal range, for example:

```
signal a:std_logic_vector(3 downto 0);
for a'range loop      -- 'range = 3 downto 0
    ...
end loop;
```

Attributes for types are, for example:

<type name>'left
returns the left array number

<type name>'right
returns the right array number

<type name>'high
returns the largest array number

<type name>'low
returns the smallest array number

<type name>'length
returns the data types length

Example:

```
subtype my_type1 is integer range 15 downto 0;
subtype my_type2 is integer range 0 to 15;
```

```
my_type1'left = 15          my_type2'left = 0
my_type1'right = 0          my_type2'right = 15
my_type1'high = 15          my_type2'high = 15
my_type1'low = 0            my_type2'low = 0
my_type1'length = 16        my_type2'length = 16
```

<type name>'succ(data value)
returns the next value in the data type of data value

<type name>'pred(data value)
returns the previous value in the data type of data value

<type name>'rightof(data value)
returns the value right of the data value

<type name>'leftof(data value)
returns the value left of the data value

Example:

```
type my_type3 is (John, Tom, Petra);
type my_type4 is my_type3 range Tom downto John;
```

```
my_type3'succ(John) = Tom        my_type4'succ(John) = Error
my_type3'pred(John) = Error      my_type4'pred(John) = Tom
my_type3'rightof(John) = Tom     my_type4'rightof(John) = Error
my_type3'leftof(John) = Error    my_type4'leftof(John) = Tom
```

4.12. Different clock descriptions in clocked processes

There are many different description methods for the clock for a flip-flop in a clocked process. The commonest methods are examined below.

```
Alt 1:  process(clk)
        begin
           if clk'event and clk='1' then
              q<=d;
           end if;
        end process;

Alt 2:  process(clk)              -- Alt 2 is not valid for asynchronous reset
        begin
           if clk='1' then
              q<=d;
           end if;
        end process;

Alt 3:  process(clk)
        begin
           if clk'event and clk='1' and clk'last_value='0' then
              q<=d;
           end if;
        end process;

Alt 4:  process
        begin
           wait until clk='1';
              q<=d;
        end process;

Alt 5:  process
        begin
           wait until prising(clk);        -- "prising" is a function
              q<=d;
        end process;
```

The choice of description method is dependent on which synthesis tool is to be used. Table 4.4 summarizes what are, perhaps, the four commonest synthesis tools and what they support.

	Synopsys	Autologic 1	Autologic 2	View Logic
Alt 1	Yes	No	Yes	Yes
Alt 2	Yes	No	Yes	No
Alt 3	No	Yes	Yes	No
Alt 4	Yes	Yes	Yes	Yes
Alt 5	No	Yes	Yes	Yes

Table 4.4: Different clock descriptions and what they support.

Only alternative 3 guarantees that the clock edge will go from '0' to '1' in the VHDL simulation. The other alternatives accept a clock edge from 'X' to '1', for example. The value 'X' does not exist in hardware, so in reality there is no trouble approving the edge 'X' to '1' as a clock edge in the VHDL simulation.

The clock descriptions which occur most frequently are alternatives 1, 2 and 4. All the alternatives produce the same synthesis result, however (Figure 4.14).

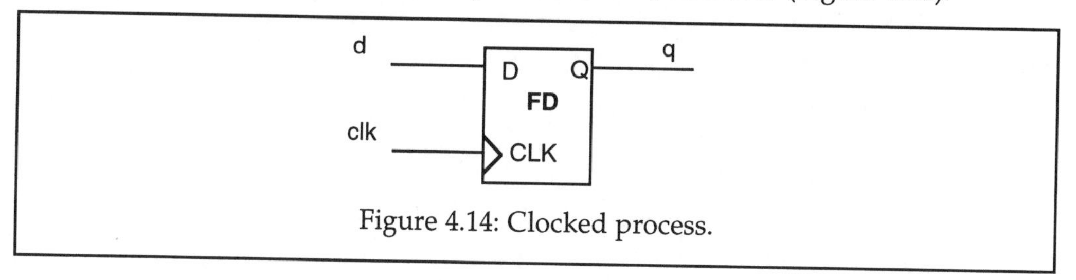

Figure 4.14: Clocked process.

Several synthesis tools contained a misunderstanding of the VHDL standard as far as clock descriptions were concerned in the early 1990s. They required the clock to be defined as:

```
wait until clk'event and clk='1';  -- Not good
```

This description is identical to the following description:

```
wait until clk='1';                -- Good
```

For the condition after **wait until** to be satisfied, there first has to be a signal event, i.e. clk'event, and then the value has to be equal to '1'. This leads to **wait until** clk'event **and** clk='1'; being unnecessarily complicated and on the verge of being incorrect (redundant information through clk'event being introduced after the wait until command). Unfortunately, this "erroneous" clock definition spread and can still be seen in books and manuals.

4.13. Asynchronous and synchronous reset

There are two different ways of resetting a flip-flop: asynchronous or synchronous reset. As previously mentioned, there are several ways of describing clocks in clocked processes in VHDL. How the clock edge is described controls how the reset for the flip-flop has to be described. The two commonest methods, which are supported by the majority of synthesis tools, are set out below.

4.13.1. Asynchronous reset

With asynchronous reset the flip-flop will be reset as soon as the reset is activated, irrespective of the clock event. This means that, if a sensitivity list is used in a clocked process which has to have an asynchronous reset, the reset must be included on the sensitivity list as well as the clock (alt 1). From a documentation point of view it is easy to see whether the clocked process has an asynchronous reset or not: you just have to check whether or not the reset is on the sensitivity list.

A general method of description which is supported by the majority of synthesis tools (e.g. Synopsys, Autologic 2) is:

Alt 1:
```
process(clk,reset)
begin
  if reset='1' then
    data<="00";
  elsif clk'event and clk='1' then
     data<=in_data;
  end if;
end process;
```

An alternative description method supported by View Logic, for example, is:

Alt 2:
```
process
begin
  wait until (prising(clk) or resetn='0');
  if resetn='0' then
    data<="00";
  else
    data<=in_data;
  end if;
end process;
```

Both description methods result in the same synthesis result (Figure 4.15). The choice should be based on which synthesis tool is to be used.

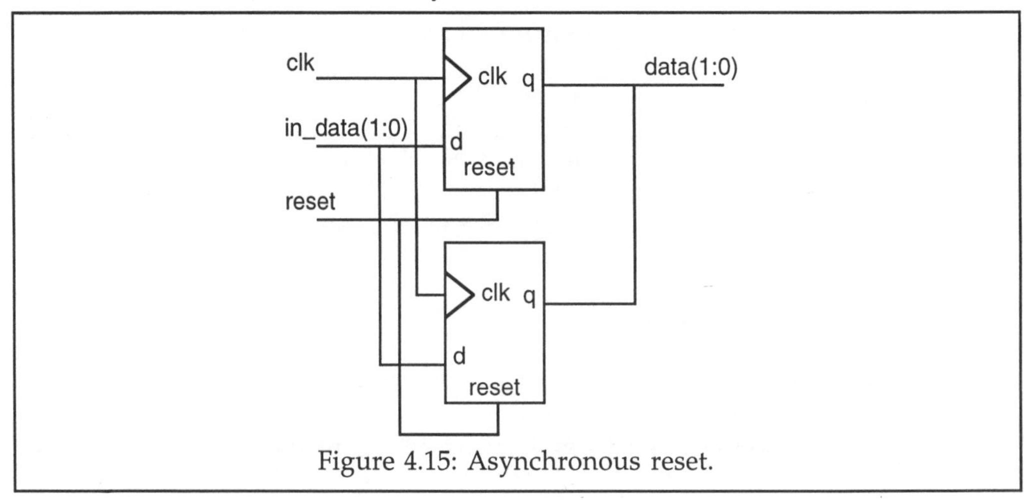

Figure 4.15: Asynchronous reset.

From the point of view of design methodology and documentation only one signal should be used for the clock or reset. Synthesis tools do not support several clocks in the same process. What is more, any gating of the clock should happen outside the process if it must happen at all (see Chapter 12, "Test methodology"). The same applies to the reset. If several signals can reset the flip-flop, they should be gated together outside the clocked process so that the above synthesis model can be followed.

4.13.2. Synchronous reset

With synchronous reset the flip-flop can only be reset at an active clock edge. This means that if a sensitivity list is used in a clocked process which has to have a synchronous reset, only the clock should be included on the sensitivity list (alt 1).

A general method of description which is supported by the majority of synthesis tools (e.g. Synopsys, Autologic 2) is:

Alt 1:

```
process(clk)
begin
   if clk'event and clk='1' then
      if reset='1' then
         data<="00";
      else
         data<=not in_data;
      end if;
   end if;
end process;
```

An alternative description method supported by View Logic, for example, is:

Alt 2:

```
process
begin
   wait until prising(clk);
   if reset='1' then
      data<="00";
   else
      data<=not in_data;
   end if;
end process;
```

The synthesis result of both methods is shown in Figure 4.16.

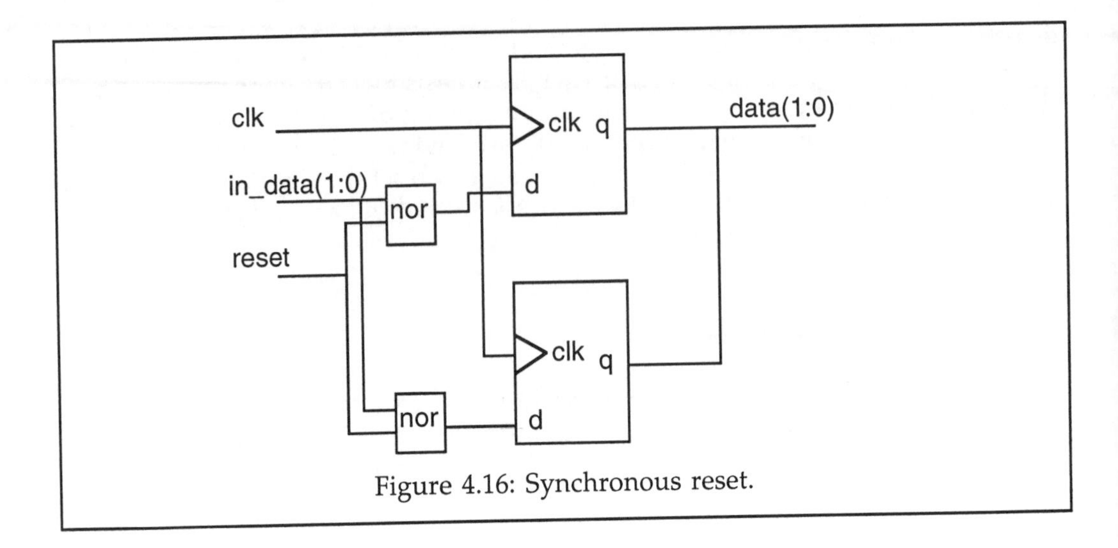

Figure 4.16: Synchronous reset.

4.14. Latches

If latches are to be described in VHDL, the process command is normally used. A normal latch (transparent latch) acts by propagating through the value from the d-input to the q-output when the enable signal is active. If the enable signal is then deactivated, the output will retain the value it had when the latch was closed. In VHDL latches are described with combinational processes. All signals which are read in the process must be on the sensitivity list. The output of the latch must only be assigned a value when its enable signal is active. This results in latches after synthesis.

Example:

```
process(enable,d_in)
begin
  if enable='1' then
    q<=d_in;
  end if;
end process;
```

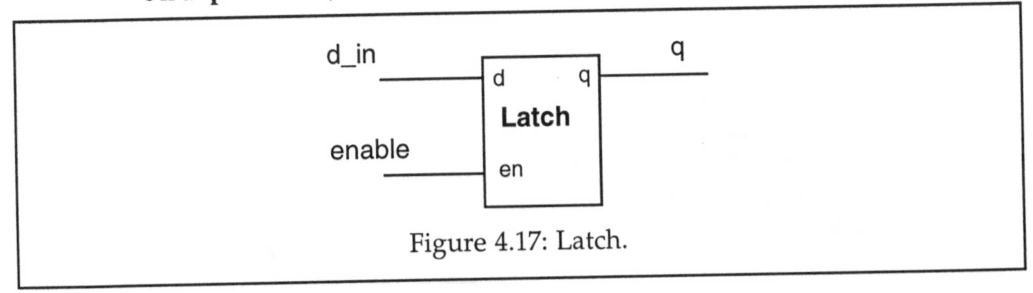

Figure 4.17: Latch.

4.15. Exercises

1. (a) What two main types of process are there?

(b) What kind of logic do they represent?

2. (a) What characterizes an incomplete combinational process?

(b) How do you avoid incomplete combinational processes?

3. (a) Why are there different methods of describing a clock edge?

(b) Give at least three alternatives.

4. What happens if you do not assign an output signal a value in a clocked process?

5. How many flip-flops does the following VHDL code give rise to?

(a)

```
Architecture rtl of ex is
signal a,b: std_logic_vector(3 downto 0);
begin
  process(clk)
  begin
    if clk='1' and clk'event then
      if q(3)/='1' then
        q<=a + b;
      end if;
    end if;
  end process;
end;
```

(b)

```
Architecture rtl of ex is
signal a,b: std_logic_vector(3 downto 0);
begin
  process(clk)
  variable int: std_logic_vector(3 downto 0);
  begin
    if clk='1' and clk'event then
      if int(3)/='1' then
        int:=a + b;
        q<=int;
      end if;
    end if;
  end process;
end;
```

6. (a) What are the two ways of zeroing a flip-flop?

   (b) If a sensitivity list is used for a clocked process, what type of reset should the reset signal have in the sensitivity list?

7. How many flip-flops will be given a clocked reset and how many an asynchronous reset in the following VHDL code?

```
process(clk,resetn)
begin
   if resetn='0' then
      q1<='0';
      q2<='0';
   elsif clk'event and clk='1' then
      q1<=a and b;
      q2<=c;
   end if;
end process;

process(clk)
begin
   if clk='1' then
      if resetn='0' then
         q3<='0';
      else
         q3<=q2;
      end if;
   end if;
end;
```

8. How many latches will the following VHDL code produce?

```
Architecture rtl of ex is
signal a,b,c,d,e: std_logic_vector(1 downto 0);
begin
  process(c,d,e,en)
  begin
     if en='1' then
        a<=c;
        b<=d;
     else
        a<=e;
     end if;
  end process;
end;
```

9. Design a component with the following inputs and outputs:

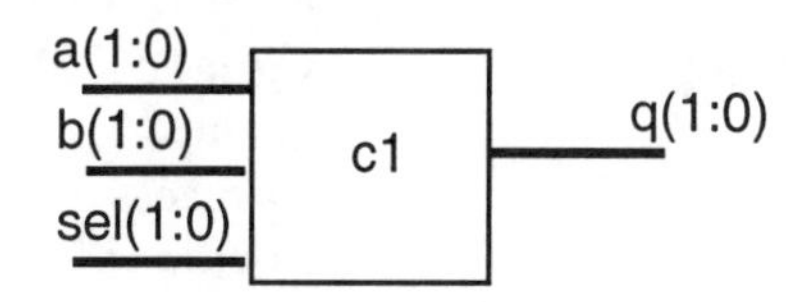

The component should have the following behaviour:

Sel	*q*
00	a **xor** b
01	a **or** b
10	a **nor** b
11	a **and** b
others	"XX"

(a) Use an **if** statement.

(b) Use a **case** statement.

(c) Use a **when else** statement.

(d) If the above exercise is synthesized, which of the alternatives will generate the least logic?

10. Suppose that the following logic is to be implemented:

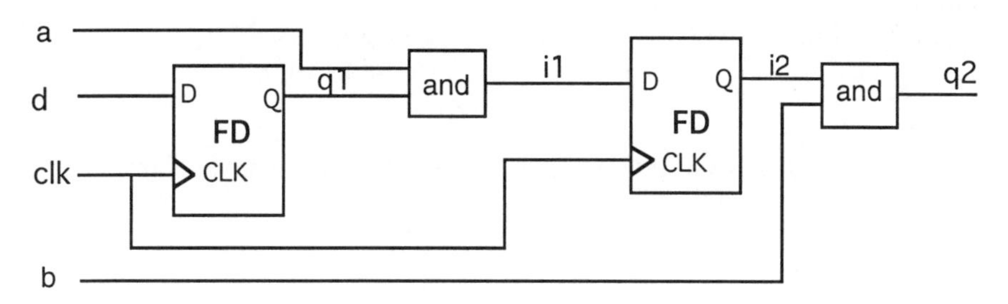

(a) Which alternative(s) is/are correct synthesizable VHDL code for the above schematic?

(1)
```
Architecture rtl of ex is
signal i1: std_logic;
begin
  process
  begin
    wait until clk='1';
    i1<=d and a;
    q2<=i1 and b;
  end process;
end;
```

(2)
```
Architecture rtl of ex is
signal i1,i2: std_logic;
begin
  process
  begin
    wait until clk='1';
    i1<=d and a;
    i2<=i1;
  end process;
  q2<=i2 and b;
end;
```

(3)
```
Architecture rtl of ex is
signal q1, i1, i2: std_logic;
begin
  process
  begin
      wait until clk='1';
      q1<=d;
      i1<=a and q1;
      i2<=i1;
      q2<=i2 and b;
  end process;
end;
```

(4)
```
Architecture rtl of ex is
signal q1, i2: std_logic;
begin
  process
  begin
     wait until clk='1';
     q1<=d;
     i2<=q1 and a;
  end process;
  q2<=i2 and b;
end;
```

(5)
```
Architecture rtl of ex is
signal q1, i1, i2: std_logic;
begin
  process
  begin
      wait until clk='1';
      q1<=d;
      i2<=i1;
  end process;
  q2<=i2 and b;
  i1<=q1 and a;
end;
```

(6)
```
Architecture rtl of ex is
signal q1, i2: std_logic;
begin
  process(clk)
  begin
     if clk'event and clk='1' then
        q1<=d;
        i2<=q1 and a;
     end if;
  end process;
  q2<=i2 and b;
end;
```

(b) Draw the synthesis result (in principle) from all the alternatives above.

5

Library, package and subprograms

When designing systems, it is possible to use ready-made components, functions, procedures, etc., which are stored in a library. For hardware designers it has been natural to use tried and tested components (type 74LSxx). When VHDL is used, this pattern can continue or, as in the world of software, it may be difficult to reuse other people's work. VHDL has several constructions which invite users to build component libraries so that the facility is available.

This chapter deals mainly with how to store components, functions, etc., and how to use them later in new design work.

5.1. Libraries

When a VHDL component is compiled, it is saved in the work **library** as default. The work library is not the name of a directory on the PC or workstation on which the compilation is being done, but a logical name. In the system there has to be a pointer which defines the physical address of the work library, i.e. which directory it points to. VHDL tools usually define the work library automatically when the tool is started up. This means that different work libraries will be obtained depending on where the VHDL compiler is started.

All compiled components are stored in a library. Packages too are usually stored in a library. A **package** can contain functions, procedures, constants and types, for example. In order to use components and packages from a particular library, the library and packages must first be specified in the VHDL code using the following commands:

```
Library <library_name>;
Use <library_name>.<package_name>.ALL;
```

The VHDL standard is defined in such a way that the work and std library are always visible. These two libraries do not, therefore, have to be specified in the VHDL code. All predefined data types and functions in the VHDL standard are available in a package called standard. This package is located in the std library. It is in this package that data types bit, bit_vector, character, time and integer are defined, for example (see "VHDL packages" in Appendix B). Package standard is always visible according to the VHDL standard and so does not have to be specified in the VHDL code either. This means that the following "invisible" lines are always included in each piece of VHDL code:

```
Library work;
Library std;
Use std.standard.ALL;
```

If other libraries and packages are to be used, they must be defined at the top of the VHDL code, i.e. before the entity. Data types std_logic, std_ulogic, std_logic_vector and std_ulogic_vector are defined in a package called std_logic_1164. This package is defined by the IEEE; the name of the library where package std_logic_1164 is located is also ieee. This means that if these data types are to be used (which is recommended), the following lines must always be included before each entity:

```
Library ieee;
Use ieee.std_logic_1164.ALL;

entity ...
```

5.2. Packages

In the case of large designs, there are often functions, data types and constants, for example, which have to be used in several components and by several designers. The constant might be the bus size in the system, for example. The function might be to calculate the checksum of the data message in the system. If the project management changes the bus size or the checksum calculation, it is preferable if the change has to be made only once. This is made possible by the use of packages. If each designer who is to use a shared constant or function for the project uses a pointer (**Use**) to a package, all the designs can be updated by just changing the shared package for the project.

Example:

```
Library project;
Use project.proj_pack.ALL;
```

All data types, constants and subprograms defined in a package can also be reused in future designs simply by defining the package using the **Use** command at the top of the VHDL code.

A package can contain, for example, functions, procedures, types, constants, attributes and components. A package consists of two parts:

- Declaration
- Body

The declaration consists of function or procedure declarations, components, constants and types which are exported. The declaration in the package can be compared to an entity, i.e. the declaration defines what is to be "outwardly" visible, exactly the same as it does in an entity. The difference is that the entity defines which signals are to be exported outside the component, while the package declaration defines which subprograms, constants and data types are to be exported outside the package.

The **body** consists of the program proper for the functions and procedures which have been declared in the package declaration. If internal subprograms are created in the body, they will be invisible outside the package, i.e. you cannot use an internal subprogram in your own VHDL code. The body can be compared to an architecture in a VHDL component. An architecture describes the behaviour proper of the component. The package body describes the behaviour for the declared subprograms from the package declaration.

Syntax for a package:

```
package <package_name> is
        [exported_subprogram_declarations]
        [exported_constant_declarations]
        [exported_components
        [exported_type_declarations]
        [attribut_declarations]
        [attribut_specifications]
end [<package_name>];

package body <package_name> is
        [exported_subprogram_bodies]
        [internal_subprogram_declarations]
        [internal_subprogram_bodies]
        [internal_constant_declarations]
        [internal_type_declarations]
end [<package_name>];
```

Example:

```
package mypack is
  function minimum (a,b:in std_logic_vector)
          return std_logic_vector;

  -- declarations of exported constant and types:
  constant maxint: integer := 16#FFFF#;
  type arithmetic_mode_type is (signed, unsigned);

end mypack;

package body mypack is
  function minimum (a,b:in std_logic_vector)
          return std_logic_vector is
  begin
    if a<b then
       return a;
    else
       return b;
    end if;
  end minimum;
 end mypack;
```

Syntax for a package (VHDL-93):

```
package <package_name> is
   ...
end [package] [<package_name>];

package body <package_name> is
   ...
end [package body] [<package_name>];
```

The **Use** command allows a user or ready-made package to be used later. This command must come before the entity declaration but after library.

Example:

```
Library ieee;
Library mylib;

Use ieee.std_logic_1164.ALL;
Use mylib.mypack.ALL;
```

```
entity <name> is
   port( a: in  std_logic;
         q: out std_logic;
   ...
```

The above example assumes that the package is compiled to library mylib. The system must contain a pointer which defines where library mylib is located, i.e. which directory it corresponds to. The example also specifies *mypack.ALL*. **ALL** means that all the subprograms in the package can be used. If only one or some of the subprograms from those included in a package are to be declared, e.g. if just function *minimum* is to be used, the following has to be written:

```
Use mylib.mypack.minimum;
```

If several different functions from different packages are to be used, the following can be written:

```
Library mylib, ieee;
Use mylib.mypack.minimum, ieee.std_logic_1164.ALL;
```

or:

```
Library mylib;
Library ieee;
Use mylib.mypack.minimum;
Use ieee.std_logic_1164.ALL;
```

5.3. Subprograms

There are two sorts of subprogram:

- Functions
- Procedures

Functions produce just one value, whereas a procedure is used to define an algorithm and can produce several values or none. Subprograms can be defined in three places in the VHDL code:

- Package
- Architecture
- Process

As it is only in a package that the same subprogram can be reused in several designs, it is recommended that all subprograms should be defined in a package.

Inside subprograms the VHDL code is sequential. This means that only sequential VHDL commands, e.g. **if-then-else** and **case** commands, can be used in subprograms. Concurrent commands such as **process** and **when else** are not permissible in subprograms. The package declaration between the subprogram and **begin** is therefore a sequential declaration. This means that only variables, not signals, can be declared inside subprograms. Note, however, that it is not permissible to use the sequential command **wait** inside a function.

5.3.1. Procedures

A **procedure** can change its argument. It accepts constant, variable and signal as an object class. These can have three different modes: **in**, **out** and **inout**. If no mode is specified, it defaults to in. If the mode is out or inout, the object class must be a variable or signal. If the object class has not been specified, it is assumed to be variable or constant, depending on the mode (in, out, inout). In other words, an input (mode in) defaults to class constant and an output (mode out or inout) to class variable. Variable parameters are not allowed in concurrent procedure call.

In the declaration part of the package:

```
    procedure <procedure_name>
        [ ( [object_class] arg_name {, arg_name}:
        [mode] type [ ( index_range [, index_range] } ];
        { [object_class] arg_name {, arg_name}:
        [mode] type [ (index_range [, index_range] ) ] } ) ];
```

In the package body:

```
    procedure <procedure_name>
        [ ( [object_class] arg_name {, arg_name}:
        [mode] type [ ( index_range [, index_range] } ];
        { [object_class] arg_name {, arg_name}:
        [mode] type [ (index_range [, index_range] ) ] } ) ] is
        [constant_declarations]
        [variable_declarations]
        [type_declarations]

    begin
        sequence_of_statements
    end [<procedure_name>];
```

Syntax for procedures (VHDL-93):

```
procedure <procedure_name> is
    ...
end [procedure] [<procedure_name>];
```

Example of a procedure:

```
procedure ff ( signal clk, data : in   std_logic;
               signal q :         out std_logic) is
begin
  loop
    wait until clk='1';
    q <= data;
  end loop;
end ff;
```

Example 2 (bad):

```
Architecture rtl of ex2 is
procedure calc  (a,b:         in   integer;
                 avg,max:     out integer) is -- Default variables
begin
   avg:=(a+b)/2;
   if a>b then
      max:=a;
   else
      max:=b;
   end if;
end calc;
begin
   calc(d1,d2,q1,q2);          -- Error, q1,q2 not variables
                               -- Concurrent procedure call
   process(d3,d4)
   variable a,b: integer;
   begin
     calc(d3,d4,a,b);          -- Good, a and b variables
                               -- Sequential procedure call
     q3<=a;
     q4<=b;
   ...
   end process;
   ...
end;
```

Example 3 (good):

```
Architecture rtl of ex3 is
procedure calc ( a,b:                in  integer;
                 signal avg,max:     out integer) is

begin
   avg<=(a+b)/2;
   if a>b then
      max<=a;
   else
      max<=b;
   end if;
end calc;

begin
   calc(d1,d2,q1,q2);            -- Concurrent procedure call

   process(d3,d4)
   begin
     calc(d3,d4,q3,q4);          -- Sequential procedure call
   ...
   end process;
   ...
end;
```

5.3.2. Functions

A **function** cannot change its argument and can only use parameters of type constant or signal and mode in. If an object class or mode is not specified, it is assumed to be of type constant and/or mode in.

```
In the declaration part of the package:

     function <function_name>
          [ ( [ object_class] arg_name {, arg_name}:
          [IN ] type [ (index_range [, index_range] ) ]
          { [object_class] arg_name {, arg_name}:
          [IN ] type [ (index_range [, index_range] ) ] } ) ]
     return type;
```

In the package body:

```
function <function_name>
      [ ( [ object_class] arg_name {, arg_name}:
      [IN ] type [ (index_range [, index_range] ) ]
      { [object_class] arg_name {, arg_name}:
      [IN ] type [ (index_range [, index_range] ) ] } ) ]
return type is
      [constant_declarations]
      [variable_declarations]
      [type_declarations]
begin
      sequence_of_statements
      return(expression);
end [<function_name>];
```

Syntax for functions (VHDL-93):

```
function <function_name> is
      ...
end [function] [<function_name>];
```

As previously mentioned, subprograms can be defined in package, architecture and process. Below are examples of how a function is defined and used in three places.

Example 1: Package

```
package mypack is
  function max (a,b:in std_logic_vector) return
     std_logic_vector;
end;

package body mypack is
  function max (a,b:in std_logic_vector) return
     std_logic_vector is
  begin
     if a>b then
        return a;
     else
        return b;
     end if;
  end;
end;
```

Function *max* in package *mypack* can then be used freely in an architecture as long as the line **Use** work.mypack.**ALL** is written at the top of the VHDL code.

```
Use work.mypack.ALL;
...
Entity ex is
 port(...
end;
Architecture rtl of ex is
begin
  ...
  q<=max(d1,d2);                 -- Concurrent function call
  process(data,g)
  begin
    data_out<=max(data,g);       -- Sequential function call
  end process;
  ...
end;
```

Example 2: Architecture

```
Architecture rtl of ex2 is
   function max (a,b: in std_logic_vector) return
   std_logic_vector is
   begin
      if a>b then
         return a;
      else
         return b;
      end if;
   end max;
   ...
begin
   q<=max(d1,d2);                -- Concurrent function call
   process(data,g)
   begin
     data_out<=max(data,g);      -- Sequential function call
   end process;
  ...
end;
```

Example 3: Process

```
Architecture rtl of ex3 is
begin
  process(data,g)
     function max (a,b: in  std_logic_vector) return
     std_logic_vector is
     begin
        if a>b then
           return a;
        else
           return b;
        end if;
     end max;
  begin                                  -- Start of process
     data_out<=max(data,g);
  end process;
  ...
end;
```

All functions and procedures are normally written without specifying the vector length for the input and output parameters. This means that the subprogram can be called with any vector length.

Example:

```
Library  ieee;
Use  ieee.std_logic_1164.ALL;
Use work.mypack.ALL;

Entity ex is
  port( a,b:  in   std_logic_vector(3  downto 0);
        c,d:  in   std_logic_vector(5  downto 0);
        q1:   out  std_logic_vector(3  downto 0);
        q2:   out  std_logic_vector(5  downto 0));
end;

Architecture rtl of ex is
begin
  q1<=max(a,b);        -- Vector length = 4 bits
  q2<=max(c,d);        -- Vector length = 6 bits
end;
```

As the above examples show, it is permissible to have several **return** statements in a function. The requirement is that only one of the **return** statements may be executed. Virtually all synthesis tools support both procedures and functions. Some simpler synthesis tools may have the restriction that only one **return** statement may be used and that it must come at the end of the function. In order to satisfy this restriction, an internal variable must be declared in the function for temporary storage of the value to be returned.

Example:

```
function max (a,b: in integer) return integer is
variable int: integer;
begin
  if a>b then
    int:=a;              -- Store return value
  else
    int:=b;              -- Store return value
  end if;
  return int;            -- Return value
end max;
```

The majority of synthesis tools support several **return** statements, however.

As with ADA, it is possible in VHDL to specify the entire search path for a function when it is called. If a full search path is specified, it may facilitate understanding of where the function is declared. The following has to be specified: <library>.<package>.function>.

Example:

```
q<=work.mypack.max(a,b);
```

If a full search path is used for the subprogram, the package does not have to be declared with **Use** library.package.**ALL**; at the top of the VHDL code.

5.3.3. Resolution functions

Data types std_logic and std_logic_vector are what are known as resolved data types. This means that data type std_logic is defined as follows in package std_logic_1164:

```
subtype std_logic is resolved std_ulogic;
```

Function resolved is called when there are several drivers driving the same std_logic signal. Function resolved will return the value which applies when there are several drivers for the signal. The function is defined as follows in package std_logic_1164:

```
function resolved (s : std_ulogic_vector ) return std_ulogic is
 variable result : std_ulogic := 'Z';       -- weakest state default
 begin
   if (length = 1) then
      return s(s'low);
   else
      for i in s'range loop
       result := resolution_table(result, s(i));
      end loop;
   end if;
   return result;
end resolved;

constant resolution_table : stdlogic_table := (
--  ------------------------------------------------
--  | U   X   0   1   Z   W   L   H   -         |   |
--  ------------------------------------------------
    ('U','U','U','U','U','U','U','U','U' ), -- | U |
    ('U','X','X','X','X','X','X','X','X' ), -- | X |
    ('U','X','0','X','0','0','0','0','X' ), -- | 0 |
    ('U','X','X','1','1','1','1','1','X' ), -- | 1 |
    ('U','X','0','1','Z','W','L','H','X' ), -- | Z |
    ('U','X','0','1','W','W','W','W','X' ), -- | W |
    ('U','X','0','1','L','W','L','W','X' ), -- | L |
    ('U','X','0','1','H','W','W','H','X' ), -- | H |
    ('U','X','X','X','X','X','X','X','X' )  -- | - |);
```

An example of a resolution function is given in Figure 5.1.

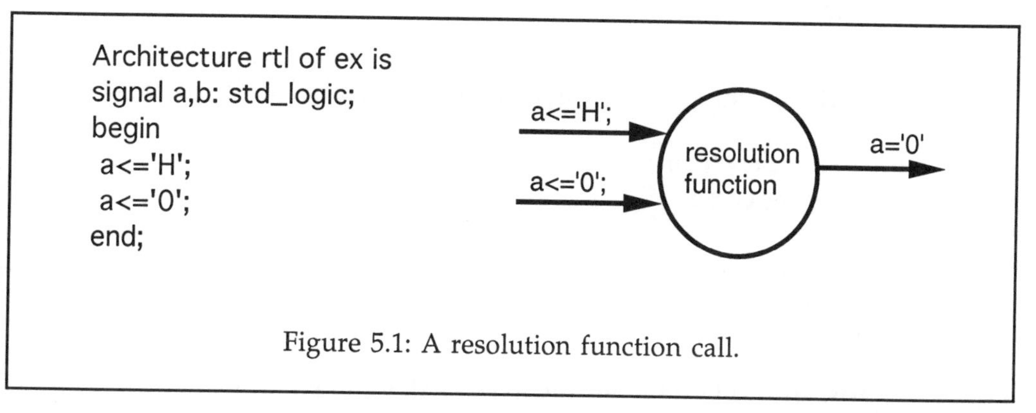

Figure 5.1: A resolution function call.

5.4. Overloading

As VHDL is a strongly typed language, it is not possible to call a function which expects bit_vector, for example, as an input parameter with a std_logic_vector. In order to solve this problem and be able to call function *max*, for example, with different data types, function *max* must be overloaded. Overloading means that the same function is defined several times with different data types as the argument for the function. The functions are also written in such a way that they can be called with any vector length.

Example:

```
Library ieee;
Use ieee.std_logic_1164.ALL;

Package ex_pack is
 function max(a,b: in  std_logic_vector) return
    std_logic_vector;
 function max(a,b: in  vlbit_vector) return vlbit_vector;
 function max(a,b: in  integer) return integer;
end;

Package body ex_pack is
  function max(a,b: in std_logic_vector) return
    std_logic_vector is
  begin
    if a>b then
       return a;
    else
       return b;
    end if;
   end;

  function max(a,b:in vlbit_vector) return vlbit_vector is
    begin
       if a>b then
          return a;
       else
          return b;
       end if;
    end;

  function max(a,b:in integer) return integer is
  begin
    if a>b then
```

```
      return a;
    else
      return b;
    end if;
  end;
end;
```

An example of calling overloaded function *max* correctly is as follows:

```
Library ieee;
Use ieee.std_logic_1164.ALL;
Use work.ex_pack.ALL;

Entity ex is
 port( a1,b1:  in   std_logic_vector(3 downto 0);
       a2,b2:  in   vlbit_vector(4 downto 0);
       a3,b3:  in   integer;
       c1:     out  std_logic_vector(3 downto 0);
       c2:     out  vlbit_vector(4 downto 0);
       c3:     out  integer);
end;

Architecture ex_beh of ex is
begin
  c1<=max(a1,b1);  -- Function max(a,b:std_logic_vector)
  c2<=max(a2,b2);  -- Function max(a,b:vlbit_vector)
  c3<=max(a3,b3);  -- Function max(a,b:integer)
end;
```

There are several very useful overloaded functions in the ieee library. The functions "+", "-", "*", "=", "<", ">", "/=", "<=" and ">=" are, for example, overloaded for data types std_logic_vector or integer in package ieee.std_logic_unsigned. Some advanced synthesis tools support these overloaded functions, which means that the code can be written very readably and effectively. The functions are defined not only with both std_logic_vectors as input arguments, but also with std_logic_vector as one input argument and integer as the other input argument.

Example:

```
function "=" (L:std_logic_vector; R:integer) return boolean;
function ">" (L:std_logic_vector; R:integer) return boolean;
```

Example:

```
Architecture rtl of ex is
signal a,b,c: std_logic_vector(15 downto 0);
begin
  process(a,b,c,d1,d2,d3)
  begin
    if a=12 or c=11 then -- Overloaded function "=", left side
                         -- std_logic_vector, right side integer
      q<=d1;
    elsif b>5 then
      q<=d2;
    else
      q<=d3;
    end if;
  end process;
end;
```

The ieee library contains two packages which have std_logic_vector as an overloaded data type: std_logic_unsigned and std_logic_signed. Depending on which of these packages has been declared in the VHDL code, the std_logic-vectors will be interpreted as unsigned or signed.

Example:

```
Library ieee;
Use ieee.std_logic_1164.ALL;
Use ieee.std_logic_unsigned.ALL;
...
Architecture rtl of ex is
begin
  q<=a + b;     -- unsigned add
end;
```

```
Library ieee;
Use ieee.std_logic_1164.ALL;
Use ieee.std_logic_signed.ALL;
...
Architecture rtl of ex is
begin
  q<=a + b;     -- signed add
end;
```

If both unsigned and signed vectors are required in the same architecture, the VHDL code can be written as follows:

```
Library ieee;
Use ieee.std_logic_1164.ALL;
Use ieee.std_logic_arith.ALL;
...
Architecture rtl of ex is
begin
  q1<=unsigned(a) + unsigned(b);   -- unsigned add
  q2<=signed(a) + signed(b);       -- signed add
end;
```

Most synthesis tools do not support packages std_logic_unsigned and std_logic_signed. The exceptions include Synopsys and Autologic 2. There are, however, several which support package std_logic_arith.

Package std_logic_unsigned has been used in some examples in this book. If the synthesis tool used by the reader does not support this package, the Use line declaring the package simply has to be exchanged for the package supported by the synthesis tool.

If package std_logic_arith is used, it is important to know the length of the vector returned by the function.

Example:

```
signal a,b: std_logic_vector(3 downto 0);
q1<=unsigned(a) + unsigned(b);
q2<=unsigned(a) + signed(b);
q3<=signed(a) + signed(b);
```

Below is a table for function "+".

	Unsigned	*Signed*
Unsigned	4 bits	5 bits
Signed	5 bits	4 bits

If package ieee.std_logic_unsigned, for example, is used to add two vectors and carry is wanted in the result, one side of the addition must be extended by one bit ('0'). Function "+" results in a vector which is the same length as the longest input argument.

Example:

```
Use ieee.std_logic_unsigned.ALL;
...
signal a,b,q2:      std_logic_vector(7 downto 0);
signal q1:          std_logic_vector(8 downto 0);
...
  q1<=('0' & a) + b;       -- carry (9 bits)
  q2<=a + b;               -- no carry bit (8 bits)
```

5.5. Type conversion

VHDL is a strongly typed language, which means that it is not permissible to assign to a signal the value of another signal if they have different data types. This problem can normally be minimized by using the same data type consistently for most of the signals in the design. Many of the commonest operators are also overloaded, which means that function "=" can, for example, be used to compare a std_logic_vector with an integer without type conversion. As many of the synthesis tools do not support such overloading and the need for type conversion cannot always be avoided, there are several ready-made functions for that purpose. There are differences between the various synthesis tools in terms of which package the tool supports for the conversion, but the principle is always the same, i.e. it is only the name of the function and package which differ if the synthesis tool is changed. There are several packages in the ieee library. Among other things, package std_logic_1164 contains type conversion functions between the following data types (see "VHDL packages" in Appendix B):

```
std_logic            <---->   bit
std_logic_vector     <---->   bit_vector
std_ulogic           <---->   bit
std_ulogic_vector    <---->   bit_vector
```

Example: In package ieee.std_logic_1164:

```
function to_stdlogicvector( s: bit_vector) return
  std_logic_vector;
```

An example of how to use a type conversion function might be:

```
Library ieee;
Use ieee.std_logic_1164.ALL;

Entity ex is
 port( a,b: in   bit_vector(3 downto 0);
          q:  out std_logic_vector(3 downto 0));
end;

Architecture rtl of ex is
begin
  q<=to_stdlogicvector(a and b);
end;
```

A more usual conversion is, perhaps, to convert from std_logic_vector to an integer and vice versa. This conversion function is contained in package ieee.std_logic_unsigned, for example:

```
function conv_integer(arg: std_logic_vector) return integer;

function conv_std_logic_vector(arg: integer; size integer)
   return std_logic_vector;
```

Example:

```
Entity ex is
  port( a,b,c: in   integer range 0 to 15;
        q:     out  std_logic_vector(3 downto 0));
end;

Architecture rtl of ex is
begin
 q<= conv_std_logic_vector(a, 4) when conv_integer(c) = 8 else
     conv_std_logic_vector(b, 4);
end;
```

The code would have been more readable if all the signals had been of data type std_logic. The code could then have been written as follows:

```
Entity ex is
 port( a,b,c: in   std_logic_vector(3 downto 0);
       q:     out  std_logic_vector(3 downto 0));
end;

Architecture rtl of ex is
begin
  q<= a when conv_integer(c) = 8 else b;
end;
```

If the synthesis tool also supports function "=" being overloaded for data type std_logic_vector and integer, the code can be made even more readable:

```
Entity ex is
 port( a,b,c: in   std_logic_vector(3 downto 0);
       q:     out  std_logic_vector(3 downto 0));
end;

Architecture rtl of ex is
begin
  q<=  a when c = 8 else b;
end;
```

From the synthesis point of view, it makes no difference whether a conversion function is used or not. The **conversion function does not take any gates.** A conversion function is only used for two reasons: to make the code easier to write and the documentation easier to understand. A good design should, however, contain a minimum of conversion functions, as the recommendation is to use data type std_logic_vector as much as possible. With a good synthesis tool which also supports std_logic_vector being compared with an integer (overloading), it is possible in principle to design highly complex ASICs in VHDL without a single conversion function having to be used in the VHDL code.

5.6. Shift operators

Shift operators were added to the standard in VHDL-93. Six different shift operators are defined for data type bit_vector:

- sll -- Shift left
- srl -- Shift right
- rol -- roll over left
- ror -- roll over right
- sla -- shift left, and keep value 'right
- sra -- shift right, and keep value 'left

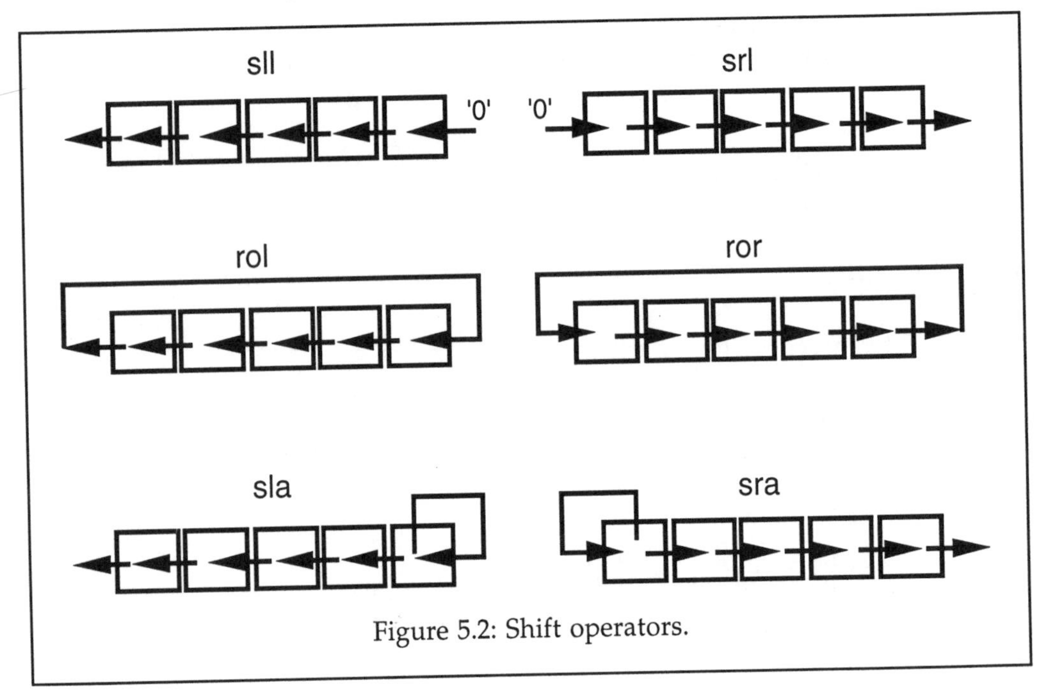

Figure 5.2: Shift operators.

Example:

```
Architecture behv of ex is
begin
    a<="01101";
    q1<= a sll 1;    -- q1 = "11010"
    q2<= a srl 3;    -- q2 = "00001"
    q3<= a rol 2;    -- q3 = "10101"
```

```
    q4<= a ror 1;   -- q4 = "10110"
    q5<= a sla 2;   -- q5 = "10111"
    q6<= a sra 1;   -- q6 = "00110"
end;
```

The majority of synthesis tools do not yet support these shift operators (with the exception of Autologic 2). However, most are planning to support them soon (1996/97), but until they do, shift functions from a package will have to be used. Two shift functions are, for example, defined in package ieee.std_logic_unsigned (see "VHDL packages" in Appendix B). The functions are called:

```
function shl(arg: std_logic_vector; count: std_logic_vector)
                        return std_logic_vector;
function shr(arg: std_logic_vector; count: std_logic_vector)
                        return std_logic_vector;
--  shl = Shift left
--  shr = Shift right
```

Example:

```
q1<=shl(data,"1");         -- Shift one step left
q2<=shr(data,"101");       -- Shift five step right
q3<=shr(data, count);      -- Shift count step right
```

It is, of course, possible to write the VHDL code without using any shift operators. For example:

```
Architecture rtl of ex is
signal data: std_logic_vector(7 downto 0);
begin
 process(clk,resetn)
 begin
   if resetn='0' then
     q1<=(others=>'1');
     q2<=(others=>'1');
   elsif clk'event and clk='1' then

     -- Shift one step right
     q1(6 downto 0)<=q1(7 downto 1);
     q1(7)<=d_in;
     -- Shift one step left
     q2(7 downto 1)<=q2(6 downto 0);
     q2(0)<=d_in;
   end if;
 end process;
end;
```

5.7. Exercises

1. (a) In which library and package is data type bit defined?

(b) Which library/libraries and package(s) are always visible?

2. Specify four things which can be stored in a package.

3. (a) Specify three places where a function can be defined.

(b) Which of these three places facilitates reuse of the function?

(c) What do you have to do to be able to use other designers' packages?

4. What is meant by data type std_logic being resolved?

5. What is the difference between a function and a procedure?

6. Why should the length of vectors in a function/procedure not be defined in the function declaration?

7. What is overloading?

8. (a) Design a package which contains two functions: *average* and *sum*. Function *average* should return the mean of two numbers (rounded down), while function *sum* should return the sum. The functions should both be defined for data types integer and std_logic_vector.

(b) Design a component (c1) which uses the functions in (a) above. The component should have four inputs *a, b, c* and *d*. Inputs *a* and *b* are of type integer(0 **to** 127) and input signals *c* and *d* are of type std_logic_vector(7 **downto** 0). The component should have four outputs (*average1, average2, sum1, sum2*), i.e. one from each function. *Average1* and *sum1* should be of type integer(0 **to** 127) and *average2* and *sum2* of type std_logic_vector (7 **downto** 0).

9. (a) What is type conversion?

(b) Does the VHDL standard contain ready-made conversion functions?

10. Which of the following functions is/are correct VHDL code?

(1)
```
function f(a,b: in std_logic) return std_logic is
begin
  if a'event then
    return b;
  else
    return a;
  end if;
end;
```

(2)
```
function f(signal a,b: std_logic) return std_logic is
begin
```

```
    if a'event then
      return b;
    else
      return a;
    end if;
  end;
```

(3)
```
  function f(constant a,b: in std_logic) return std_logic is
  begin
    if a'event then
      return b;
    else
      return a;
    end if;
  end;
```

(4)
```
  function f(a: in std_logic) return std_logic is
  begin
    wait on a;
    return a;
  end;
```

(5)
```
  function f(signal a: in std_logic) return std_logic is
  begin
    wait on a;
    return a;
  end;
```

11. (a) Which shift operators are defined in the VHDL-93 standard?

(b) Is it possible to design shift registers without using these operators?

6
Structural VHDL

Structural VHDL can be compared to a netlist which binds different components (entities) together. An entity normally contains several components, and it is their interconnection which is described with structural VHDL (in hierarchical design). Each component also forms a new hierarchical level, i.e. if a component consists of four inputs and three outputs, but contains 1000 gates, only the component shell (the entity's inputs and outputs) is visible at the top level, while the gates are on a hierarchical level below the component. Structural VHDL can therefore be used for hierarchical design in VHDL. The ability to design hierarchically is a must for handling large designs. It is not possible to say where exactly the border for hierarchical design lies - it depends on the design. A good guideline, however, is to design hierarchically when the design is larger than 2000-3000 gates. As recommended in the chapter on design methodology, each hierarchical level (entity/architecture) should contain approximately 500-6000 gates.

Structural VHDL is simply the interconnection of a number of components (entities). This interconnection can be done in schematic form, as opposed to code, enabling an overview of the design to be gained, if the design is not too large. A schematic representation can say a lot about the principles, while text can say a great deal more about the details. In a schematic representation you are not usually interested in all the signals, but just the principle. Drawing a block diagram which shows only the interconnection of the blocks and does not contain all information (all signals) is therefore a good supplement. The block diagram can easily be drawn using pen and paper or a graphics VHDL tool. The disadvantage of using a graphics tool is that the block diagram becomes platform-, version- and tool-dependent. Structural VHDL code, on the other hand, is completely platform-, tool- and version-independent. This means that, although structural VHDL is a little difficult to read, it is very useful. If structural VHDL is used for hierarchical design, the whole design will be described in VHDL, making it platform-, version-, technology- and tool-independent.

The following example illustrates the principle of structural VHDL. The synthesis result is shown in Figure 6.1. A multiplexor does not normally have to be described in this way, however. The rest of the chapter describes each element of structural VHDL.

Example:

```
Entity mux is
  port( d0,d1,sel: in    std_logic;
        q:         out   std_logic);
end;

Architecture struct_mux of mux is
  -- Component declaration.
  component and_comp
    port( a,b: in  std_logic;
          c:   out std_logic);
  end component;

  component or_comp
    port( a,b: in  std_logic;
          c:   out std_logic);
  end component;

  component inv_comp
    port( a: in  std_logic;
          b: out std_logic);
  end component;

  signal i1,i2,sel_n:std_logic;

  --Component specification.
  for U1 : inv_comp Use Entity work.inv_comp(rtl);
  for U2,U3 : and_comp Use Entity work.and_comp(rtl);
  for U4 : or_comp Use Entity work.or_comp(rtl);

  begin
    -- Component instantiation.
    U1 : inv_comp port map(sel,sel_n);
    U2 : and_comp port map(d0,sel,i1);
    U3 : and_comp port map(sel_n,d1,i2);
    U4 : or_comp port map(i1,i2,q);
  end;
```

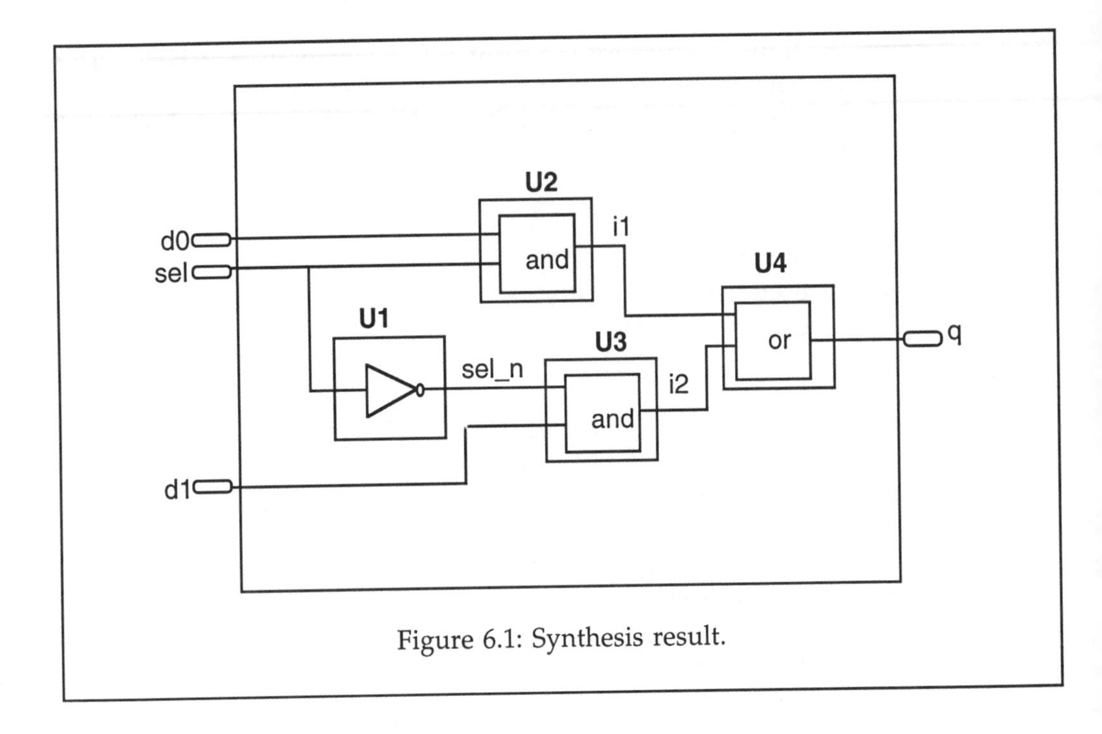

Figure 6.1: Synthesis result.

6.1. Component declaration

For a VHDL **component** to be instanced with structural VHDL, it first has to be declared. This is done in the concurrent declaration part of VHDL, i.e. between **architecture** and **begin**.

```
Syntax:
    component <entity_name>
    [generic (<generic-association-list>);]
        port(<port-association-list>);
    end component;
```

```
Syntax(VHDL-93):
    component <entity_name>
        ...
    end component [<entity_name>];
```

Port_list in the component declaration must be identical to that in the component's entity. The best way of guaranteeing this is to copy the entity to the component declaration or vice versa. **Generic** parameters do not need to be declared in the component declaration. Suppose that the component below (ex) is to be instanced with structural VHDL. The VHDL code for the component declaration would be:

```
Entity ex is
   port(  a,b:  in     std_logic_vector(2 downto 0);
            q:  out    std_logic_vector(2 downto 0));
end;

Architecture rtl of ex is
begin
   ...
end;


Entity top_level is
   port( ...
         ...);
end;

Architecture rtl of top_level is
   component ex               -- Component declaration
     port( a,b: in     std_logic_vector(2 downto 0);
             q: out    std_logic_vector(2 downto 0));
   end component;
   ...
   begin
     ...
   end;
```

If the order of the signals in the **port** command does not agree with the entity ex, an error will occur in the VHDL compilation of the code. The following component declaration is therefore incorrect:

Example (bad):

```
Architecture bad of top_level is
 component ex                    -- Component declaration
  port( b,a:  in    std_logic_vector(2 downto 0); -- Error
          q:  out   std_logic_vector(2 downto 0));
end component;
```

As it is normal for several components to be instanced in hierarchical design, the top_level architecture will contain several component declarations.

Example:

```
Architecture rtl of top_level is
  component c1
    port( a,b:  in   std_logic_vector(2 downto 0);
          q:    out  std_logic_vector(2 downto 0));
  end component;

  component c2
    port( a,b:  in   std_logic_vector(4 downto 0);
          q:    out  std_logic_vector(1 downto 0));
  end component;
  ...
```

6.2. Component specification

If the VHDL simulator is to find the declared component, the component must be specified. This is done with a **Use** command.

> Syntax:
>
> **For** <label>: <component_name> **Use entity** <library>.<entity_name>(<architecture_name>);

This specification is used to choose which library the component is to be compiled in and which architecture is to be simulated. A VHDL component can have several different architectures linked to it. They can represent different implementation methods or description levels. Specifying the architecture name in the component specification selects an architecture for simulation (compilation). Suppose that the component below (ex2) has two different architectures linked to it: behv and rtl. If the rtl architecture is to be simulated, the component specification should be written as follows:

```
Entity ex2 is                          -- Component ex2
  port(  ...
         ...);
end;

Architecture behv of ex2 is    -- Architecture behv of ex2
begin
  ...
end;
```

```
Architecture rtl of ex2 is     -- Architecture rtl of ex2
begin
  ...
end;

Entity top_level is
  component ex2
    port ...
            ...);
  end component;
  For U1: ex2 Use entity  work.ex2(rtl);
  -- Use architecture rtl of ex2
  begin
    ...
  end;
```

Component specification is not supported by synthesis tools. Synthesis tools require the declared component either to be entered in the synthesis tool's memory or to be located in one of the synthesis tool's set search paths. Synthesis tools do accept the presence of the component specification in the VHDL code, they just ignore it though.

6.3. Port map command

What we have done so far is to declare and specify the component to be instanced. The component is instanced and "linked" to other components using the **port map** command.

> Syntax:
>
> **port map**(<port-association-list>);

Suppose that the following component (ex3) has to be instanced in component top_level.

```
Entity ex3 is
  port( a,b:  in   std_logic;
           q:  out std_logic);
end;

Architecture rtl of ex3 is
begin
  ...
end;
```

The component can then be instanced as follows in component top_level:

```
Entity top_level is
  port( d1,d2 in     std_logic;
          q1:     out std_logic);
end;

Architecture rtl of top_level is
  component ex3
    port( a,b: in     std_logic;
            q:   out std_logic);
  end component;

  For U1: ex3 Use entity  work.ex3(rtl);
  begin
    U1: ex3 port map(d1, d2, q1);
  end;
```

It is the order which decides how the interconnection takes place. In the preceding example input signal *d1* in component top_level will be connected to input *a* on component ex3, etc. The **port map** command can also be written in such a way as to make it clear what is connected to what, as the order of the signals is unimportant. The **port map** command can therefore also be written as follows:

```
U1: ex3 port map( a=>d1, b=>d2, q=>q1);
```

The way in which the **port map** command is written has no effect on either simulation or the synthesis result. Any label name (U1) can be chosen. In synthesis the component will normally be given both the component name (ex2) and the label name U1 as the component name and instance name respectively.

6.3.1. Unconnected outputs

When a component is instanced, one of the outputs sometimes has to be unconnected. This can be done using the VHDL word **open**. Suppose that output *q2* in component ex4 has to be unconnected for instantiation. In this case the VHDL code is written as follows:

```
Architecture rtl of top_level is
  component ex4
    port( a,b:       in   std_logic;
            q1,q2;    out std_logic);
  end component;
```

```
For U1: ex4 Use entity  work.ex4(rtl);
begin
  U1: ex4 port map(a=>a, b=>b, q1=>dout, q2=>open);
end;
```

It is also permissible to omit the unconnected output from the **port map** command, i.e. the code could also have been written as follows:

```
U1: ex4 port map(a=>a, b=>b, q1=>dout);
```

If an order-controlled port map is used, the omitted signal must be the last one in the port map declaration. If, for example, signal *q1* has to be unconnected, the VHDL code cannot be written as follows:

Example (bad):

```
U1: ex4 port map(a, b, d2);
```

If the code is written in this way, the VHDL compiler and synthesis tool will assume that signal *q2* is unconnected (the last signal). The VHDL code must be written using => in the **port map** command instead.

Example (good):

```
U1: ex4 port map(a=>a, b=>b, q1=>open, q2=> d2);
```

Some ASIC suppliers do not accept unconnected signals from a hierarchical block. For such suppliers the VHDL code has to be rewritten in such a way that signal *q2* is not an output from component ex4. This problem usually arises when a component is reused in a new design.

6.3.2. Unconnected inputs

No silicon supplier will accept an input to a hierarchical block which is "floating" unconnected. If an input on a component is not to be used, the signal should be connected to VCC or GND (+5 V / 3.3 V or 0 V). It is not permissible to map the input directly in the **port map** command: an internal signal must be used. This signal is assigned the value '0' or '1', depending on what function is required. The unused input is then mapped to the internal signal. Example 1 below is an example of how an unconnected input should be mapped. Examples 2 and 3, as described above, are not permitted in the VHDL-87 standard.

Example 1 (good):

```
Architecture rtl of top_level is
  component ex1
    port( a,b:     in   std_logic;
          q1,q2;   out  std_logic);
  end component;

  For U1: ex1 Use entity  work.ex4(rtl);
  signal gnd: std_logic;
  begin
    gnd<='0';
    U1: ex1 port map(a=>gnd, b=>b, q1=>dout, q2=>d2);
  end;
```

Example 2 (bad):

```
Architecture bad of top_level is
  component ex1
    port( a,b:     in   std_logic;
          q1,q2:   out  std_logic);
  end component;

  For U1: ex1 Use entity  work.ex4(rtl);
  begin
    U1: ex1 port map(a=>open, b=>b, q1=>dout, q2=>d2);
                                  -- Error

  end;
```

Example 3 (VHDL-93):

```
Architecture bad of top_level is
  component ex1
    port( a,b:     in   std_logi;
          q1,q2;   out  std_logic);
  end component;

  For U1: ex1 Use entity  work.ex4(rtl);
  begin
    U1: ex1 port map( a=>'0', b=>b, q1=>dout, q2=>d2);
                            -- Error (VHDL-87), valid in VHDL-93

  end;
```

The VHDL-93 standard permits mapping directly to '0' and '1'. The VHDL code can therefore be written as shown in example 3 above if VHDL-93 is used. As previously mentioned, most synthesis tools do not yet support VHDL-93, so alternative 1 still has to be used for design.

6.4. Generic map command

If generic components have been specified in the component to be instanced, their value can be changed during instantiation using the command **generic map**.

```
Syntax:
  generic map(<generic_values>);
```

By using generics, it is possible to design components which can be parameterized. Below is an example of a component which uses generics both for component delay and for the number of inputs, i.e. when the component is instanced, the component's delay and the number of inputs are determined:

```
Entity and_comp is
     generic ( Tdelay:time:=10 ns;
               n:positive:=2);
   port( a: in    bit_vector(n-1 downto 0);
         c: out   bit);
end;

Architecture and_beh of and_comp is
begin
  p0:process(a)
  variable int:bit;
  begin
    int:='1';
    for i in a'length-1 downto 0 loop
      if a(i)='0' then
          int:='0';
      end if;
    end loop;
    c<=int after Tdelay;
  end process;
end;
```

If you want a three-input **and** component with a 12 ns delay and a two-input **and** with an 8 ns delay, the syntax is as follows:

```
Entity ex is
 port( d1,d2,d3,d4,d5:  in   bit;
       q1,q2:           out  bit);
end;

Architecture ex_beh of ex is
 component and_comp
   generic ( Tdelay:time;
             n:positive);
   port( a:  in   bit_vector(n-1 downto 0);
         c:  out  bit);
 end component;
 For U1,U2 : and_comp Use entity  work.and_comp(and_beh);
begin
 U1: and_comp generic map(n=>2,  Tdelay=>8 ns)
      port map(a(0)=>d1, a(1)=>d2, c=>q1);
 U2: and_comp generic map(n=>3, Tdelay=>12 ns)
      port map(a(0)=>d3, a(1)=>d4, a(3)=>d5, c=>q2);
end;
```

The synthesized result is shown in Figure 6.2.

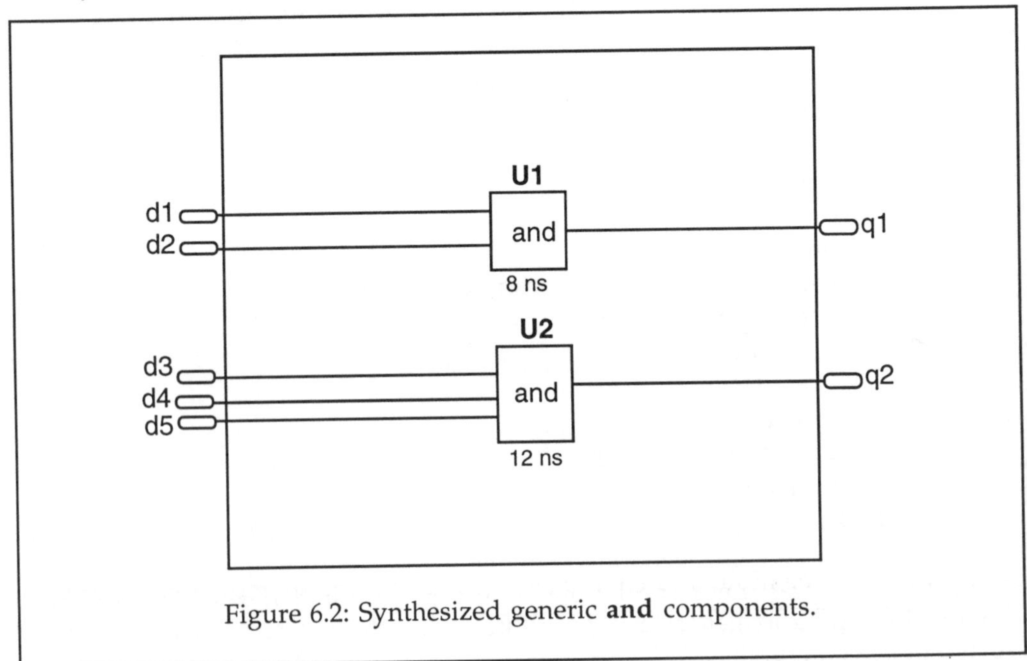

Figure 6.2: Synthesized generic **and** components.

As already mentioned, the component delays (8 and 12 ns respectively) are ignored by synthesis tools. The delays will only be valid during VHDL simulation.

6.5. Generate command

If the same component has to be instanced several times in the same architecture, it may be effective to include **port map** commands in a loop.

Syntax:

```
[label:]For <loop_index> <iteration scheme> generate
      <label>: <entity> port map(<port-association-list>);
end generate [label];
```

Suppose that component c1 has to be instanced five times in component top. This is done using the **generate** command.

Example:

```
Entity top is
  port( a,b: in    std_logic(vector(4 downto 0);
           q:  out  std_logic_vector(4 downto 0));
end;

Architecture rtl of top is
  component c1
    port( a,b: in   std_logic;
             q:  out std_logic);
  end component;
For U1: c1 Use entity  work.c1(rtl);
begin
  c_gen:  For i in 0 to 4 generate
    U: c1 port map(a(i),  b(i),  q(i));
  end generate  c_gen;
end;
```

Some of the simpler synthesis tools do not support the generate command. Most do, however. Using the generate command, the code for instancing the same component several times can be made very general and flexible. All synthesis tools which support generate require <iteration scheme> to be calculable, i.e. it may not be configurable.

6.6. Configuration

A **configuration** binds an **architecture** to an **entity**. The simplest form of configuration is not to have any at all. In this case the architecture simulated will be the last to be compiled. The next step is a default configuration. A default configuration can be used if the component does not contain any other component. Below is an example of entity mux, which has two different architectures: rtl and behv. The configuration specifies which is to be simulated. The name of the configuration which results in the rtl architecture being simulated is mux_rtl, while that for the behv architecture is mux_behv.

Example:

```
Entity mux is
  port( a,b,c,d: in   std_logic_vector(3 downto 0);
        sel:     in   std_logic_vector(1 downto 0);
        q:       out  std_logic_vector(3 downto 0));
end;

Architecture behv of mux f
begin
  process(a,b,c,d,sel)
  variable int: std_logic_vector(3 downto 0);
  begin
    case sel is
      when "00"    ⇒ int<=a;
      when "01"    ⇒ int<=b;
      when "10"    ⇒ int<=c;
      when "11"    ⇒ int<=d;
      when others ⇒ int<=(others=>'X');
    end case;
    q<=int after 10 ns;
  end process;
end;

Architecture rtl of mux is
begin
  process(a,b,c,d,sel)
  begin
    case sel is
      when "00"    ⇒ q<=a;
      when "01"    ⇒ q<=b;
      when "10"    ⇒ q<=c;
```

```
            when others=> q<=d;
         end case;
      end process;
   end;


   -- Default configuration for architecture behv
   Configuration  mux_behv of mux is
      For behv
      end for;
   end mux_behv;

   -- Default configuration for architecture rtl
   Configuration  mux_rtl of mux is
      For rtl
      end  for;
   end mux_rtl;
```

Depending on which of the configurations above is compiled, architecture behv or rtl will be simulated. Unless other components are instanced inside the component, it is normal to omit the configuration in design, i.e. to enter the architecture to be simulated. **Synthesis tools normally ignore all configurations and synthesize the architecture last entered on the tool instead.**

If the component to be simulated contains other components, additional information has to be included in the configuration. A simplified form of configuration is that illustrated earlier in the chapter, namely component specification. The component specification is placed in the declaration part of the architecture for the component to be simulated.

Example:

```
   Entity ex is
     port( a,b:  in    std_logic;
           c:    out   std_logic);
   end;

   Architecture behv of ex is
   begin
     ...
   end;


   Entity top_level is
     component c1
```

```
    port( a,b:  in    std_logic;
          c:    out   std_logic);
  end component;
  For U1: ex Use entity work.ex(rtl); -- Simple form of
  begin                               -- configuration
    ...
  end;
```

The configuration can also be placed in a separate file, just as in the case of default configuration. This configuration can define both the underlying configurations to be used for the components which are going to be instanced and the architecture to be used for the respective entity.

Example:

```
Entity c1 is
  port( a,b: in   std_logic;
        c:   out  std_logic);
end;
Architecture behv of c1 is
begin
  c<=a nor b after 10 ns;
end;

Configuration  c1_conf is
  For behv
  end for;
end;

Entity c2 is
  port( a,b:  in    std_logic;
        c:    out   std_logic);
end;
Architecture behv of c2 is
begin
  c<=a xor b after 12 ns;
end;

Configuration  c2_conf  is
  For behv
  end for;
end;
```

```
Entity top_level is
  port( d1,d2,d3: in   std_logic;
        q1,q2:    out std_logic);
end;

Architecture top_behv of top_level is
  component c1
    port( a,b: in   std_logic;
          c:   out std_logic);
  end component;

  component c2
    port( a,b: in   std_logic;
          c:   out std_logic);
  end component;

  signal i1: std_logic;

  begin
    U1: c1 port map(d1,d2,i1);
    U2: c2 port map(d3,i1,q2);

    q1<=i1;
  end;


Configuration  top_conf is
  For top_behv
    For U1: c1 Use configuration work.c1_conf;
    end for;

    For U2: c2 Use configuration work.c2_conf;
    end for;
  end for
end top_conf;
```

Configuration top_level stipulates that configurations c1_conf and c2_conf are to be used as the configurations for components c1 and c2 respectively. It also stipulates that architecture top_behv is to be used for simulation of component top_level.

In summary it can be said that configuration is used more for designing simulatable models, and less for designing hardware.

6.7. Direct instantiation (VHDL-93)

Direct instantiation of components is permitted in the VHDL-93 standard. Direct instantiation means that neither component declaration nor configuration are needed to instance a component. The examples below show instantiation according to the VHDL-87 standard, followed by the new method according to the VHDL-93 standard. Note that it is still permissible to instance in accordance with VHDL-87 in the new VHDL-93 standard.

Example (VHDL-87):

```
Architecture rtl of top_level is
  component c1
    port( a,b:  in   std_logic;
           q:   out  std_logic);
  end component;
For U1: c1 Use entity  work.c1(rtl);
begin
  U1: c1 port map(a,b,q);
end;
```

Example (VHDL-93):

```
Architecture rtl of top_level is
begin
  U1: entity work.c1(rtl) port map(a,b,q);
end;
```

Specification of the architecture name is not obligatory in the VHDL-93 example. If the architecture name is not specified, the default configuration rules will apply.

Unfortunately, most synthesis tools do not yet support this simple and quick method of instancing a component.

6.8. Components in package

It is possible to define **component** in a **package**. One of the advantages of doing this is that no component declaration is needed in the architecture when the component is instantiated.

Example:

```
package mypack is
  function minimum (a,b:in std_logic_vector)
            return std_logic_vector;

  component mycomp1
    port( clk,resetn,din: in  std_logic;
          q1,q2:          out std_logic);
  end component;

  component mycomp2
    port( a,b:  in  std_logic;
          q:    out std_logic);
  end component;
end mypack;

package body mypack is
  function minimum (a,b:in std_logic_vector)
            return std_logic_vector is
  begin
    if a<b then
       return a;
    else
       return b;
    end if;
  end minimum;
end mypack;
```

The component mycomp1 and 2 can then be used by declare the package (mypack) at the top of the VHDL-code. As the component is declared in the package and the package is stored in library work there is no need for the component declaration or the component specification when the component is instantiated.

Example:

```
Library ieee;
Use ieee.std_logic_1164.ALL;
Use work.mypack.ALL;

entity ex is
    port( clk,resetn: in  std_logic;
          d1,d2:      out std_logic;
          a,b:        in  std_logic_vector(3 downto 0);
          q1,q2,q3:   out std_logic;
```

```
                q4:         out std_logic_vector(3 downto 0));
end;

Architecture rtl of ex is
begin
  U1: mycomp1  port map(clk,resetn,d1,q1,q2);
  U2: mycomp2  port map(d1,d2,q3);

  q4<=minimum(a,b);
end;
```

6.9. Exercises

1. What is hierarchical design?

2. Describe the following block diagram using structural VHDL, i.e. instance component c1 (twice) and component c2 in component top.

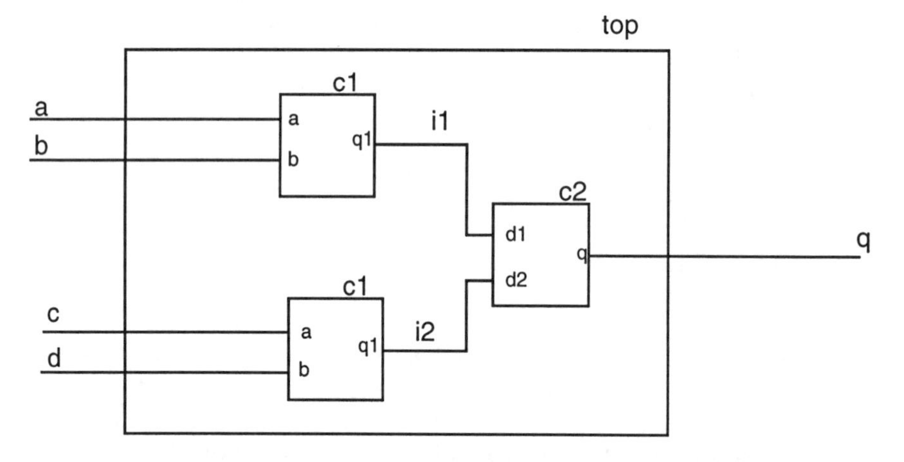

3. (a) Is it permissible to have several entities linked to an architecture?

(b) Is it permissible to have several architectures linked to an entity?

4. If there are several architectures, how do you define which is to be simulated/synthesized?

5. What command can be used to change a generic parameter during instantiation?

6. (a) Design a generic OR gate which has N inputs and a delay of Tdelay ns. N should be equal to 3 default and Tdelay equal to 3 ns.

(b) Instance the generic component twice in component top. Change the generic parameters to N=4 and Tdelay=2 ns in one component and to N=5 and Tdelay=3 ns in the other.

7. Instance the following component (c1) 10 times in component top.

```
Entity c1 is
   port( a,b:  in   std_logic;
         q:    out  std_logic);
end;
```

Component top should have the following entity:

```
Entity top is
   port( a,b:  in   std_logic_vector(9 downto 0);
         q:    out  std_logic_vector(9 downto 0));
end;
```

Hint: Use the **generate** command.

7
RAM and ROM

Both RAM and ROM can be included in ASIC and FPGA designs. This chapter examines how the VHDL code should be written.

7.1. ROM

There are two ways of defining ROM (Read Only Memory) in VHDL code:

- Using an array constant
- Instancing a technology-specific ROM

7.1.1. Using an array constant

Describing a ROM with an array is only area-efficient if the ROM is relatively small. If the ROM is large, the area overhead will be much larger than for alternative 2. It is suggested that the ROM to be declared using an array constant should be placed in a package. This facilitates reuse of the ROM. The package should define the size of the ROM using constants. Below is an example of a ROM of 4 x 8 bits (4 bytes):

```
Library ieee;
Use ieee.std_logic_1164.ALL;

Package rom is
  constant   rom_width:integer:=8;
  constant   rom_length:integer:=4;
  subtype   rom_word is std_logic_vector
              (rom_width-1 downto 0);
  type rom_table is array (0 to rom_length-1) of rom_word;
```

```
    constant rom: rom_table:= rom_table'( "00101111",
                                          "11010000",
                                          "01101010",
                                          "11101101");
end;
```

The ROM can then be used in a VHDL design as follows:

```
Library ieee;
Use ieee.std_logic_1164.ALL;
Use ieee.std_logic_unsigned.ALL;
Use work.rom.ALL;

Entity waveform is
 port( addr: in   std_logic_vector(1 downto 0);
       q:    out  rom_word);
end;

Architecture testbench of waveform is
begin
  q<=rom(conv_integer(addr));      -- Read data from ROM
end;
```

The synthesis tool will construct the ROM using combinational logic. Figure 7.1 shows the synthesis result of the above example.

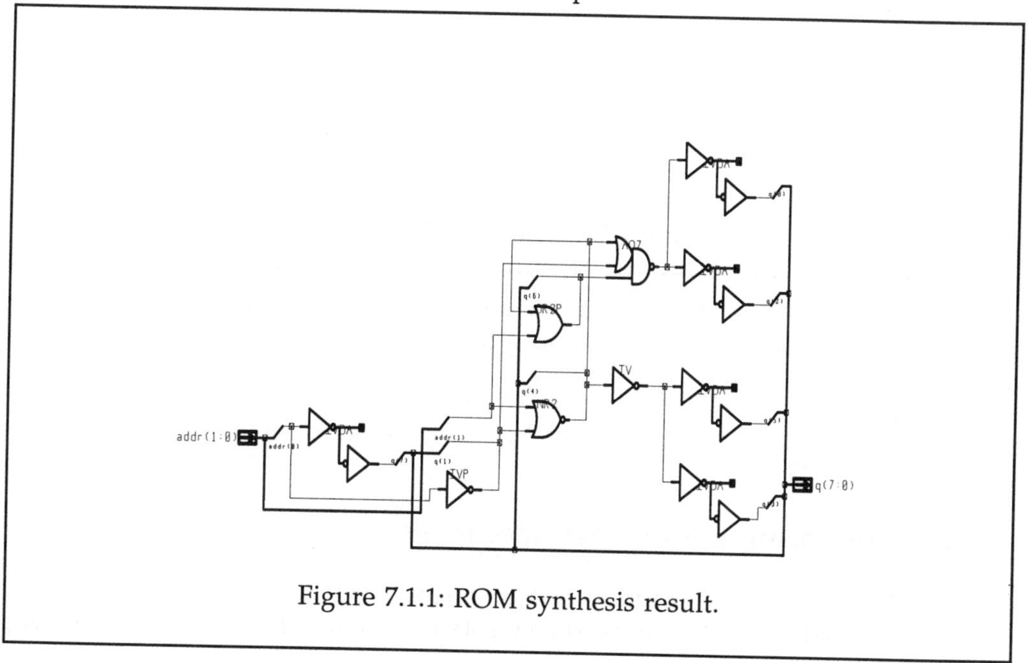

Figure 7.1.1: ROM synthesis result.

In the above example the ROM address (addr) was not calculable in advance. If the address could be calculated, output signal *q* would be constant, which means that the synthesis tool would not generate any logic for such a design

An example of calculable ROM address might be:

```
Architecture testbench of waveform is
begin
   q<=rom(2);              -- Read data from ROM
end;
```

If the above example is synthesized, Figure 7.2 results.

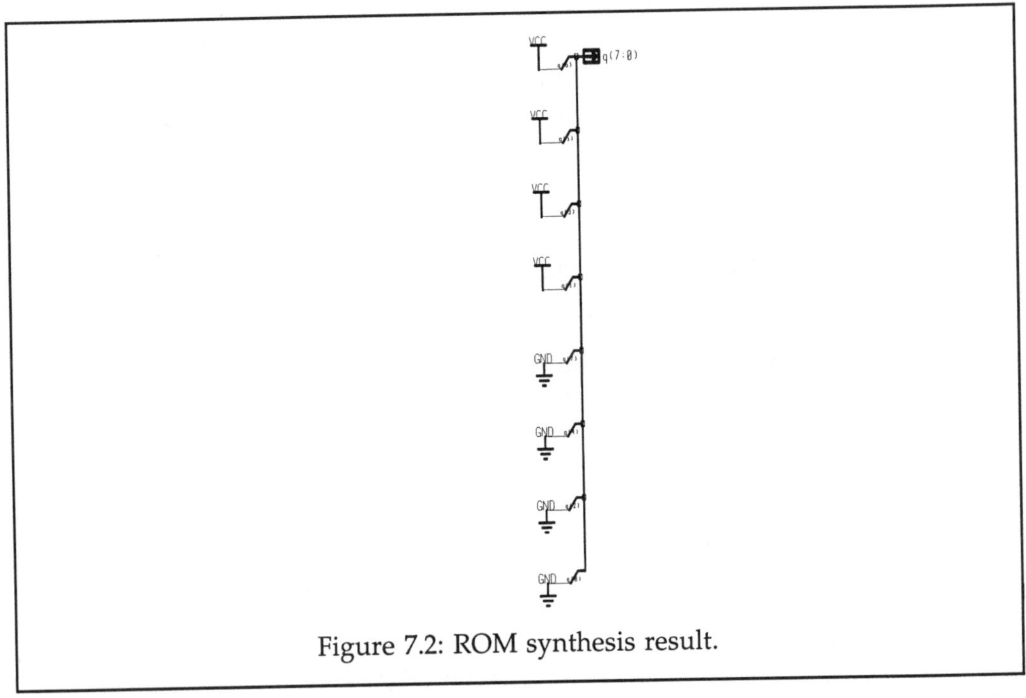

Figure 7.2: ROM synthesis result.

As the address to the ROM could be calculated, no logic was generated. The VHDL code only assigned output signal *q* the value of rom(2), which is defined as a constant with the value "01101010".

Different sizes of ROM can be obtained by changing the constants in the package.

7.1.2. Instancing a technology-specific ROM

Larger ROMs should be instanced, which means that the selected technology must have ROM in its library. Instancing the ROM has several disadvantages.

Among other things, the VHDL code becomes technology-dependent, as there is a reference to a specific ROM macro in the library. The other disadvantage is that problems can arise with the VHDL simulation if there is no VHDL model for the ROM. Alternatively, a mix level (both VHDL and gate level) simulator can be used. The ROM is instanced just like any other component (for information on instantiation, see Chapter 6, "Structural VHDL").

7.1.3. Summary

The two alternatives for introducing ROM are summarized in Table 7.1.

	Area efficient	Easy to simulate	Technology independent
alt 1	NO	YES	YES
alt 2	YES	NO	NO

Table 7.1: ROM.

7.2. RAM

Just as with ROM, there are two alternative procedures when it comes to introducing RAM (Random Access Memory) in the VHDL code:

- Using registers
- Instancing RAM

7.2.1. Using registers

Registers can be used for very small RAM. With larger RAM, the area overhead will be much larger for using registers than for instancing RAM. As explained in Chapter 4, registers are described using a clocked process:

```
process(clk,resetn)
begin
  if resetn='0' then
    q<=(others=>'0');
  elsif clk'event and clk='1' then
    if wr='1' then
      q<=data;
    end if;
  end if;
end process;
```

7.2.2. Instancing RAM

It is not possible to obtain a good synthesis result for RAM unless the RAM is instanced. Just as with ROM, the disadvantages are that the VHDL code becomes technology-dependent and problems can arise with the VHDL simulation. The RAM on offer varies according to the choice of manufacturer and technology.

Several RAMs can be instanced in order to achieve the required size. The VHDL **generate** command may be useful for such instantiation. If, for example, a 4 x 1-bit RAM is instanced four times, a 4 x 4-bit RAM is obtained:

```
Architecture rtl of RAM4 is
component RAM4x1
  port( d,a0,a1,we:  in   std_logic;
           q:         out std_logic);
end component;
--for U1:RAM4x1 use Entity work.RAM4x1(rtl);
                   -- Only used for simulation
begin
  for i in 0 to 3 generate
    RAM_b: RAM4x1 port map (d_in(i),a0,a1, write,d_out(i));
  end generate;
end;
```

RAM can also be introduced automatically by some behavioural synthesis tools. This is the simplest, fastest and best way of introducing RAM, as it is area-efficient, easy to simulate and technology-independent. The VHDL code uses an array which is read and written like any other signal. The behavioural tool can then replace this array with any RAM from the selected technology (see Chapter 17, "Behavioural synthesis").

7.3. Exercises

1. What are the advantages and disadvantages of the two ways of introducing ROM in VHDL code?

2. What are the advantages and disadvantages of the three ways of introducing RAM in VHDL code?

3. Design a component c1 which instances a 4 x 8-bit RAM. The library only contains a 4 x 1-bit RAM:

```
Entity RAM4x1 is
   port( d,a0,a1,we:    in   std_logic;
            q:          out  std_logic);
end;
```

8

Testbench

Once the designer has described the design, the design must be verified to check that the specification has been followed.

The commonest verification method is to apply input stimuli signals during a simulation and then "read" the output signals from the design (type logic analyser). A major disadvantage of using the simulator's input stimuli language is that it varies from simulator to simulator.

Another method of verification is to write the test model generation and the output signal check in VHDL. This means that you design a testbench which both provides input signals and tests the output signals from the design. As the testbench too can generate errors, it has to be tested as well. The recommendation is to use both the VHDL testbench and to verify the components using models of the environment, known as system simulation (see Chapter 11, "Design methodology").

The advantage of system simulation is that the stimuli for the component are generated by models which are intended to represent reality. The disadvantage of system simulation is that it takes quite a long time. The advantages of the VHDL testbench are its speed and the fact that it is platform-independent, as it is described in VHDL. The disadvantage is that the VHDL testbench can contain the same logical errors as the design, i.e. the designer thinks the test is correct when it is in fact incorrect.

In some cases, however, it may be beneficial to build a test environment in full or in part. Figure 8.1 shows the logical structure of a testbench. The model of the design receives input stimuli (input signals) and gives a response (output signals) to the testbench.

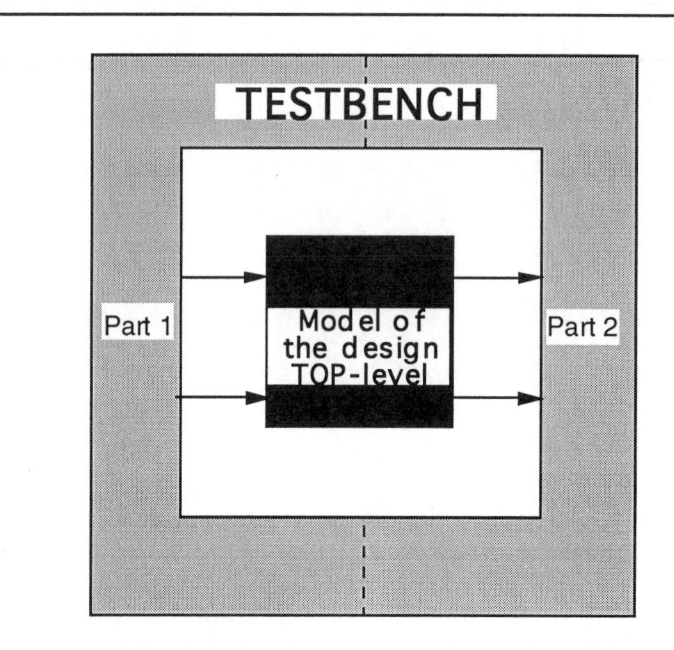

Figure 8.1: Model of the design and the test environment (testbench).

The testbench is made up of two parts. One part generates the input signals for the model to be tested, while the other part checks the output signals from the model.

Part 1
Below is an example of how input signals for the model can be generated:

```
in_model <=  '1' after 5 ns,
             '0' after 100 ns,
             '1' after 200 ns,
             '0' after 300 ns + 4 ns;

clk <= not clk after 50 ns;
```

Signal *in_model* provides a square-topped signal which inverts at 5, 100, 200 and 304 ns. Signal *clk* will oscillate with a period of 100 ns.

The clock assignment statement, clk<=not clk after 50 ns, can lead to problems if *clk* is not declared as data type bit. As the recommendation for VHDL design is to use data type std_logic as much as possible, *clk* in the testbench should also be of the same type to avoid type conversion. The problem is that all signals in VHDL are initiated to signal'LEFT. In the case of std_logic this means the value 'U' (unknown). The inversion of 'U' is still 'U'. This problem can easily be solved by initiating *clk* to the value '0', for example, in the declaration of *clk*:

```
signal clk:std_logic:='0';
```

If the component to be tested contains state machines, it may be practical to use the **wait** command to assign the input signals *a* value in the respective states:

```
process
begin
  in_signal1<='1';                -- Input signals value at reset
  in_signal2<='0';
  wait until resetn='1';

  wait until clk='1';
  in_signal1<='1';
  in_signal2<='0';

  wait until clk='1';
  in_signal1<='1';
  in_signal2<='0';

  wait until clk='1';
  in_signal1<='1';
  in_signal2<='0';
end process;
```

As an alternative to waiting for the clock's edge, it is possible to declare the period for *clk* as a constant so that it can be used in the wait command. For example:

```
constant period:time:=50 ns;

wait for period;
```

The advantage of this method is that a behavioural model may not contain a clock. It is also possible to wait for several clock cycles on one line:

```
wait for 2 * period;
```

If the wait for command is used in several clock periods, it is recommended that the time should be declared as a constant:

```
constant period:time:=50 ns;
constant period2:time:=2 * period;
  ..
wait for period2;
```

This is because the simulator might otherwise need to calculate the value of 2 * period several times during the same simulation if the line is executed several times. This, of course, takes up simulation time unnecessarily.

Part 2

Part 2 of the testbench checks the output signals from the model. The assert statement is often used for this stage. An **assert** command is executed if the statement is false.

An example of the **assert** command is:

```
process(clk)
begin
assert now < 900 ns
  report "stopping simulator (max simulation time 900 ns)"
  severity Failure;
end process;
```

The simulator stops after 900 ns.

```
process(out_model, in_model)
begin
if out_model = 1 and in_model = 1 then
  assert false
  report "The signals out and in are '1' at the same time"
  severity Error;
end process;
```

If the output signals and input signals are the same, the simulator stops and prints out the error code. The above example contains an **if** statement which determines whether the assert statement is to be executed or not. The assert condition in the example is always false ("assert false").

It is also possible to use an output signal which is deactivated if the test does not go well:

```
process(out_model, in_model)
begin
  if out_model = 1 and in_model = 1 then
    test_ok<='0';
  end if;
end process;
```

The technique of using the **wait until** clk command can also be used to check the output signals from a state machine, for example. The testbench should be written in such a way that it can also be used at gate level. At gate level the output signals will not change immediately after a clock edge, which is the norm in VHDL simulation. Allowance has to be made for this in the testbench. A simple method is to use the command **wait until** clk='0';. The output signals must, however, also be stable at gate level after half the clock period. Alternatively, verification of the output signals can take place just before or simultaneously with the clock edge.

The model of the design can be a VHDL code, netlist or another format. This means that the same testbench can be used during different phases of the design work (Figure 8.2). The design model is verified using the testbench. It is not until a behavioural model of the design and above that the model is refined to an RTL model, after which the netlist model is verified. Finally, a model with delay from the routing is verified.

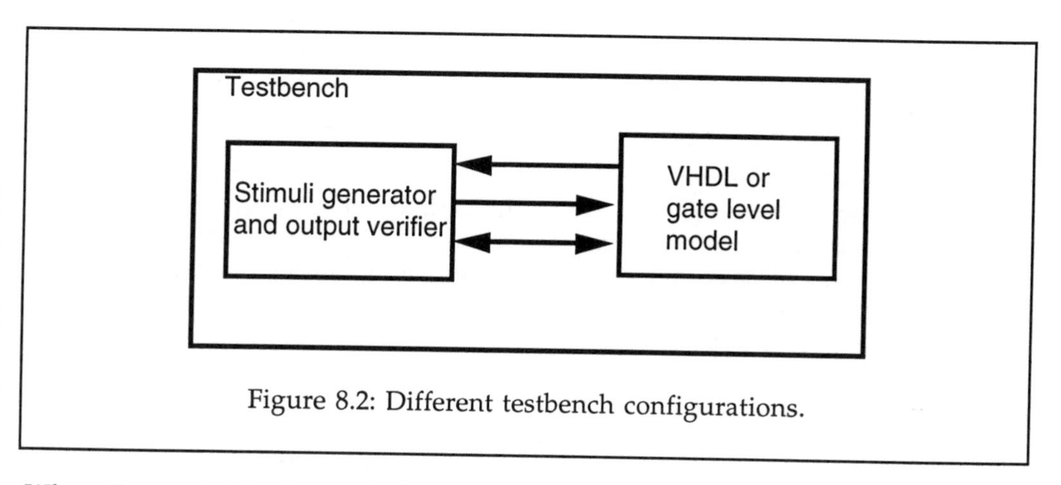

Figure 8.2: Different testbench configurations.

When the model is changed, the I/O interface with the testbench also has to be changed in some cases (Figure 8.3).

At behavioural level it is possible to work with more abstract data types than at netlist level. This produces different interfaces between testbench and model. An example would be using a floating point number at behavioural level, but translating the floating point number to a number of interconnections (vector) at a lower level of abstraction.

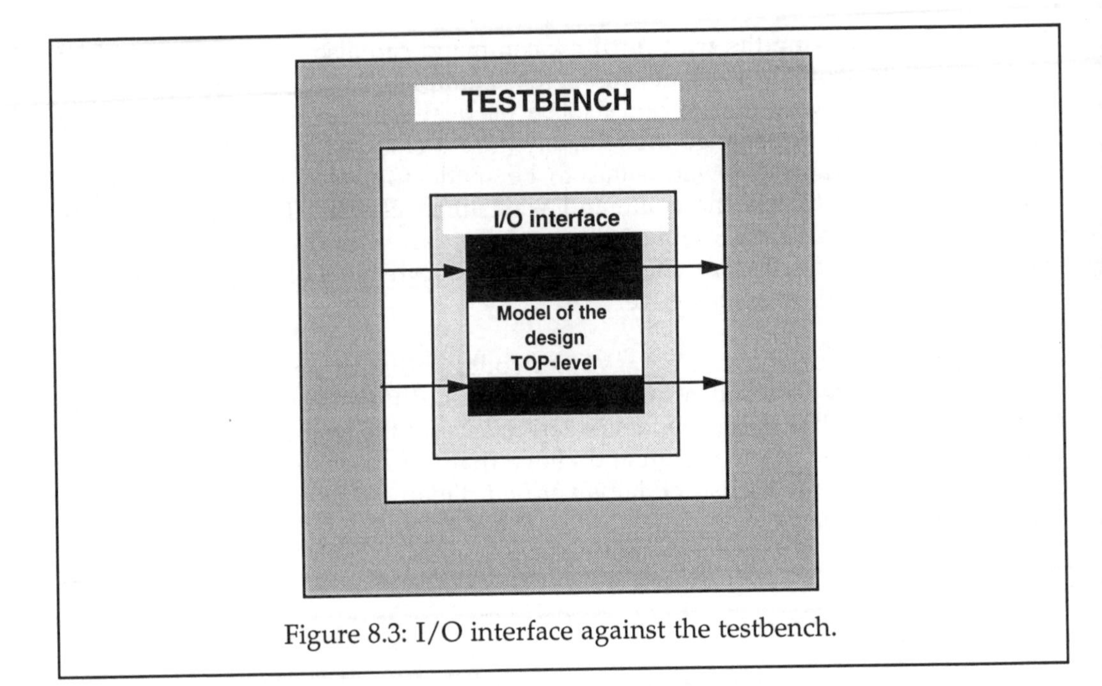

Figure 8.3: I/O interface against the testbench.

It is normally easier to have a uniform interface (entity) in all models. This can be accomplished by also using vectors in the entity at behavioural level and then using more abstract data types in the architecture. One advantage of this is that it becomes very easy to swap between the various models. The only thing that has to be done is to change the component specification in the testbench to the component to be simulated.

Example:

Behavioural level:	**For** U1: **Use entity** work.my_comp(behv);
RTL VHDL model:	**For** U1: **Use entity** work.my_comp(rtl);
Gate level:	**For** U1: **Use entity** work.my_comp(gate);

8.1. Different levels of testbench

The testbenches can be divided into three different classes, with the first category being the least complex and third the most complex (see Table 8.1). The difference between the various testbenches lies in whether they check the output signals' value and time. All the testbench variants generate the input signals for the model to be tested.

	Input stimuli	Test of output signals	Timing test
Class 1	Testbench	Manual	Manual
Class 2	Testbench	Testbench	Manual
Class 3	Testbench	Testbench	Testbench

Table 8.1: Different types of testbench.

In a normal testbench the component to be tested is instanced with structural VHDL. As it is the testbench that generates the input stimuli for the component, the testbench does not have to contain any input signal in the entity. Suppose that we have the following VHDL component for testing with a testbench.

```
Library ieee;
Use ieee.std_logic_1164.ALL;
Use ieee.std_logic_unsigned.ALL;

Entity my_comp is
 port( clk,resetn,d_in:     in  std_logic;
       a,b:                 in  std_logic_vector(2 downto 0);
       d_out,en,overflow:   out std_logic;
       q:                   out std_logic_vector(2 downto 0));
end;

Architecture rtl of my_comp is
begin
  Controller:Block
  type state_type is (s0,s1,s2);
  signal state:state_type;
  begin
    process(clk,resetn)
    begin
      if resetn='0' then
        state<=s0;
        d_out<='1';
        en<='0';
      elsif clk'event and clk='1' then
        case state is
          when s0 => if d_in='1' then
                       state<=s1;
                     end if;
                     d_out<='1';
                     en<='0';
```

```
                when s1 =>  if d_in='0' then
                              state<=s2;
                            end if;
                            d_out<='0';
                            en<='1';

                when s2 =>  state<=s0;
                            d_out<='0';
                            en<='0';

                when others  => state<=s0;
                                d_out<='1';
                                en<='0';

              end case;
            end if;
          end process;

        end block;

        check:Block
        signal q_b: std_logic_vector(2 downto 0);
        begin
          q<=q_b;
          q_b<=a + b;
          overflow<='1' when q_b>2 else '0';
        end block;
      end;
```

Class 1

Testbenches of type class 1 only generate the input signals for the model. The output signals must be verified manually.

```
Library ieee;
Use ieee.std_logic_1164.ALL;

Entity test_my_comp is
 port( d_out,en,overflow:out std_logic;
       q:                out std_logic_vector(2 downto 0));
end;

Architecture testbench of test_my_comp is
 component my_comp
   port( clk,resetn,d_in: in  std_logic;
         a,b:             in  std_logic_vector(2 downto 0);
         d_out,en:        out std_logic;
```

```
            overflow:       out std_logic;
            q:              out std_logic_vector(2 downto 0));
    end component;
    signal clk:     std_logic:='0';
    signal resetn:  std_logic:='0';
    signal d_in:    std_logic;
    signal a,b:     std_logic_vector(2 downto 0);
    For U1:my_comp Use Entity work.my_comp(rtl);
    begin
     U1:my_comp port map( clk,resetn,d_in,a,b,d_out,en,
                          overflow,q);
     clk<=not clk after 50 ns;
     resetn<='1' after 125 ns;
     a<=  "000",
          "010" after 125 ns,
          "100" after 175 ns;
     b<=  "000",
          "100" after 125 ns,
          "011" after 175 ns;
    process
    begin
      d_in<='0';
      wait until resetn='1';
      d_in<='1';
      wait until clk='1';
      d_in<='0' after 10 ns;
      wait;                    -- Wait for end of simulation
    end process;
  end;
```

The class 1 testbench above generates the input signals for the component, while the output signals have to be verified manually. In the example the output signals have been included in the entity of the testbench. This is not necessary, however, as it is possible in the simulation to take the output signals from component my_comp a hierarchical level below verification. From the documentation point of view, however, it is an advantage to have the signals that are to be verified manually in the entity of the testbench. If this rule is always followed, it is easy to decide which of the output signals have to be verified manually.

Class 2
Testbenches of this class not only generate the input signals for the model but also verify that the output signals from the model have the correct value. The timing has to be verified manually, however.

An example of a testbench of type class 2 might be:

```
Library ieee;
Use ieee.std_logic_1164.ALL;

Entity test2_my_comp is
 port(test_ok:   out   std_logic:='H');
end;

Architecture testbench of test2_my_comp is
 component my_comp
  port( clk,resetn,d_in:      in  std_logic;
        a,b:                  in  std_logic_vector(2 downto 0);
        d_out,en,overflow:    out std_logic;
        q:                    out std_logic_vector(2 downto 0));
 end component;

 signal clk:                  std_logic:='0';
 signal resetn:               std_logic:='0';
 signal d_in:                 std_logic;
 signal a,b:                  std_logic_vector(2 downto 0);
 signal d_out,en,overflow:    std_logic;
 signal q:                    std_logic_vector(2 downto 0);

 For U1:my_comp Use Entity work.my_comp(rtl);

 begin
   U1:my_comp port map( clk,resetn,d_in,a,b,d_out,en,
                        overflow,q);

   clk<=not clk after 50 ns;
   resetn<='1' after 125 ns;

   a<= "000",
       "010" after 125 ns,
       "100" after 175 ns;

   b<= "000",
       "100" after 125 ns,
       "011" after 175 ns;
```

```
p0:process
begin
  wait for 125 ns;
  if q/="000" or overflow/='0' then
    test_ok<='0';
  end if;

  wait for 50 ns;
  if q/="110" or overflow/='1' then
    test_ok<='0';
  end if;

  wait for 50 ns;
  if q/="111" or overflow/='1' then
    test_ok<='0';
  end if;

  wait;
end process;

process
begin
  d_in<='0';
  wait until resetn='1';
  d_in<='1';

  wait until clk='1';
  d_in<='0' after 10 ns;
  wait;
end process;

p1:process
begin
  wait for 30 ns;
  if en/='0' or d_out/='1' then
    test_ok<='0';
  end if;

  wait until resetn='1';

  wait until clk='1';
  if en/='0' or d_out/='1' then
    test_ok<='0';
  end if;

  wait until clk='1';

  wait until clk='1';
```

```
            if en/='1' or d_out/='0' then
              test_ok<='0';
            end if;
            wait until clk='1';
            if en/='0' or d_out/='0' then
              test_ok<='0';
            end if;
            wait;
          end process;
        end;
```

The class 2 testbench above tested the output signals by using a *Test_ok* signal. *Test_ok* is set to the value 'H' (weak 1) in the entity declaration.

The *Test_ok* signal is declared as data type std_logic. Std_logic is a resolved data type. This means that if several sources are driving the signal, a predefined resolution function will be called. This function decides which value the output signal will be given in the simulation. The above testbench drives signal *Test_ok* continuously to the value 'H'' and, in the event of an error, also to the value '0'. As *Test_ok* is driven by several drivers simultaneously, the resolution function with input parameters 'H' and '0' is called, with value '0' winning. If, on the other hand, signal *Test_ok* still has value 'H' after the simulation, the testbench has not found any errors and the component has passed the test.

In the above example *Test_ok* has to be initiated to 'H' in the entity and not '1'. This is because *Test_ok* has two drivers linked to it. Processes p0 and p1 both drive *Test_ok*. This means that the resolution function will be called with input arguments 'H' and 'H' from the start. If one of the processes finds an error, *Test_ok* is set to '0' by the driver of that process, i.e. the resolution function is called with input arguments '0' and 'H', with '0' winning. If *Test_ok* had been initiated to '1' instead of 'H', the resolution function would have been called with input arguments '0' and '1' in the event of an error, the result of which would have been 'X'. If, on the other hand, the testbench had been written in such a way that there was just one concurrent command (in this case a process) driving *Test_ok*, then *Test_ok* could have been initiated to '1' without a problem.

The test of the output signals and d_out is carried out simultaneously with the clock edge. Neither the VHDL models nor the gate level has managed to update its outputs, i.e. the outputs still have the old value. The VHDL models' outputs will be updated 1 delta after the test has been performed. At gate level the change in the outputs probably waits for a number of nanoseconds (depending on the choice of technology).

As an alternative to signal *Test_ok*, the command **assert** could have been used to report errors. In the **assert** command it is possible to set severity level, so that the simulation either stops or continues when an error is encountered. As default, most VHDL simulators stop at severity level error, although it is normally possible to set the level at which the simulator is to stop. It is, of course, possible to combine the *Test_ok* output signal with the **assert** command. The advantage of this method is that it is easy to see whether the test has been passed using the *Test_ok* signal, but the **assert** command also prints out in plain text which of the tests have not been passed.

Example:

```
if en/='0' or d_out/='0' then
   test_ok<='0';
   assert false
   report "en/='0' or d_out/='0' "
   severity warning;
end if;
```

Class 3

Testbenches of this class generate the input signals for the model and check the value and timing of the output signals.

An example of a testbench of type class 3 might be:

```
Library ieee;
Use ieee.std_logic_1164.ALL;

Entity test3_my_comp is
 port(test_ok:    out   std_logic:='H');
end;

Architecture testbench of test3_my_comp is
 component my_comp
  port( clk,resetn,d_in:       in  std_logic;
        a,b:                   in  std_logic_vector(2 downto 0);
        d_out,en,overflow:     out std_logic;
        q:                     out std_logic_vector(2 downto 0));
 end component;

 signal clk:                   std_logic:='0';
 signal resetn:                std_logic:='0';
 signal d_in:                  std_logic;
 signal a,b:                   std_logic_vector(2 downto 0);
```

```
signal d_out,en,overflow:    std_logic;
signal q:                    std_logic_vector(2 downto 0);

For U1:my_comp Use Entity work.my_comp(rtl);
begin
  U1:my_comp port map(clk,resetn,d_in,a,b,
                      d_out,en,overflow,q);

  clk<=not clk after 50 ns;
  resetn<='1' after 125 ns;

  a<= "000",
      "010" after 125 ns,
      "100" after 175 ns;

  b<= "000",
      "100" after 125 ns,
      "011" after 175 ns;

  process
  begin
    wait for 125 ns;
    if q/="000" or not q'stable(15 ns) then
      test_ok<='0';
    elsif overflow/='0' or not overflow'stable(15 ns) then
      test_ok<='0';
    end if;

    wait for 50 ns;
    if q/="110" or not q'stable(15 ns) then
      test_ok<='0';
    elsif overflow/='1' or not overflow'stable(15 ns) then
      test_ok<='0';
    end if;

    wait for 50 ns;
    if q/="111" or not q'stable(15 ns) then
      test_ok<='0';
    elsif overflow/='1' or not overflow'stable(15 ns) then
      test_ok<='0';
    end if;

    wait;
  end process;

  process
  begin
```

```
    d_in<='0';
    wait until resetn='1';
    d_in<='1';
    wait until clk='1';
    d_in<='0' after 10 ns;

    wait;
  end process;

  process
  begin
    wait for 30 ns;
    if en/='0' or d_out/='1' then
      test_ok<='0';
    end if;

    wait until resetn='1';
    wait until clk='1';
    if en/='0' or d_out/='1' then
      test_ok<='0';
    elsif not en'stable(80 ns) or not d_out'stable(80 ns) then
      test_ok<='0';
    end if;

    wait until clk='1';
    wait until clk='1';
    if en/='1' or d_out/='0' then
      test_ok<='0';
    elsif not en'stable(80 ns) or not d_out'stable(80 ns) then
      test_ok<='0';
    end if;

    wait until clk='1';
    if en/='0' or d_out/='0' then
      test_ok<='0';
    elsif not en'stable(80 ns) or not d_out'stable(80 ns) then
      test_ok<='0';
    end if;

    wait;
  end process;
end;
```

The class 3 testbench above also checks that the output signals are stable for 15 ns and 80 ns respectively. Addition q<=a + b is checked with command q'stable(15) after 50 ns, i.e. the addition must take max. 50-15 = 35 ns for the test to go through. At a positive clock edge output signals en and d_out are tested to see if they have been stable for 80 ns, i.e. they have 100-80 = 10 ns from the internal flip-flop to the circuit's output signal. This timing constraint is satisfied without difficulty in the VHDL simulation. It is only in gate level simulation that a timing violation can occur.

8.2. Pull up/down

If an input or output signal of the component to be designed has to have a pull up or pull down in the IO cell, the pull up/down can be described in two ways:

- In the VHDL component
- In the VHDL testbench

Suppose we have the following component which is to be given a pull up at bidirectional output io.

```
Library ieee;
Use ieee.std_logic_1164.ALL;

Entity my_comp is
  port( a,b: in    std_logic;
        io:  inout std_logic;
        q:   out   std_logic);
end;

Architecture rtl of my_comp is
begin
    q<=a and io;
    io<='1' when a='1' and b='0' else 'Z';
end;
```

The way to model this is by assigning the bidirectional signal value 'H' (weak '1'). The signal must have data type std_logic. As previously mentioned, std_logic is a resolved data type. As signal io is driven continuously to value 'H' and sometimes also to other values ('0' or '1') at the same time, the resolution function will be called. If the bidirectional signal is driven by the testbench or internally by the circuit to value '0', for example, the resolution function is called with input arguments 'H' and '0', with value '0' winning. This is also in accordance with reality.

If we put the description of the pull up in the component, the following architecture is obtained:

```
Architecture rtl of my_comp is
begin
  io<='H';
  q<=a and io;
  io<='1' when a='1' and b='0' else 'Z';
end;
```

If we only want to initiate the signal in the component, the following entity is obtained:

```
Library ieee;
Use ieee.std_logic_1164.ALL;
Entity my_comp is
  port( a,b: in     std_logic;
        io:  inout  std_logic:='H';
        q:   out    std_logic);
end;
```

Note that as soon as the signal is assigned to a value in the architecture the initiated value is "disconnected" from the signal.

The synthesis tool will ignore the fact that port *io* has been pulled up/initiated to 'H'. This pull up must be put in when connecting and choosing what sort of IO cell the component is to have. In VHDL simulation, however, port *io* will be given the correct value ('H') if there is no driver (internal or external) connected to the port.

As an alternative to the above, account can be taken of the pull up in the testbench. Below is a simple class 1 testbench in which account has also been taken of a pull up.

```
Library ieee;
Use ieee.std_logic_1164.ALL;
Entity tb is
  port(  io:  inout std_logic;
         q:   out   std_logic);
end;
Architecture testbench of tb is
begin
  io<='H';                  -- Pull up
  a<= '0',
       '1' after 100 ns,
       '0' after 50 ns;
```

```
    b<=  '0',
         '1' after 150 ns,
         '0' after 50 ns;

    io<= '0',
         'Z' after 100 ns,        -- io internally driven
         '0' after 50 ns;
end;
```

8.3. Several components in the same testbench

If several components which are to be connected to each other are designed at the same time, all can be tested in the same testbench. In this case some of the input signals are controlled by the components and some by the testbench. In this test configuration a mixture of system simulation and testbench simulation is being tested. Suppose that we have the logical interconnection of two components shown in Figure 8.4.

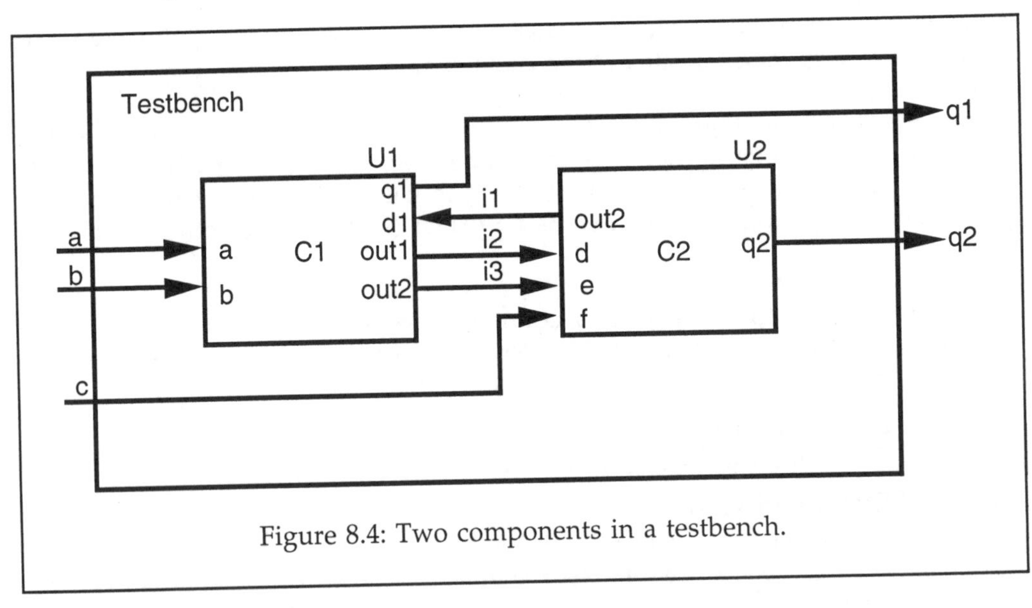

Figure 8.4: Two components in a testbench.

The testbench can then be written as follows:

```
Library ieee;
Use ieee.std_logic_1164.ALL;

Entity tb2comp is
 port( q1,q2: out std_logic);
end;
```

```
Architecture testbench of tb2comp f
 component cl
  port( a,b,d1:          in     std_logic;
        q1,out1,out2:    out    std_logic);
 end component;

 component c2
  port( d,e,f:           in     std_logic;
        q2,out2:         out    std_logic);
 end component;

 For U1:cl Use entity work.c1(rtl);
 For U2:c2 Use entity work.c2(rtl);

 signal a,b,c,i1,i2,i3: std_logic;
 begin
  U1:c1 port map(a,b,i1,q1,i2,i3);
  U2:c2 port map(i2,i3,c,q2,i1);

  a<=   '1',
        '0' after 50 ns,
        '1' after 125 ns;

  b<=   '0',
        '1' after 70 ns,
        '0' after 125 ns,
        '1' after 225 ns;

  c<=   '1',
        '0' after 50 ns,
        '1' after 150 ns,
        '0' after 250 ns;

end;
```

8.4. Waveform generators

Another way of generating input stimuli in the testbench is to use what is known as a waveform generator. The waveform generator makes use of ROM which contains information on the stimuli values.

The preceding chapter explained how ROM can be described using arrays. It is advisable to put the definition and content of the ROM in a package.

Suppose that we have the testbench shown in Figure 8.5.

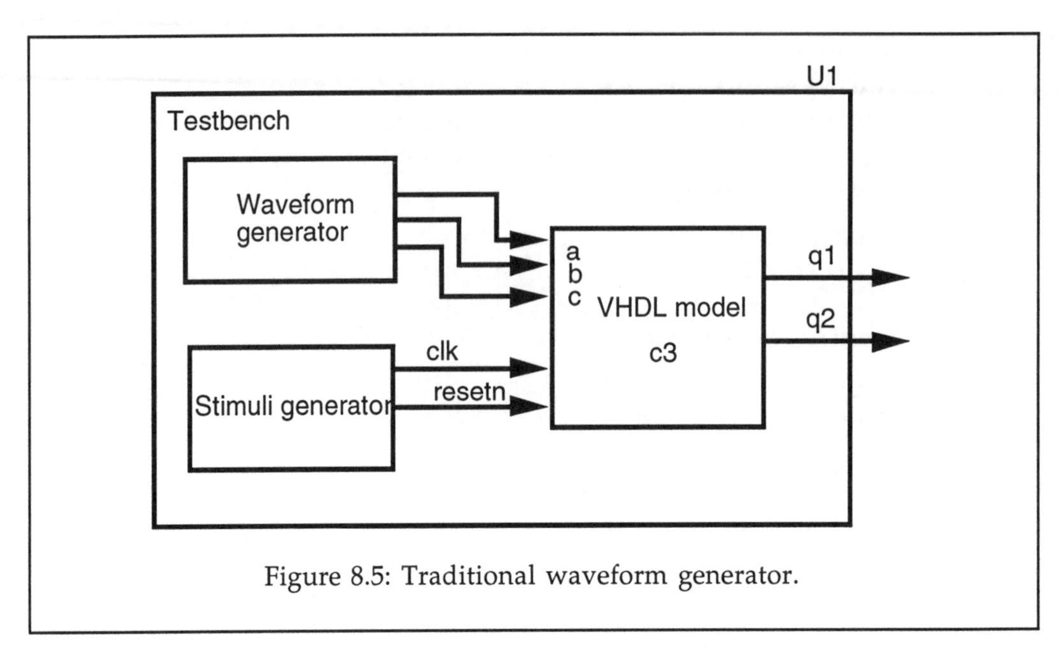

Figure 8.5: Traditional waveform generator.

Below is an example of a testbench which uses a waveform generator to generate stimuli for component c3.

```
Library ieee;
Use ieee.std_logic_1164.ALL;

Package rom is
  constant   rom_width:integer:=3;
  constant   rom_length:integer:=10;

  subtype rom_word is std_logic_vector(rom_width-1 downto 0);

  type  rom_table is array (0 to rom_length-1) of rom_word;

  constant rom: rom_table:= rom_table'( "001",
                                         "110",
                                         "011",
                                         "111",
                                         "101",
                                         "000",
                                         "101",
                                         "001",
                                         "101",
                                         "010");
end;
```

```
Library ieee;
Use ieee.std_logic_1164.ALL;
Use ieee.std_logic_unsigned.ALL;
Use work.rom.ALL;

Entity waveform is
 port( q1,q2: out std_logic);
end;

Architecture testbench of waveform is
component c3
   port( clk,resetn,a,b,c:  in   std_logic;
         q1,q2:             out  std_logic);
end component;

For U1:c3 Use entity work.c3(rtl);

signal clk:  std_logic:='0';
signal resetn: std_logic;
signal step:std_logic_vector(3 downto 0);
signal waves:rom_word;
begin
  resetn<= '0',
           '1' after 25 ns;

  clk<=not clk after 50 ns;

  process(clk,resetn)
  begin
    if resetn='0' then
      step<=(others=>'0');             -- Restart the waveform

    elsif clk'event and clk='1' then
      if step/=rom_length-1 then
        step<=step+1;                  -- Increment the waveform
      else
        step<=(others=>'0');           -- Restart the waveform
      end if;
    end if;
  end process;

  waves<=rom(conv_integer(step));
            -- Read waveform value from ROM

  U1:c3 port map( clk,resetn,waves(2),waves(1),waves(0),
                  q1,q2);
end;
```

The test pattern shown in Figure 8.6 will be obtained from the waveform generator during simulation (the first four clock cycles).

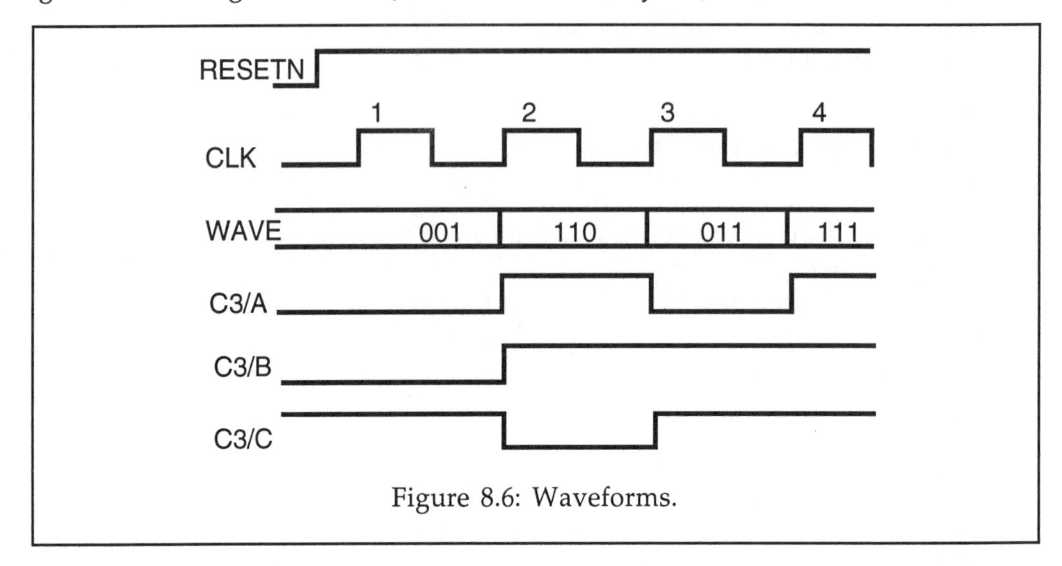

Figure 8.6: Waveforms.

As the waveform generator is based on synthesizable ROM, this technique can also be used in design. The method can be much more area-efficient in some applications than using a state machine to generate an output signal sequence.

If the input signals from the waveform generator do not change value often, gates are wasted when the traditional method of describing a waveform generator is used. In the case of design, and also in the case of testbenches, it may be effective just to specify the ROM value at change and the number of clock cycles that the value should be stable for. A ROM model is used for this too. Suppose that two output signals are to have the waveform shown in Figure 8.7, which is to be repeated every 16 clock cycles.

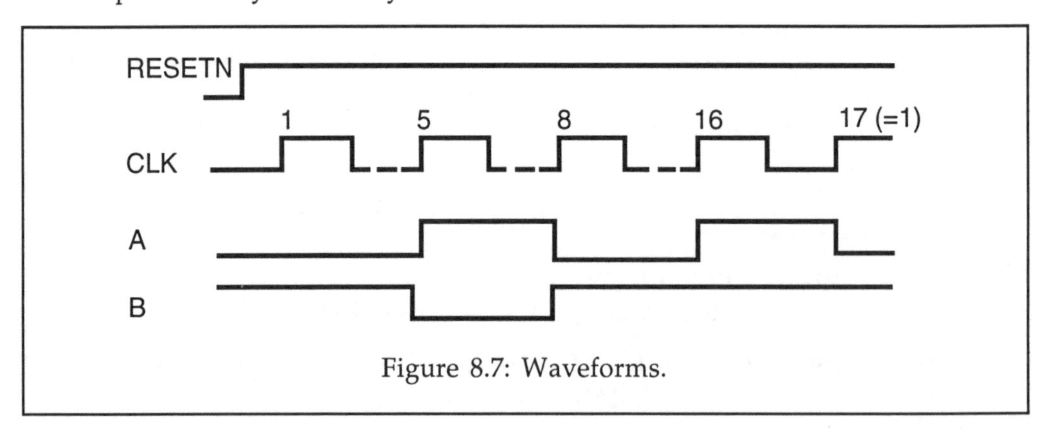

Figure 8.7: Waveforms.

If this is described using the traditional waveform method, the following VHDL code is obtained:

```
Library ieee;
Use ieee.std_logic_1164.ALL;

Package rom is
 constant   rom_width:integer:=2;
 constant   rom_length:integer:=16;
 subtype rom_word is  std_logic_vector(rom_width-1  downto 0);

 type  rom_table is array (0 to  rom_length-1)  of  rom_word;

 constant rom:  rom_table:=  rom_table'("01",
                                        "01",
                                        "01",
                                        "01",
                                        "10",
                                        "10",
                                        "10",
                                        "01",
                                        "01",
                                        "01",
                                        "01",
                                        "01",
                                        "01",
                                        "01",
                                        "01",
                                        "11");

end;


Library ieee;
Use ieee.std_logic_1164.ALL;
Use ieee.std_logic_unsigned.ALL;
Use work.rom.ALL;

Entity waveform is
 port( clk,resetn: in     std_logic;
       waves:      out    rom_word);
end;

Architecture testbench of waveform is
signal start_up:std_logic;
signal step:std_logic_vector(3 downto 0);
```

```
  begin
    process(clk,resetn)
    begin
      if resetn='0' then
        step<=(others=>'0');        -- Restart the waveform
        start_up<='1';

      elsif clk'event and clk='1' then
        start_up<='0';

        if start_up='0' then         -- Delay one clock cycle after reset

          if step/=rom_length-1 then
            step<=step+1;            -- Increment the waveform
          else
            step<=(others=>'0');     -- Restart the waveform
          end if;
        end if;
      end if;
    end process;

    waves<=rom(conv_integer(step));
              -- Read waveform from ROM

  end;
```

If we use a more advanced waveform generator instead, the following must be declared:

- The value of the output signals
- The number of clock cycles for which the output signals have to be stable

Example:

```
Library ieee;
Use ieee.std_logic_1164.ALL;

Package rom is
 constant rom_width:integer:=2;
 constant rom_length:integer:=4;
 subtype rom_word is std_logic_vector(rom_width-1 downto 0);

 type rom_table is array (0 to rom_length-1) of rom_word;

 constant rom: rom_table:= rom_table'( "01",
                                       "10",
                                       "01",
                                       "11");
```

```
 subtype max_cycle is integer range 0 to 15;
 type cycle_length_table is array (0 to rom_length-1)
      of max_cycle;

 constant cycle_length: cycle_length_table
          :=cycle_length_table'(4,3,8,1);

end;


Library ieee;
Use ieee.std_logic_1164.ALL;
Use ieee.std_logic_unsigned.ALL;
Use work.rom.ALL;

Entity wave_adv is
   port( clk,resetn:  in    std_logic;
         waves:       out   rom_word);
end;

Architecture testbench of wave_adv is
signal step:std_logic_vector(3 downto 0);
begin
   process(clk,resetn)
   variable count:max_cycle;
   begin
     if resetn='0' then
       step<=(0=>'1',others=>'0');      -- step=1
       count:=cycle_length(0);

     elsif clk'event and clk='1' then
       if count=0 then
         count:=cycle_length(conv_integer(step));

         if step/=rom_length-1 then
           step<=step+1;            -- Increment the waveform
         else
           step<=(others=>'0');     -- Restart the waveform
         end if;
       end if;

       count:=count-1;

     end if;
   end process;
```

```
        waves<=rom(conv_integer(step));
                                        -- Read waveform from ROM
    end;
```

In the above package the ROM which contains data only has to be 4 x 2 bits in size instead of the 16 x 2 in the previous example. The number of clock cycles for which the output signals should be stable is declared in a separate ROM:

```
constant rom: rom_table:= rom_table'(  "01",
                                       "10",
                                       "01",
                                       "11");

cycle_length:cycle_length_table:=cycle_length_table'(4,3,8,1);
```

The above constants should be interpreted as follows:

Value "01" should be sent out in 4 clock cycles
Value "10" should be sent out in 3 clock cycles
Value "01" should be sent out in 8 clock cycles
Value "11" should be sent out in 1 clock cycle

If we compare the area after synthesis for these two small examples, the traditional waveform generator is smaller in this case. If, on the other hand, there had been more than two signals or if the sequence had been longer, the area result would have been the opposite. The choice of waveform generator should be based on the waveform that you want to generate.

8.5. TextIO

TextIO routines are built into the VHDL standard. Files can be read and written with them. The routines used are as follows:

```
read(..)
readline(..)
write(..)
writeline(..)
```

In order to be able to use these routines, the following line has to be included at the beginning of the VHDL code:

```
Use std.textio.ALL;
```

One area of application for textIO is in testbenches. The input signals for the model can be put in an external file, and the value of the expected output signals can be stored in another file.

Suppose that we have defined the output signals' value in a file (my_inputs.vec). The testbench has to connect these input signals to input signals *d1, d2* and *d3* in component c4:

```
Library ieee;
Use ieee.std_logic_1164.ALL;
Use ieee.std_logic_arith.ALL;
Use std.textio.ALL;

Entity text_io is
 port(q1,q2:   out   std_logic_vector(2 downto 0));
end;

Architecture testbench of text_io is
 component c4
   port( clk,resetn:   in    std_logic;
         d1,d2,d3:     in    std_logic_vector(3 downto 0);
         q1,q2:        out   std_logic_vector(2 downto 0));
 end component;

 signal clk:        std_logic:='0';
 signal resetn:     std_logic:='0';
 signal d1,d2,d3:   std_logic_vector(3 downto 0);

 For U1:c4 Use Entity work.c4(rtl);

 begin
 U1:c4 port map(clk,resetn,d1,d2,d3,q1,q2);

 clk<=not clk after 50 ns;
 resetn<='1' after 125 ns;

 process
 variable text_line:line;      -- Stores a complete line
 variable i1:integer;          -- Used for instimuli for signal d1
 variable i2:integer;          -- Used for instimuli for signal d2
 variable i3:integer;          -- Used for instimuli for signal d3
 file my_file: text is in  "my_inputs.vec";  -- File where the input
 begin                                       -- stimuli are stored
   while not endfile(my_file) loop
     readline(my_file,text_line); -- Read one line from file
                                  -- my_inputs.vec into text_line

     read(text_line,i1);          -- Read the first value into i1
     d1<=conv_std_logic_vector(i1,d1'length);
```

```
      read(text_line,i2);           -- Read the second value into i2
      d2<=conv_std_logic_vector(i2,d2'length);

      read(text_line,i3);
      d3<=conv_std_logic_vector(i3,d3'length);

      wait until clk='1';
    end loop;
    wait;
  end process;
end;
```

Suppose that file my_inputs.vec has the following contents:

my_inputs.vec

```
3   6   8
5   7   2
12  1   9
```

The testbench will then give signals *d1*, *d2* and *d3* the following values:

Time	*Signal*	*Value*
0	d1	"0011"
0	d2	"0110"
0	d3	"1000"
50	d1	"0101"
50	d2	"0111"
50	d3	"0010"
150	d1	"1100"
150	d2	"0001"
150	d3	"1001"

Unfortunately, VHDL-93 is not backwards compatible with the VHDL-87 standard when it comes to TextIO. The above example can be simulated with the VHDL-87 standard, but if VHDL-93 is to be used, the following line in the code has to be replaced:

```
file my_file: text is in  "my_inputs.vec";                -- VHDL-87
file my_file: text open read_mode is "my_inputs.vec"  -- VHDL-93
```

In VHDL-93, read_mode is default and so does not have to be specified in the above example, i.e. the line could have been written as follows:

```
file my_file: text is "my_inputs.vec"                     -- VHDL-93
```

Another difference between the 1987 and 1993 VHDL standards is that the same file can be both read and written in the same simulation in VHDL-93, though not simultaneously.

This method of using TextIO in testbenches is very flexible and fairly simple. The disadvantage is that textIO calls are normally quite slow to simulate compared with ordinary VHDL code.

It is also important to point out that a TextIO call is not synthesizable, but can only be used for simulation.

8.6. Exercises

1. What are the different parts of a testbench?

2. What are the level alternatives of the three different testbenches and what is the difference between them?

3. (a) What is a waveform generator in VHDL?
   (b) What is the difference between a simple and an advanced waveform generator?

4. (a) What is TextIO?
   (b) How can TextIO be used in testbenches?

5. Design a testbench to test the following component:

```
Library ieee;
Use ieee.std_logic_1164.ALL;
Use ieee.std_logic_unsigned.ALL;

Entity my_comp is
 port( clk,resetn,a,b:  in    std_logic;
       c,d:             in    std_logic_vector(2 downto 0);
       q1,q2:           out   std_logic;
       q3:              out   std_logic_vector(5 downto 0));
end;

Architecture rtl of my_comp is
type state_type is (s0,s1,s2);
signal state:state_type;
signal q3_b: std_logic_vector(5 downto 0);
begin
  q3<=q_b;
  q3_b<=c * d;
```

```
  process(clk,resetn)
  begin
    if resetn='0' then
         state<=s0;
         q1<='0';
         q2<='1';
    elsif clk'event and clk='1' then
         case state is
           when s0 =>  if a='1' then
                         state<=s1;
                       end if;
                       q1<='1';
                       q2<='0';
           when s1 =>  if b='0' then
                         state<=s2;
                       end if;
                       q1<='0';
                       q2<='1';
           when s2 =>  state<=s0;
                       q1<='0';
                       q2<='0';
           when others  => state<=s0;
                           q1<='1';
                           q2<='0';
      end case;
    end if;
  end process;
end;
```

(a) Testbench according to class 1 (generation of stimuli for input signals only).

(b) Testbench according to class 2.

(c) Testbench according to class 3. The output signals should be stable 30 ns after each clock edge.

(d) Testbench with a simple waveform generator. The testbench should only generate stimuli for the input signals.

(e) Testbench with an advanced waveform generator. The testbench should only generate stimuli for the input signals.

(f) Testbench with TextIO. The testbench should only generate stimuli for the input signals.

9

State machines

Using state machines is an effective means of implementing control functions. In a von Neumann structure (CPU), for example, operations such as fetch and execution phases, data paths, ALU registers, etc., are required. A state machine's performance is much greater than a CPU's, as the code for behaviour is stored in the machine in a gate network which can be compared to Boolean equations; the state is stored in a number of flip-flops.

The following code can be executed both in a CPU and in a state machine described in VHDL:

```
if a>37 and c<7 then
   state <= alarm;
   out_a <= '0';
   out_b <= '0';
   out_analog <= a+b;
else
   state <= running;
end if;
```

If the code is implemented in a CPU, it is translated into approximately 10-20 machine instructions. Execution of the program requires a different number of assembler instructions depending on which path it chooses in the **if** statement. This means that it is not possible to determine an exact execution time, with minimum and maximum execution time being used instead. If the same code is implemented in gates and flip-flops, it is executed in one clock cycle. **The conclusion is that performance and time determinism are considerably better in state machines implemented in gates and flip-flops than in a CPU.**

The state machine works in two phases. In the first phase the new state is calculated, and in the second phase the new state is sampled into a register. What decides the highest clock frequency that the state machine can cope with is the maximum time it takes to "calculate" the next state. In the previous example we showed that the condition "a>37 and c<7" determines whether the state is to be "alarm" or "running" next time.

Figure 9.1 shows this reasoning graphically. At each line a new value is sampled into the register. This means that the internal signals can be unstable between these register samplings. They are unstable because the input signal to the logic is changed after every clock pulse, following which it takes a certain amount of time before all the gates are set. The signals to the state machine must be stable when the next clock pulse arrives.

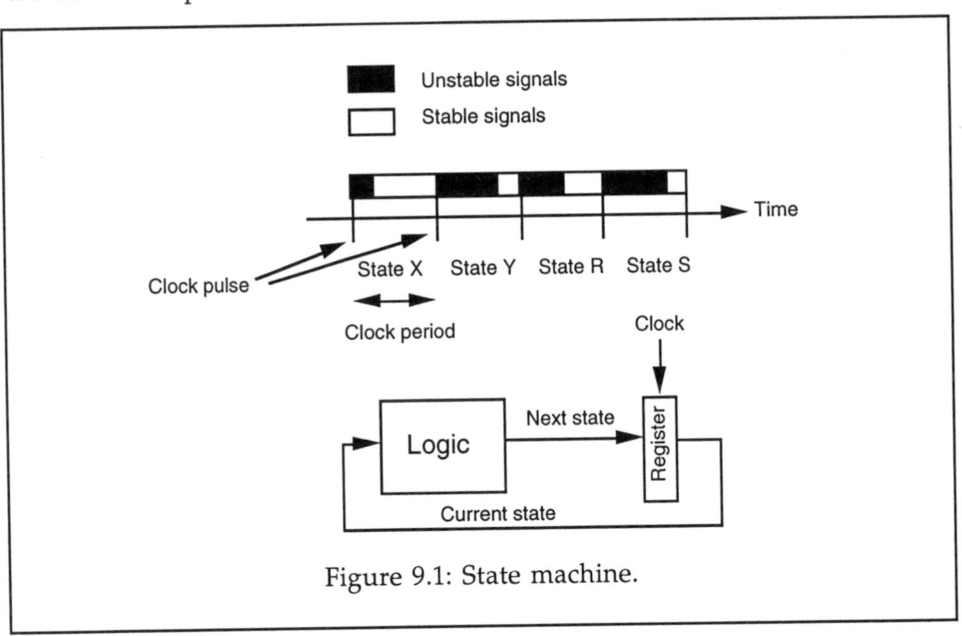

Figure 9.1: State machine.

There are two basic types of machine: Moore and Mealy. The difference between Mealy and Moore machines lies in the outputs.

- **Mealy machine:**

A Mealy machine's outputs are a function of the present state and all the inputs.

- **Moore machine:**

A Moore machine's outputs are a function of the present state only.

Furthermore, a Mealy machine always works one clock cycle in advance of a Moore machine, as a Mealy machine changes its outputs immediately when the inputs change. The Moore machine has to change state first, i.e. wait a clock cycle before the output values can be changed.

The following example explains how a state machine is constructed. The example is a simple sense-of-rotation indicator for detecting sense of rotation (POS and NEG) in a motor, for example. The sense of rotation is detected by two pulse sensors which produce '1' for a white background and '0' for a black background. The design uses a state machine and Karnaugh map for translation into gates (NAND) and minimization. The measuring equipment can be studied in Figure 9.2.

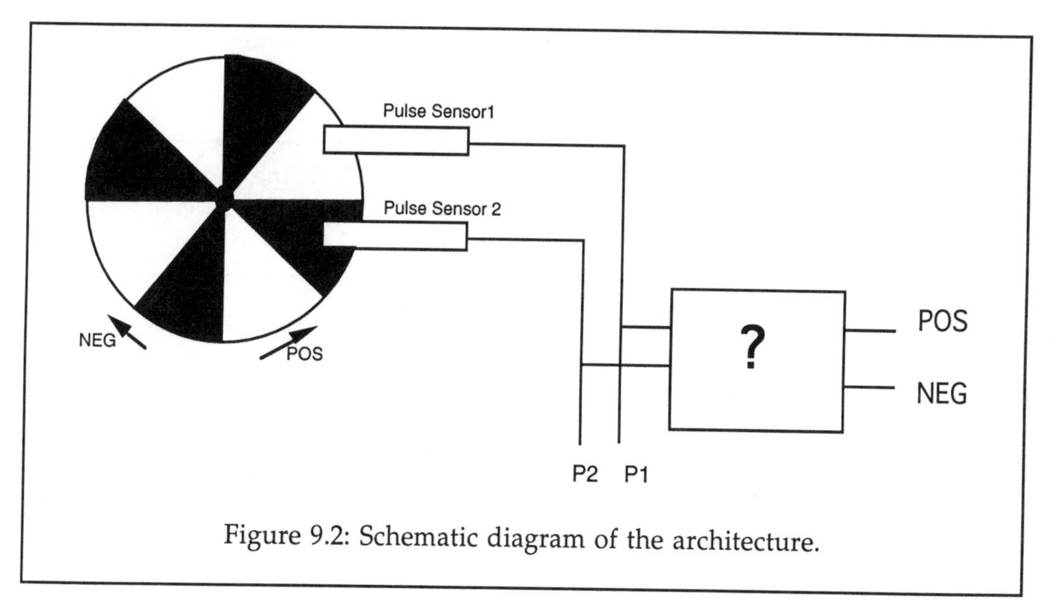

Figure 9.2: Schematic diagram of the architecture.

The pulse sensor produce two pulses, phase-shifted by 90 degrees, in lines P1 and P2. The sense of rotation is determined as positive (POS-1, NEG-0) or negative (POS=0, NEG=1) depending on which pulse comes first. It should be possible to change the sense of rotation at any time (see Figure 9.3). The sense of rotation is only determined when the input signals P1 and P2 go from "00" to "01" or "10".

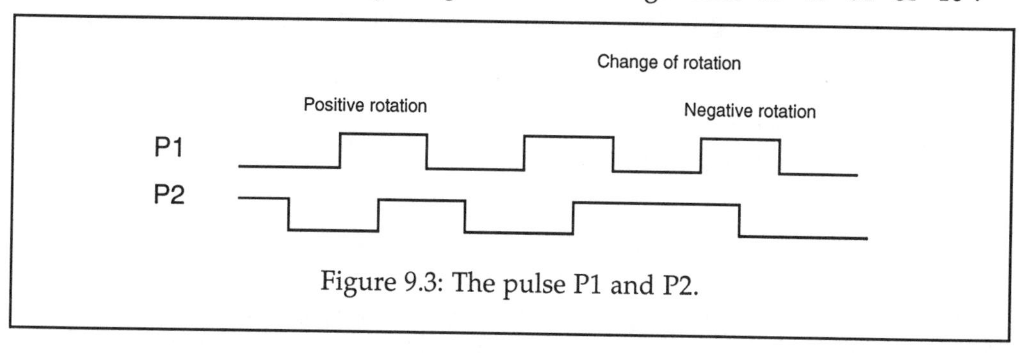

Figure 9.3: The pulse P1 and P2.

Design description

A state machine of the Mealy type with four states is adequate to describe the control unit (see Figure 9.4). Four states can be defined with two bits.

State

The state decoding is in brackets:

(00) +	Positive sense indicated
(11) -	Negative sense indicated
(01) Pos	Positive sense indicated
(10) Neg	Negative sense indicated

State "+" (see Figure 9.4) can change to state "Pos" if the input signals become "01" (P2,P1) or to "Neg" if the input signals become "10". At the same time as the state changes, the output signal is changed to "0" (Pos Neg). This is characteristic of a Mealy machine.

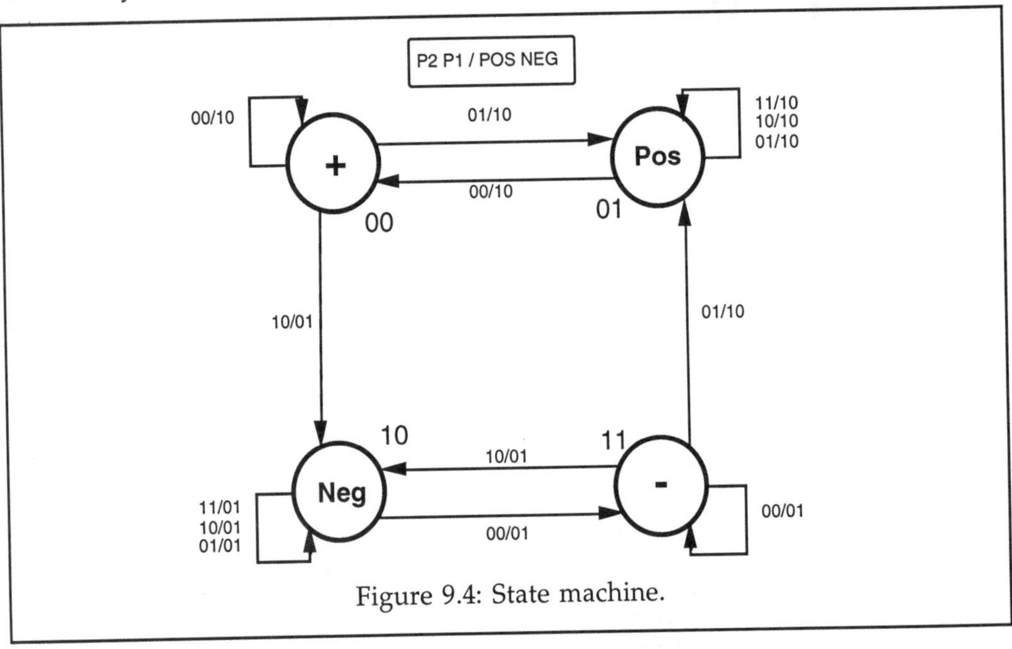

Figure 9.4: State machine.

A Karnaugh map is used for logic minimization and logic description. Karnaugh maps are described in all books on basic electronics. The Boolean equations for the state machine are as follows:

$$D2 = q_2 \overline{p}_1 + q_2 \overline{q}_1 + \overline{q}_1 p_2 = ((q_2 \overline{p}_1)' \cdot (q_2 \overline{q}_1)' \cdot (\overline{q}_1 p_2)')'$$

$$D1 = \overline{q}_2 p_1 + q_1 p_1 + q_2 \overline{p}_2 \overline{p}_1 + \overline{q}_2 q_1 p_2 = ((\overline{q}_2 p_1)' \cdot (q_1 p_1)' \cdot (q_2 \overline{p}_2 \overline{p}_1)')' \cdot \overline{q}_2 q_1 p_2 =$$
$$= ((((\overline{q}_2 p_1)' \cdot (q_1 p_1)' \cdot (q_2 \overline{p}_2 \overline{p}_1)')')' \cdot (\overline{q}_2 q_1 p_2)')'$$

$$POS = \overline{q}_2 \overline{p}_2 + q_1 p_1 + \overline{q}_2 q_1 = ((\overline{q}_2 \overline{p}_2)' \cdot (q_1 p_1)' \cdot (\overline{q}_2 q_1)')'$$

$$NEG = q_2\overline{p}_1 + q_2\overline{q}_1 + \overline{q}_1 p_2 = ((q_2\overline{p}_1)' \cdot (q_2\overline{q}_1)' \cdot (\overline{q}_1 p_2)')'$$

D1 and D2 describe the next state and are inputs to two flip-flops. q1 and q2 describe the present state and are output signals from the flip-flops. In the equations, D2 and NEG are equal, so one and the same subcircuit is used for both the signals. POS should also be the inverse of NEG, so POS should not need its own subcircuit. The NEG signal is divided and inverted instead.

The Boolean statements result in the gate schematic shown in Figure 9.5.

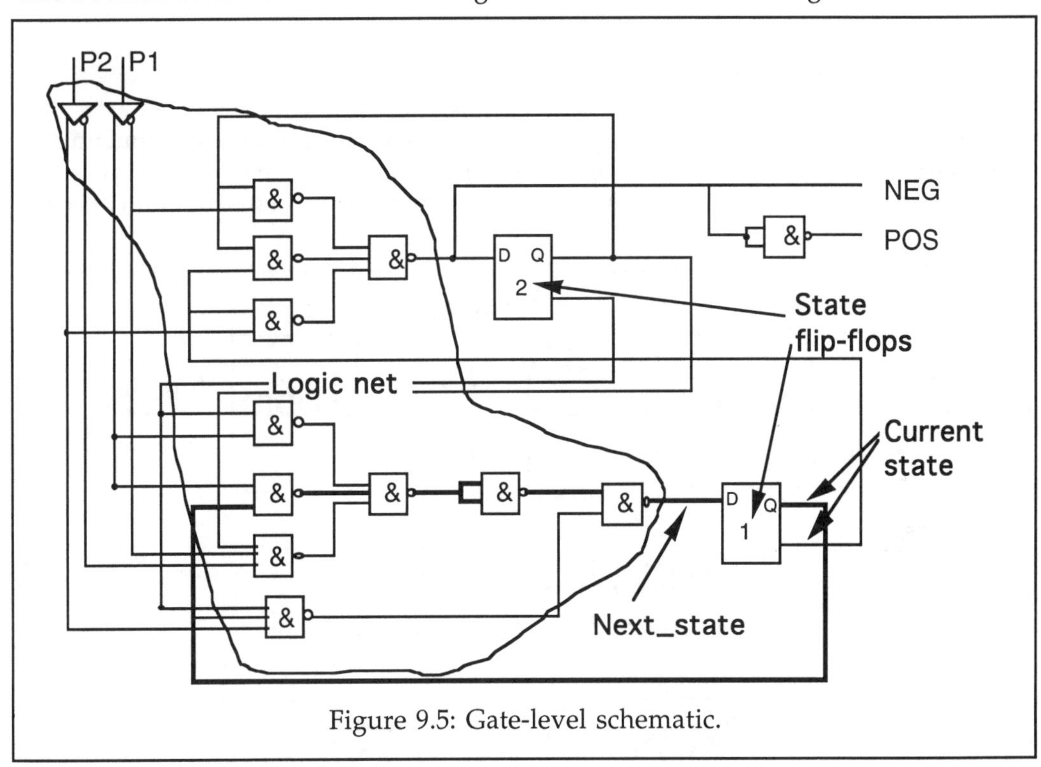

Figure 9.5: Gate-level schematic.

Flip-flops 1 and 2 are state flip-flops. The clock and reset are not drawn in for the flip-flops, but they are connected. The logic network for determining the next state is circled with a line. Output signals NEG and POS are connected to the logic network. This means that spikes can occur in these signals. The longest gate network is four gates (to the D1 flip-flop, marked with bolder lines), so if we assume a 2.5 ns delay at each gate, the fastest clock frequency will be 100 MHz (10 ns).

The design in Figure 9.5 would probably be accepted as a translation exercise from RTL to gate-level description, but it would be rejected as a design at RT level. The comments on the solution would be as follows:

- A change of direction in the state machine can only be indicated if the preceding input signal was "00".
- The outputs are not spike-free.

These points are so serious that redesign is needed.

Today's designers do not have to be able to use a Karnaugh map, minimize, translate to technology-dependent gates or draw schematics. It is all done automatically using the RTL synthesis tool.

9.1. Moore machine

As we can see from the block diagram in Figure 9.6 the outputs are a function of the state vectors, i.e. combinational logic between the state vectors and the outputs. A Moore machine can also be represented with a state diagram (Figure 9.7).

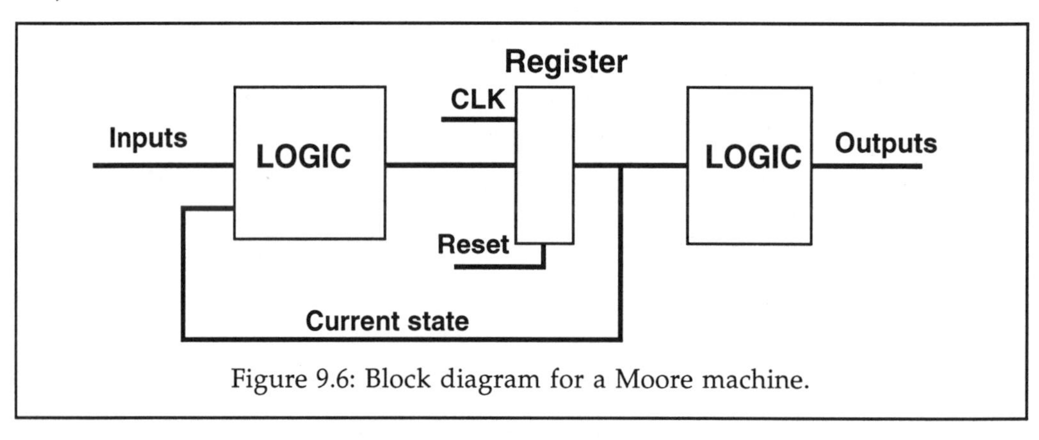

Figure 9.6: Block diagram for a Moore machine.

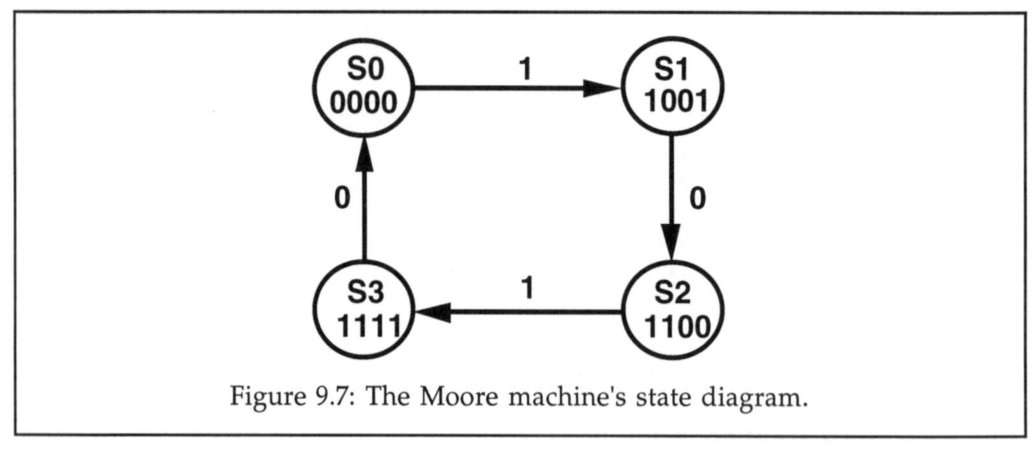

Figure 9.7: The Moore machine's state diagram.

As Figure 9.6 shows, the output signals are only dependent on which state the machine is in, which means that the output signals are usually drawn inside the state bubbles. The noughts and ones inside the "bubbles" in Figure 9.7 are the output signals' values, i.e. in state S0 the output signal has the value "0000" and in state S1 the value "1001", etc. This should not be confused with the state coding for S0, S1, S2 and S3. Figure 9.7 does not say anything about how they have been coded. The state coding does not affect the behaviour of the state machine and does not therefore have to be described in the VHDL code. Advanced synthesis tools often include a state machine optimizer. This makes it possible to select state coding in the synthesis tool to achieve the desired performance and area.

An example of a Moore machine might be:

```
Entity demo is
 port( clk,in1,reset:  in   std_logic;
       out1:           out std_logic);
end demo;

Architecture moore of demo is
type state_type is (s0,s1,s2,s3);     -- State declaration
signal state: state_type;

begin
  demo_process: process(clk,reset) -- Clocked process
 begin
 if reset = '1' then
    state<=s0;                      -- Reset state

 elsif clk'event and clk='1' then
    case state is
      when s0=> if in1='1' then
                  state<=s1;
                end if;
      when s1=> if in1='0' then
                  state<=s2;
                end if;
      when s2=> if in1='1' then
                  state<=s3;
                end if;
      when s3=> if in1='0' then
                  state<=s0;
                end if;
   end case;
 end if;
end process;
```

```
    output_p:process(state)         -- Combinational process
    begin
      case state is
        when s0=> out1<="0000";
        when s1=> out1<="1001";
        when s2=> out1<="1100";
        when s3=> out1<="1111";
      end case;
    end process;
  end moore;
```

The state machine consists of three parts: a declaration, a clocked process and a combinational process.

Declaration
The declaration declares the state. Any name can be chosen for the state with what is known as an enumerated data type. From the documentation point of view, it is an advantage to choose an explanatory name for the state.

Example:

```
type state_type is (start_state, run_state, error_state);
signal state: state_type;
```

The name also facilitates simulation. State vector state will assume values defined for its data type, i.e. start, idle, wait, run and error. Its values will be printed out in the waveform window during simulation, just like the value of all other signals.

Clocked process
The clocked process decides when the state machine should change state. This process is activated by the state machine's clock. Depending on the present state and the value of the input signals, the state machine can change state at every active clock edge. After checking that an active clock edge has taken place, the appropriate **case** command is used to check which state the machine is in. The **if-then-else** command is very suitable for checking the value of the input signals. Using a **case** command followed by an **if-then-else** command usually produces easy-to-read and well structured VHDL code. The state vector in the above example is stored in the signal state which has been declared in the declaration in the architecture.

Combinational process
The combinational process assigns the output signals their value depending on the present state. In the example, only the state vectors are enumerated in the

sensitivity list for process output_p, which agrees with the definition of a Moore machine's output signals.

The general model is just one of many ways of describing a state machine in VHDL. Other models use three different processes: one to decode next state, one to assign state vector current_state and one for the output signals. The following VHDL code is slightly more like the defined block diagram for a Moore machine (see Figure 9.8). The disadvantage is that it takes slightly more VHDL code to write and is slightly slower to simulate. It does not really matter which of these models is used. It may, however, be an advantage from the documentation point of view for all the designers in a firm to use the same model. There is no difference between the synthesis results from the model below with three processes and the above model with two processes for describing a Moore machine.

Example of a Moore machine (three processes):

```
Entity demo is
 port( clk,in1,reset:  in  std_logic;
       out1:           out std_logic);
end demo;

Architecture moore of demo is

type state_type is (s0,s1,s2,s3); -- State declaration
signal current_state, next_state: state_type;
begin
 P0:process(state,in1)
 begin
   case state is                 -- Combinational process
     when s0=> if in1='1' then
                  next_state<=s1;
               end if;

     when s1=> if in1='0' then
                  next_state<=s2;
               end if;

     when s2=> if in1='1' then
                  next_state<=s3;
               end if;

     when s3=> if in1='0' then
                  next_state<=s0;
               end if;
```

```
      end case;
    end if;
  end process;

  P1: process(clk,reset)            -- Clocked process
  begin
    if reset = '1' then
      state<=s0;                    -- Reset state
    elsif clk'event and clk='1' then
      current_state<=next_state;
    end if;
  end process;

  P2:process(current_state)         -- Combinational process
  begin
    case state is
      when s0=> out1<="0000";
      when s1=> out1<="1001";
      when s2=> out1<="1100";
      when s3=> out1<="1111";
    end case;
  end process;

end moore;
```

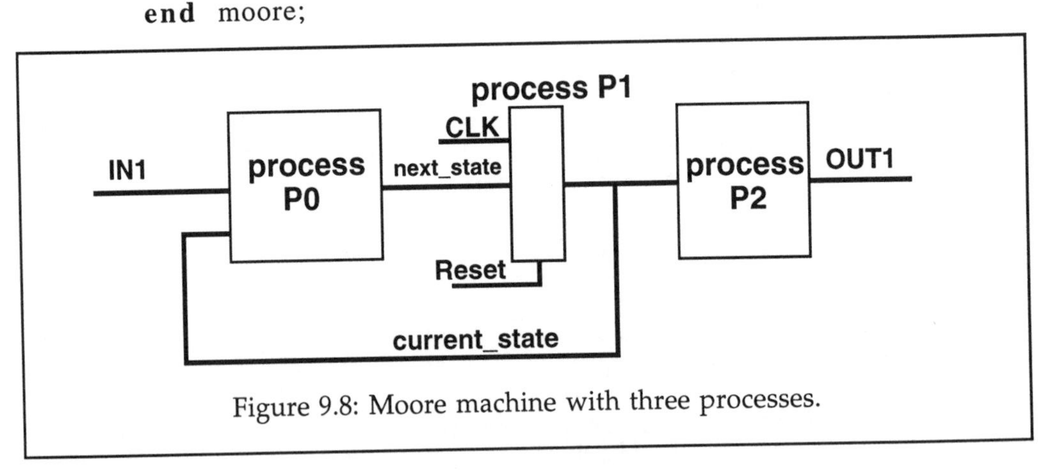

Figure 9.8: Moore machine with three processes.

Some simpler synthesis tools do not support the state vector being defined with an enumerated data type. These tools require the state coding to be done in VHDL code. In some of these simpler synthesis tools there is also limited support for the **case** command. The ViewLogic PC synthesis tool is an example of where these restrictions are currently found. In this case the VHDL code for a Moore machine can be written as follows:

```
Entity demo is
 port( clk,in1,reset:  in   vlbit;
      out1:            out  vlbit_vector(3 downto 0));
end demo;

Architecture moore of demo is
 type state_type is array (1 downto 0) of vlbit;
 constant s0: state_type :="00";   -- State encoding.
 constant s1: state_type :="01";
 constant s2: state_type :="10";
 constant s3: state_type :="11";
 signal state: state_type;

 begin
  demo_process: process
 begin
   wait until prising(clk);
   if reset = '1' then
     state<=s0;

   else
     if state=s0 then
       if in1='1' then
         state<=s1;
       end if;

     elsif state=s1 then
       if in1='0' then
         state<=s2;
       end if;

     elsif state=s2 then
       if in1='1' then
         state<=s3;
       end if;

     else
       if in1='0' then
         state<=s0;
       end if;
     end if;
   end if;
  end process;
```

```
    output_p:process(state)
    begin
      if state=s0 then
        out1<="0000";
      elsif state=s1 then
        out1<="1001";
      elsif state=s2 then
        out1<="1100";
      else
        out1<="1111";
      end if;
    end process;
  end moore;
```

In the above example data type vlbit (ViewLogic bit) has also been used. The principles of the general model and what ViewLogic supports in this case are identical. Both have a clocked process which assigns the state vector its value and a combinational process for the output signals' value. If the synthesis tool supports the general model (most do), it is recommended that the state machine be written on the basis of that model. The designer will then avoid having to do the state coding - this work is left to the synthesis tool.

9.2. Mealy machine

A Mealy machine is always one clock cycle in advance of an equivalent Moore machine. Figures 9.9 and 9.10 show a state diagram and block diagram for a Mealy machine.

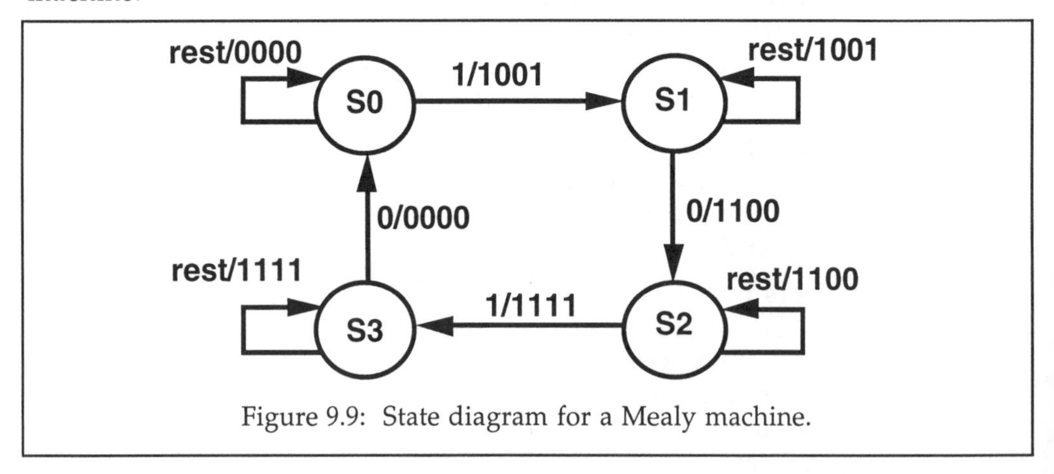

Figure 9.9: State diagram for a Mealy machine.

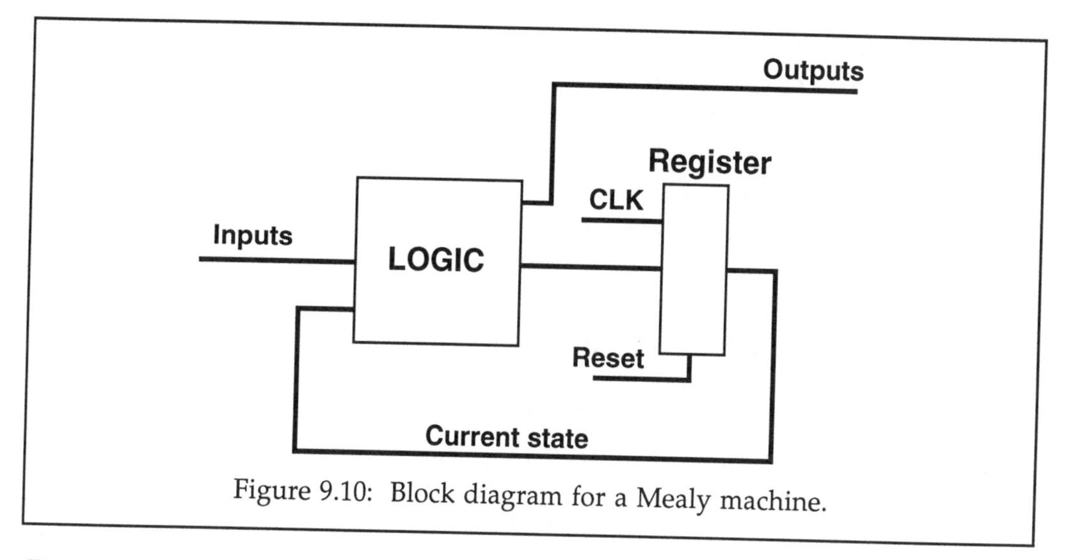

Figure 9.10: Block diagram for a Mealy machine.

Figure 9.10 shows that the output signals are dependent on both present state and all the input signals. This means that the output signals change immediately if the input signals change or if the state is changed. The state diagram cannot define the value of the output signals unambiguously on the basis of the present state, with the value of the output signals also having to be taken into account. This results in the output signals normally being drawn on the state transitions (arrows) in the state diagram, e.g. in state S0 the output signals will assume the value "1001" if the input signal is equal to '1' and the value "0000" for any other input signal value.

A Mealy machine is constructed in exactly the same way as a Moore machine. The difference is the combinational output signal process, which, according to the definition of a Mealy machine, should be a function of the state vector and all the inputs. The clocked process, on the other hand, is identical, regardless of whether a Mealy machine or a Moore machine is being designed.

Example of a Mealy machine:

```
Entity demo is
port( clk,in1,reset:  in    std_logic;
      out1:           out   std_logic_vector(3 downto 0));
end demo;

Architecture mealy of demo is
type state_type is (s0,s1,s2,s3);
signal state: state_type;

begin
  demo_process: process(clk,reset)
```

```
begin
 if reset = '1' then
   state<=s0;
 elsif clk'event and clk='1' then
  case state is
       when s0= if in1='1' then
                      state<=s1;
                end if;
       when s1=> if in1='0' then
                      state<=s2;
                 end if;
       when s2=> if in1='1' then
                      state<=s3;
                 end if;
       when s3=> if in1='0' then
                      state<=s0;
                 end if;
  end case;
 end if;
end process;

output_p:process(state,in1)
begin
  case state is
    when s0 => if in1='1' then
                  out1<="1001";
               else
                  out1<="0000";
               end if;
    when s1=> if in1='0' then
                  out1<="1100";
               else
                  out1<="1001";
               end if;
    when s2=> if in1='1' then
                  out1<="1111";
               else
                  out1<="1001";
               end if;
```

```
      when s3=> if in1='0' then
                   out1<="0000";
                else
                   out1<="1111";
                end if;

    end case;
  end process;
end mealy;
```

As already mentioned in the Moore machine examples, there are simpler synthesis tools which do not support the general Mealy machine model shown above. The commonest problem is that the state coding has to be included in the VHDL code. In this case the VHDL code can be written as follows:

```
Architecture mealy of demo is
 type state_type is array (1 downto 0) of vlbit;
 constant s0: state_type :="00";
 constant s1: state_type :="01";
 constant s2: state_type :="10";
 constant s3: state_type :="11";
 signal state: state_type;

 begin
  demo_process: process
 begin
   wait until prising(clk);
   if reset = '1' then
     state<=s0;
   else
     if state=s0 then
       if in1='1' then
         state<=s1;
       end if;

     elsif state=s1 then
       if in1='0' then
         state<=s2;
       end if;

     elsif state=s2 then
       if in1='1' then
         state<=s3;
       end if;
```

```
      else
        if in1='0' then
          state<=s0;
        end if;
      end if;
    end if;
  end process;

  output_p:process(state, in1)
  begin
    if state=s0 then
      if in1='1' then
        out1<="1001";
      else
        out1<="0000";
      end if;
    elsif state=s1 then
      if in1='0' then
        out1<="1100";
      else
        out1<="1001";
      end if;
    elsif state=s2 then
      if in1='1' then
        out1<="1111";
      else
        out1<="1100";
      end if;
    else
      if in1='0' then
        out1<="0000";
      else
        out1<="1111";
      end if;
    end if;
  end process;
end;
```

9.3. Mealy and Moore variants

Both Mealy and Moore machines can have spikes in the outputs. This is because the output signals come from combinational logic. The state machine can have spike-free outputs at a certain temperature and supply voltage, only to have spikes at another temperature or supply voltage. Normally such spikes are not important. In synchronous design it is only at the active clock edge that the data signal must be stable. The data signals can have spikes shortly after the clock edge without their having any effect. If, on the other hand, the output signals from a state machine are to be used for three-state enable or as a clock, for example, the Mealy or Moore machine's outputs cannot be used, as they are not guaranteed to be spike-free.

If spike-free outputs are required, there are three different state machine variants that can be used:

- Output=state machine
- Moore machine with clocked outputs
- Mealy machine with clocked outputs

9.4. Output=state machine

In an output=state machine the output signals are assigned the value of the state vector direct. This means that the state coding must be determined in the VHDL code, as it affects the function of the output signals. An output=state machine is a special instance of a Moore machine in which the combinational process for the outputs has been optimized out. If a state diagram has to be drawn for an output=state machine, it will be drawn exactly like an ordinary Moore machine (Figure 9.11). There is no good way of representing this machine with a state diagram. In the block diagram (Figure 9.12) on the other hand, it is clear that the output signals come from the state vector direct, i.e. the state flip-flops, which means that the outputs are guaranteed spike-free.

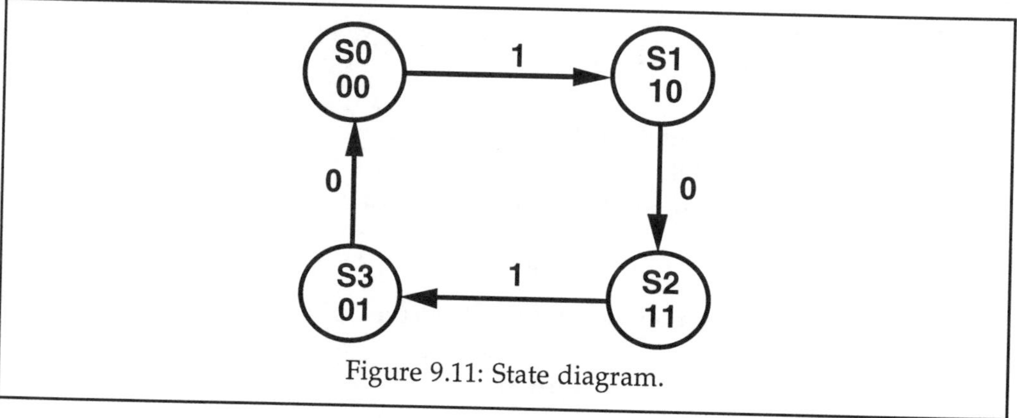

Figure 9.11: State diagram.

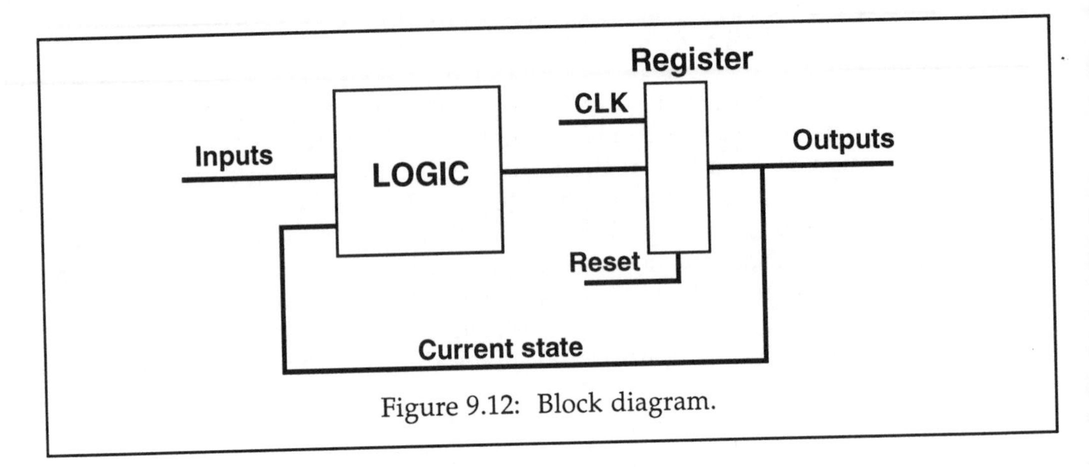

Figure 9.12: Block diagram.

An example of an output=state machine might be:

```
Entity demo is
  port( clk,in1,reset:  in    std_logic;
           out1:        out   std_logic_vector(1  downto 0));
end demo;

Architecture moore of demo is
 type state_type is array (1 downto 0) of  std_logic;

  -- State encoding.
 constant s0: state_type :="00";
 constant s1: state_type :="10";
 constant s2: state_type :="11";
 constant s3: state_type :="01";
 signal state: state_type;

 begin
  demo_process: process(clk,reset)
  begin
   if reset = '1' then
      state<=s0;

   elsif clk'event and clk='1' then
      case state is
         when s0=> if  in1='1' then
                      state<=s1;
                   end if;

         when s1=> if in1='0' then
                      state<=s2;
                   end if;
```

```
            when s2=>  if in1='1' then
                           state<=s3;
                       end if;
            when s3=>  if in1='0' then
                           state<=s0;
                       end if;
        end case;
      end if;
    end process;

    out1<=state;              -- Assign state vector to output
  end;
```

9.5. Moore machine with clocked outputs

The difference between a Moore machine with clocked outputs and an ordinary Moore machine is that all the outputs have been synchronized with an extra D-type flip-flop to obtain spike-free outputs. This means that the outputs are a clock cycle behind the ordinary Moore machine's outputs. There is no really good way of representing this machine in a state diagram: it will be identical to that of the ordinary Moore machine (Figure 9.13). In the block diagram (Figure 9.14), on the other hand, it is clear that the output signals are synchronized with an extra D-type flip-flop, which means that the outputs are guaranteed spike-free. In the VHDL code this means that the assignment of the output signals has to be done inside the clocked process. This will cause all the output signals to be given the extra D-type flip-flop, as all signals assigned in a clocked process result in a flip-flop.

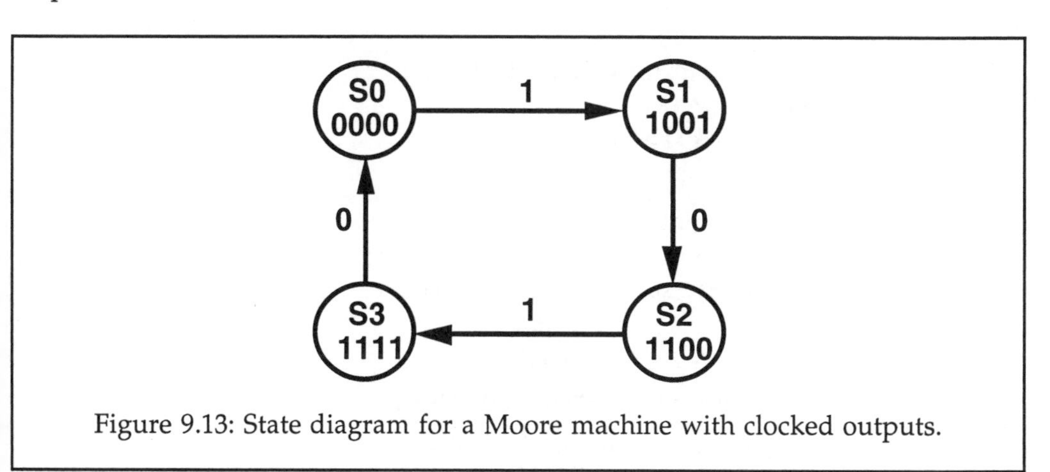

Figure 9.13: State diagram for a Moore machine with clocked outputs.

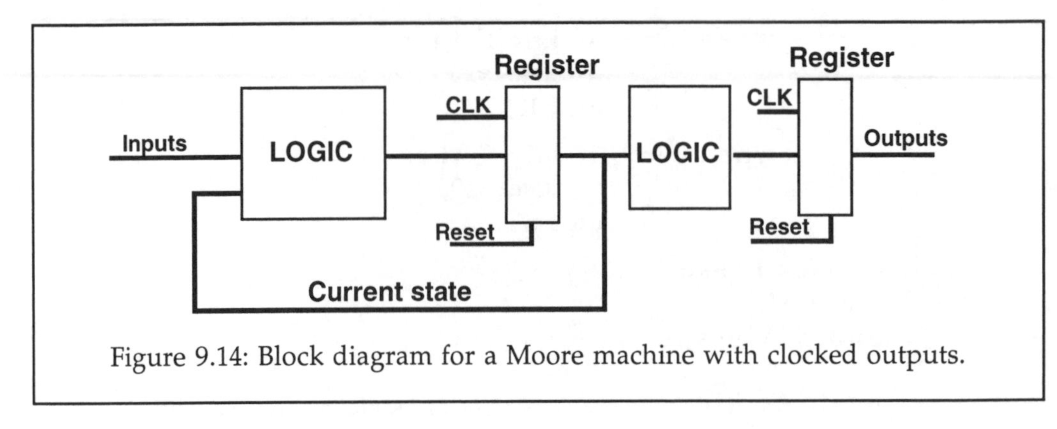

Figure 9.14: Block diagram for a Moore machine with clocked outputs.

An example of a Moore machine with clocked outputs might be:

```
Entity demo is
  port( clk,in1,reset: in     std_logic;
            out1:           out    std_logic);
end demo;

Architecture moore of demo is

 type state_type is (s0,s1,s2,s3);
 signal state: state_type;

begin
  demo_process: process(clk,reset)
 begin
  if reset = '1' then
      state<=s0;
      out1<=(others=>'0');

  elsif clk'event and clk='1' then
    case state is
      when s0=> if in1='1' then
                    state<=s1;
                end if;
                out1<="0000";

      when s1=> if in1='0' then
                    state<=s2;
                end if;
                out1<="1001";

      when s2=> if in1='1' then
```

```
                        state<=s3;
                    end if;
                    out1<="1100";

        when s3=>   if in1='0' then
                        state<=s0;
                    end if;
                    out1<="1111";

      end case;
    end if;
  end process;

end;
```

9.6. Mealy machine with clocked outputs

The difference between a Mealy machine with clocked outputs and an ordinary Mealy machine is that all the outputs have been synchronized with an extra D-type flip-flop to obtain spike-free outputs. This means that the outputs on a clocked Mealy machine are a clock cycle behind the ordinary Mealy machine's outputs. There is no really good way of representing this machine in a state diagram: it will be identical to that of the ordinary Mealy machine (Figure 9.15). In the block diagram (Figure 9.16), on the other hand, it is clear that the output signals are synchronized with an extra D-type flip-flop per output, which means that the outputs are guaranteed spike-free.

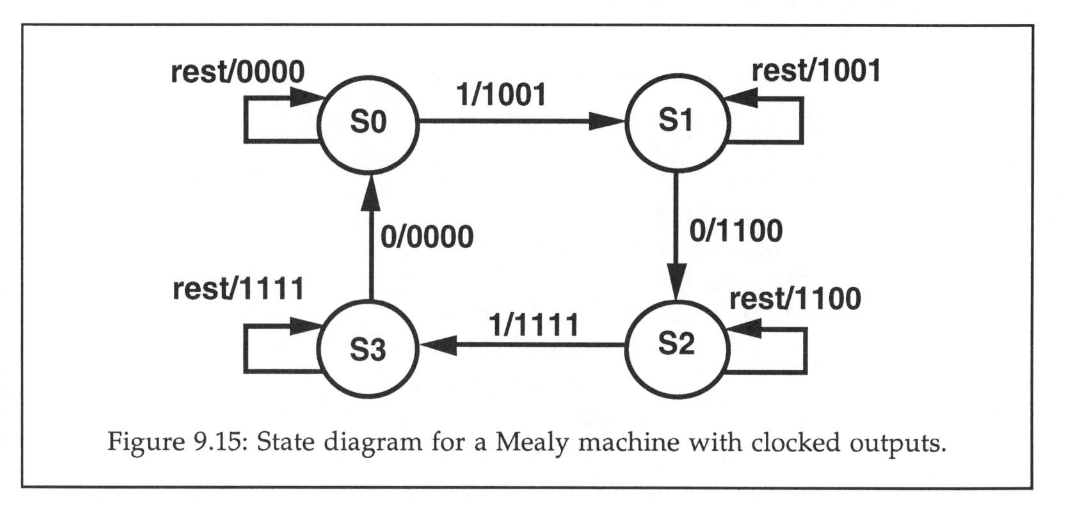

Figure 9.15: State diagram for a Mealy machine with clocked outputs.

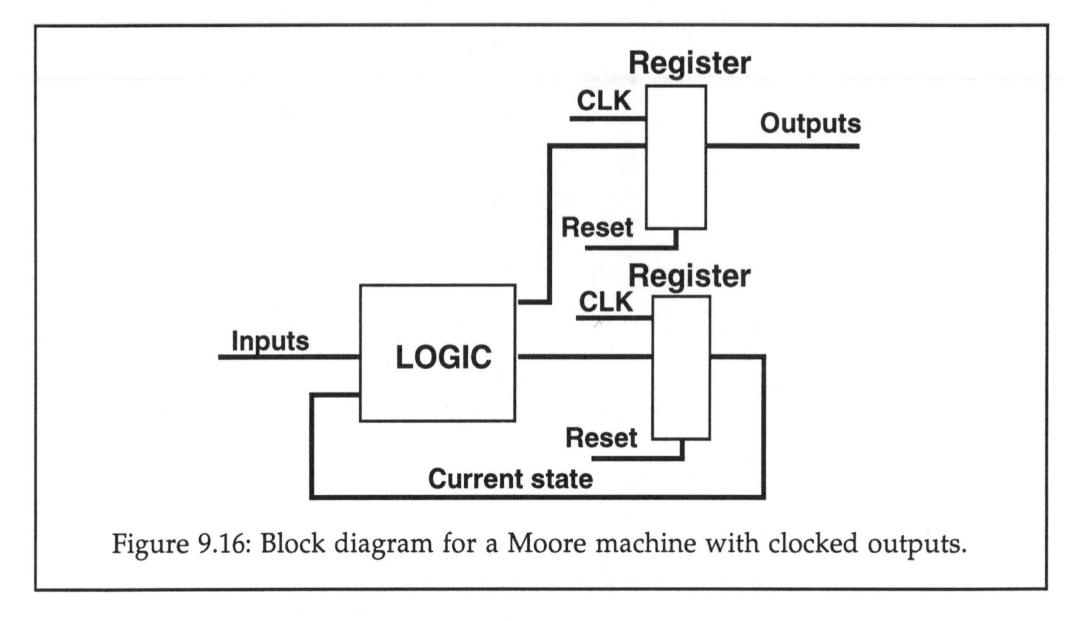

Figure 9.16: Block diagram for a Moore machine with clocked outputs.

An example of Mealy machine with clocked outputs might be:

```
Entity demo is
 port( clk,in1,reset:  in     std_logic;
       out1:           out    std_logic_vector(3 downto 0);
end demo;

Architecture mealy of demo is
type state_type is (s0,s1,s2,s3);
signal state: state_type;

begin
 demo_process: process(clk,reset)
 begin
  if reset = '1' then
    state<=s0;
    out1<=(others=>'0');
  elsif clk'event and clk='1' then
    case state is
      when s0=> if in1='1' then
                   state<=s1;
                   out1<="1001";
                else
                   out1<="0000";
                end if;
```

```
          when s1=> if in1='0' then
                       state<=s2;
                       out1<="1100";
                    else
                       out1<="1001";
                    end if;
          when s2=> if in1='1' then
                       state<=s3;
                       out1<="1111"
                    else
                       out1<="1100";
                    end if;
          when s3=> if in1='0' then
                       state<=s0;
                       out1<="0000";
                    else
                       out1<="1111";
                    end if;
      end case;
    end if;
  end process;
end;
```

9.7. State coding

As already mentioned, the state coding does not affect the function of the state machine and can therefore be selected in the synthesis tool. The only exception is the output=state machine. Some synthesis tools therefore include a state optimizer which makes it possible to determine the state coding when synthesizing the VHDL code. The commonest types of state coding are:

- Sequential
- Gray
- One-hot
- Random
- Auto (area minimization)

Suppose that the state is defined as follows in the VHDL code (Figure 9.17):

```
type state_type is (s0, s1, s2, s3);
```

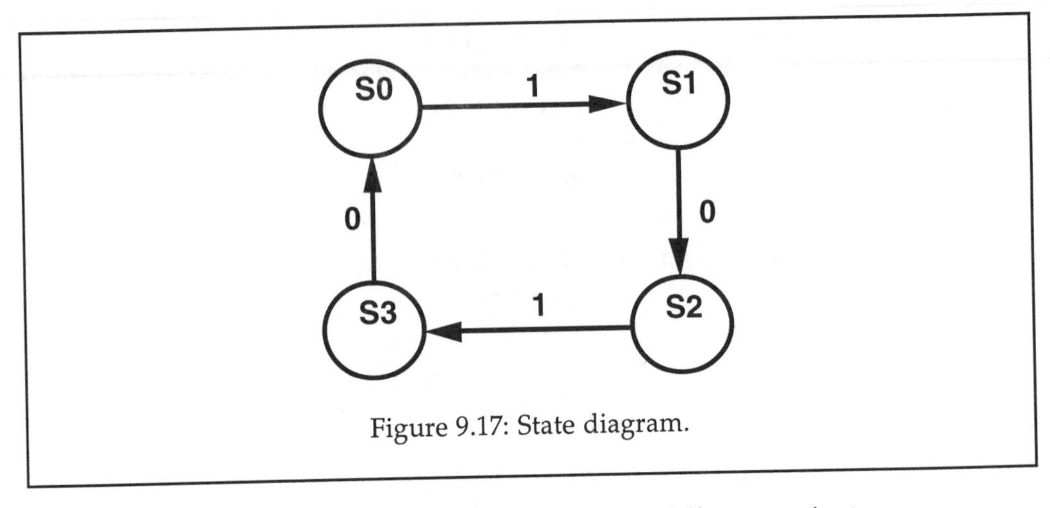

Figure 9.17: State diagram.

The following state coding is then obtained for the different variants:

Sequential	*Gray*	*One-hot*
s0="00";	s0="00"	s0="0001"
s1="01"	s1="01"	s1="0010"
s2="10"	s2="11"	s2="0100"
s3="11"	s3="10"	s3="1000"

If the state machine is optimized without using a special state optimizer, sequential state coding is obtained. Gray coding means that just one bit in the state vector changes value when the state machine changes state. One-hot has as many flip-flops as it has states. This means that the number of gates becomes larger, but there are a lot of flip-flops in some FPGA architectures (type Xilinx). This means that one-hot decoding can be effective when synthesizing to FPGA.

9.8. Residual states

If one-hot state coding is not used, the maximum number of states is equal to 2**N, where N is equal to the vector length of the state vector. In state machines where not all the states are used, there are three options:

1. Let "chance" decide what is to happen if you land in an undefined valid state.
2. Define what is to happen if you land in an undefined state by ending the case statement with when others.
3. Define all possible states in the VHDL code.

The first option normally requires the least logic, as no decoding is required for the undefined states. This can, however, lead to the state machine getting stuck in an undefined state, with the only way out being to activate the reset signal. The second option is safer but requires more logic. Unfortunately, not all synthesis tools support this alternative, which means that option 3 then has to be used, i.e. defining all possible states in the VHDL code.

An example of a state machine is given in Figure 9.18.

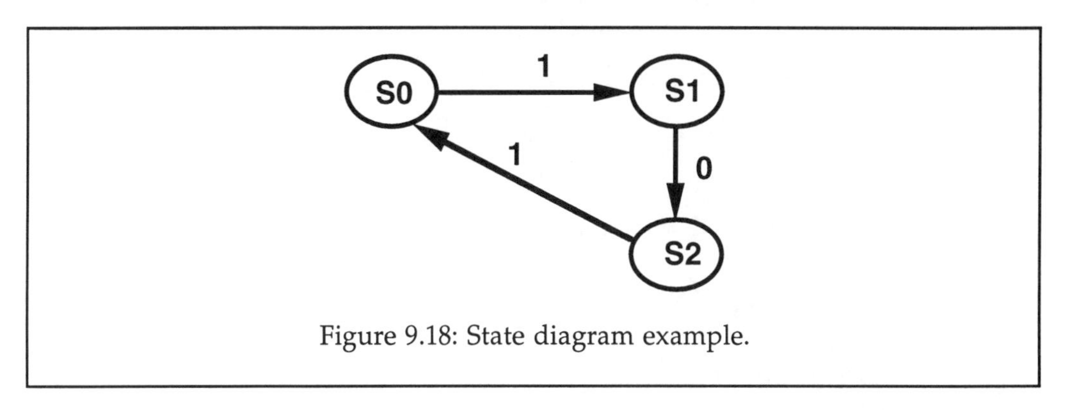

Figure 9.18: State diagram example.

In the figure the state machine has only three states. If the state machine is optimized as ordinary logic (without the state optimizer), sequential state coding is obtained, i.e.

S0="00",
S1="01"
S2="10"

State "11" is not defined. There now follows a look at the three alternative solutions.

```
Library ieee;
Use ieee.std_logic_1164.ALL;

Entity ex is
 port( clk,resetn,a: in      std_logic;
       q:            out     std_logic);
end;
```

Alternative 1:

```
Architecture alt1 of ex is
type state_type is (s0,s1,s2)
signal state: state_type;
begin
```

```
process(clk,resetn)
begin
 if resetn='0' then
    state<=s0;
    q<='0';
 elsif clk='1' and clk'event then
    case state is
        when s0 =>  state<=s1;
                    q<='0';
        when s1 =>  if a='1' then
                        state<=s2;
                    end if;
                    q<='1';
        when s2 =>  state<=s0;
                    q<='0';
    end case;
 end if;
end process;
end;
```

Alternative 2:

```
Architecture alt2 of ex is
type state_type is (s0,s1,s2);
signal state: state_type;
begin
 process(clk,resetn)
 begin
  if resetn='0' then
     state<=s0;
     q<='0';
  elsif clk='1' and clk'event then
    case state is
        when s0 =>  state<=s1;
                    q<='0';
        when s1 =>  if a='1' then
                        state<=s2;
                    end if;
                    q<='1';
        when s2 =>  state<=s0;
```

```
                q<='0';
        when others => state<=s0;
                       q<='0';

      end case;
     end if;
    end process;
   end;
```

Alternative 3:

```
Architecture alt3 of ex is
type state_type is (s0,s1,s2,s3)
signal state: state_type;
begin
 process(clk,resetn)
 begin
  if resetn='0' then
     state<=s0;
     q<='0';
  elsif clk='1' and clk'event then
     case state is
        when s0 =>  state<=s1;
                    q<='0';

        when s1 =>  if a='1' then
                       state<=s2;
                    end if;
                    q<='1';

        when s2 =>  state<=s0;
                    q<='0';

        when s3 =>  state<=s0;
                    q<='0';

     end case;
  end if;
 end process;
end;
```

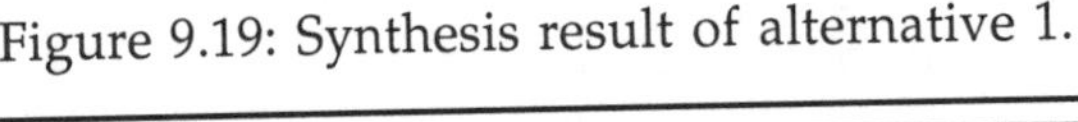

Figure 9.19: Synthesis result of alternative 1.

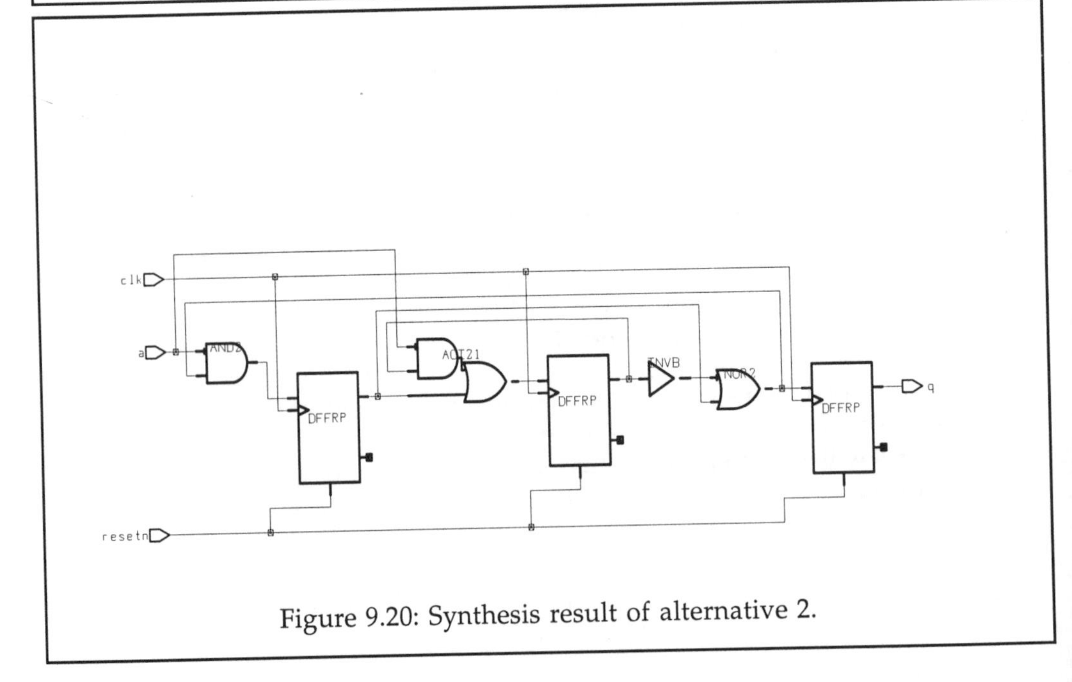

Figure 9.20: Synthesis result of alternative 2.

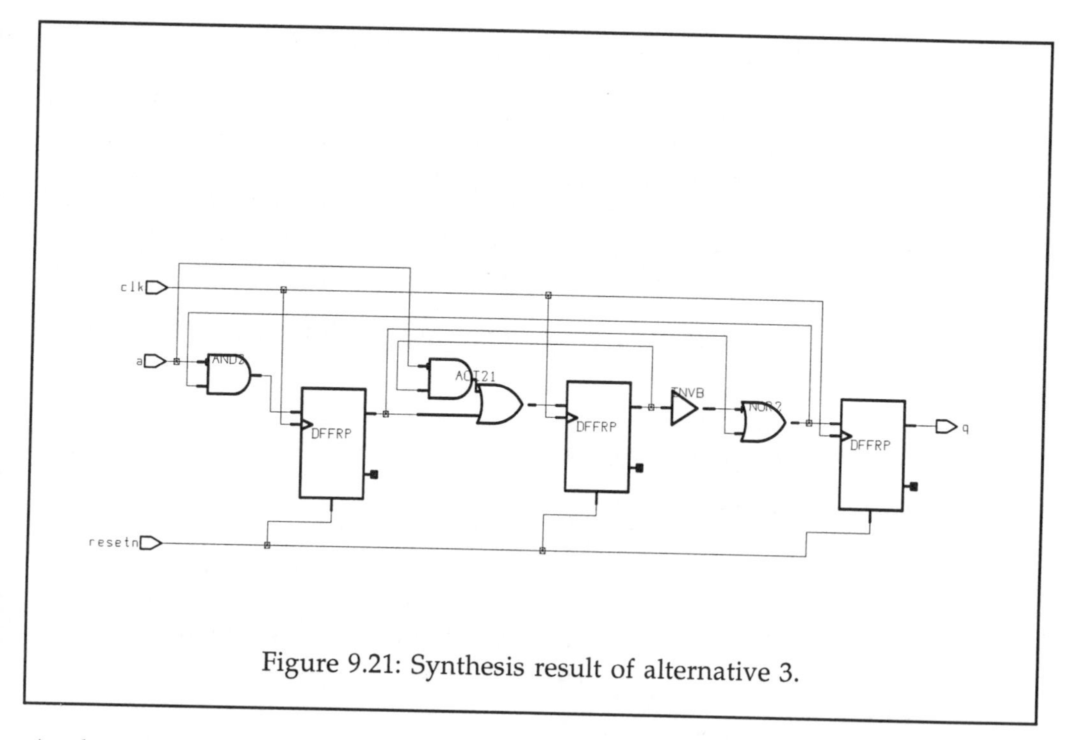

Figure 9.21: Synthesis result of alternative 3.

As the synthesis results in Figure 9.20-9.21 show, alternative 2 and 3 are identical. The state machine from alternative 1 can "get stuck" in undefined state "11" if, for example, a malfunction causes the machine to jump there. The machine will then stay in state "11" as long as input signal *a* is equal to '1'. This behaviour could only be produced after synthesis. The result is also more or less random, depending on how the synthesis tool optimizes the logic. This alternative cannot be chosen, of course, if it is not acceptable for there to be a risk of the state machine getting stuck in the event of a malfunction. It is recommended that the undefined states should always be covered up so that the state machine cannot get stuck. Alternative 1 should be considered only if alternative 2 or 3 results in too much logic.

The choice of alternative is dependent on both design and tool. If the requirement is that the state machine must not land in an undefined state (e.g. in the event of a malfunction or timing problem), alternative 2 or 3 should be chosen, depending on which synthesis tool is to be used. Synopsys is one of the synthesis tools which support alternative 2. If you use Synopsys and alternative 2, you should avoid using the state optimizer, however, as it can optimize the **others** line out. Instead, optimize the VHDL code like any other VHDL code.

9.9. How to write an optimum state machine in VHDL

We have now examined the various state machines and state coding. The choice of state machine is completely dependent on the design (spike-free, area, timing, etc.). Sometimes a combination of Mealy, Moore, clocked Mealy or clocked Moore may be the optimum solution; i.e. you only synchronize the outputs which need to be spike-free with a register (Mealy/Moore with clocked outputs) and let the other outputs be either "pure" Mealy or "pure" Moore outputs.

An example of a combination of Mealy, Moore, clocked Mealy and clocked Moore might be:

```
Library ieee;
Use ieee.std_logic_1164.ALL;

Entity mix is
 port( clk,resetn,a,b: in   std_logic;
       q1:             out  std_logic; -- Mealy output
       q2:             out  std_logic; -- Moore output
       q3:             out  std_logic; -- S. Mealy output
       q4:             out  std_logic); -- S. Moore output
end;

Architecture rtl of mix is
type state_type is (s0,s1)
signal state: state_type;
begin
 process(clk,resetn)
 begin
  if resetn='0' then
    state<=s0;
    q3<='0';
    q4<='0';

  elsif clk'event and clk='1' then
    case state is
      when s0 => if a='1' then
                   state<=s1;
                   q3<='0';
                 else
                   q3<='1';
                 end if;
                 q4<='1';
```

```
           when s1 => if b='1' then
                        state<=s2;
                        q3<='1';
                      else
                        q3<='0';
                      end if;
                      q4<='0';

        end case;
       end if;
      end process;

      process(state,a,b)
      begin
       case state is
         when s0 => if a='1' then
                      q1<='0';
                    else
                      q1<='1';
                    end if;
                    q2<='0';

         when s1 => if b='1' then
                      q1<='1';
                    elsif a='0' then
                      q1<='0';
                    else
                      q1<='1';
                    end if;
                    q2<='1';

       end case;
      end process;
    end;
```

If it is possible to choose output=state as the state machine type, this should always be done, as it is the fastest in terms of timing and the smallest in terms of area. The only exception is that a Mealy machine is faster to change its output signals. A Mealy machine's outputs change immediately when the value of the input signals changes, while an output=state machine waits for a clock cycle before the outputs change.

It is often only possible to use output=state for relatively small state machines. Output=state requires the output signals not to be equal in two or more states. If the output signals are equal, the state coding would be identical in these states, which is not permitted, of course.

The following example shows a state machine in which the output signals are identical in two of the states (S1 and S2). Sometimes it may be worth adding an extra D-type flip-flop to the state vector to differentiate between the two and make it possible to choose output=state machine.

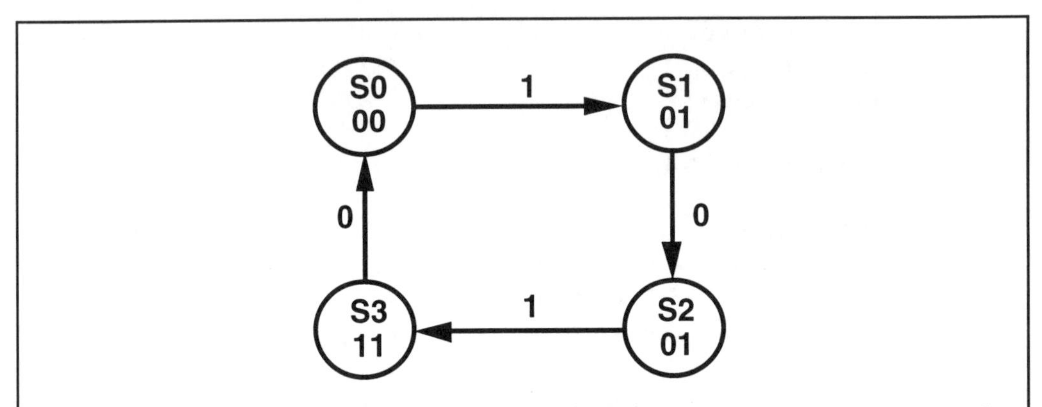

Figure 9.22: State diagram in which the output in two of the states is identical.

As can be seen from Figure 9.22, the output signals have the same value in states S1 and S2.

In the following example an extra D-type flip-flop has been added to the state vector to make output=state possible for the machine in Figure 9.22.

```
Architecture rtl of ex is
constant s0: std_logic_vector(2 downto 0):="000";
constant s1: std:logic_vector(2 downto 0):="001";
constant s2: std_logic_vector(2 downto 0):="101";
constant s3: std:logic_vector(2 downto 0):="011";
signal state: std_logic_vector(2 downto 0);

begin
 process(clk,resetn)
 begin
  if resetn='0' then
    state<=s0;
  elsif clk'event and clk='1' then
    case state is
```

```
          when s0 =>  if a='1' then
                          state<=s1;
                      end if;
          when s1 =>  if b='1' then
                          state<=s2;
                      end if;
          when s2 =>  if a='0' then
                          state<=s3;
                      end if;
          when others  =>  state<=s0;
      end case;
    end if;
  end process;
  q_out<=state(1 downto 0);
end;
```

When the state machine is designed, it is important to define the value of all the output signals in each clock cycle. An example of this is given in Figure 9.23.

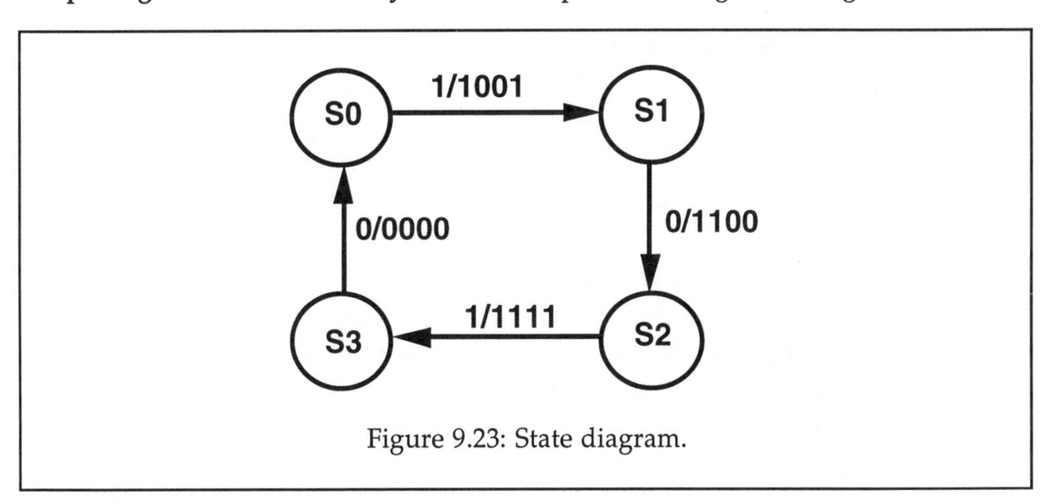

Figure 9.23: State diagram.

From a functional point of view the following description is correct. However, the output signals are not assigned a value at every active clock edge. If, for example, the state machine is in state S1 and input signal *a* is '1', the output signal is not assigned a value. This is interpreted (quite correctly) by the synthesis tool to the effect that the output signals should retain the old value.

```
Architecture bad of ex is
type state_type is (s0,s1,s2,s3)
signal state: state_type;

begin
 process(clk,resetn)
 begin
  if resetn='0' then
    state<=s0;
    q<=(others=>'0');

  elsif clk'event and clk='1' then
    case state is
      when s0 => if a='1' then
                   state<=s1;
                   q<="1001";
                 end if;

      when s1 => if a='0' then
                   state<=s2;
                   q<="1100";
                 end if;

      when s2 => if a='0' then
                   state<=s3;
                   q<="1111";
                 end if;

      when s3 => if a='0' then
                   state<=s0;
                   q<="0000";
                 end if;

    end case;
  end if;
 end process;

end;
```

In the above VHDL code the output signals are not assigned a value if the state machine does not change state. The state diagram in Figure 9.23 illustrates this behaviour. If the state diagram is redrawn instead and all the outputs are assigned a value in all states (Figure 9.24), without the functionality being changed, its code will be as follows:

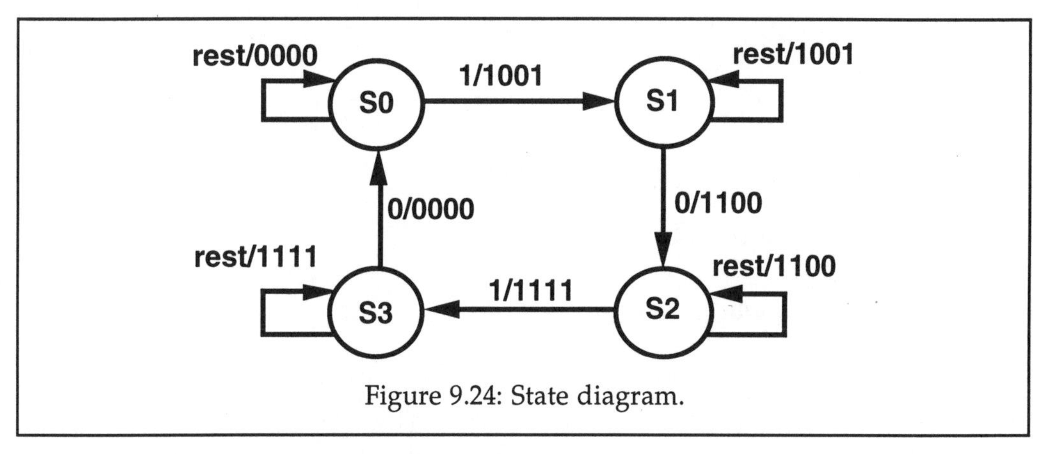

Figure 9.24: State diagram.

```
Architecture good of ex is
type state_type is (s0,s1,s2,s3)
signal state: state_type;
begin
 process(clk,resetn)
 begin
  if resetn='0' then
     state<=s0;
     q<=(others=>'0');
  elsif clk'event and clk='1' then
     case state is
       when s0 => if a='1' then
                    state<=s1;
                    q<="1001";
                  else
                    q<="0000";
                  end if;
       when s1 => if a='0' then
                    state<=s2;
                    q<="1100";
                  else
                    q<="1001";
                  end if;
       when s2 => if a='0' then
                    state<=s3;
                    q<="1111";
                  else
                    q<="1100";
                  end if;
```

```
                when s3 => if a='0' then
                              state<=s0;
                              q<="0000";
                           else
                              q<="1111";
                           end if;
             end case;
          end if;
        end process;
      end;
```

As previously mentioned, the behaviour will be identical both in the VHDL simulation and after synthesis at gate level. The disadvantage of the first description is that it will be approximately 12 gates larger because the output signals are not assigned a value in every clock cycle. This means that the hardware is required to retain the old value for the output signal. Synthesis tools solve this by feeding back the output value to a multiplexor (Figure 9.25). The multiplexor sends the value back to the output in those clock cycles where the output signals have not been assigned a value. This is how the value for the output is retained. If the output signal had been assigned a value instead, the multiplexor and output signal feedback would not have been needed (Figure 9.26), saving approximately 3 gates per output. Note that the state vector does not have to be assigned a value in each clock cycle, as it should be fed back and so does not entail extra logic.

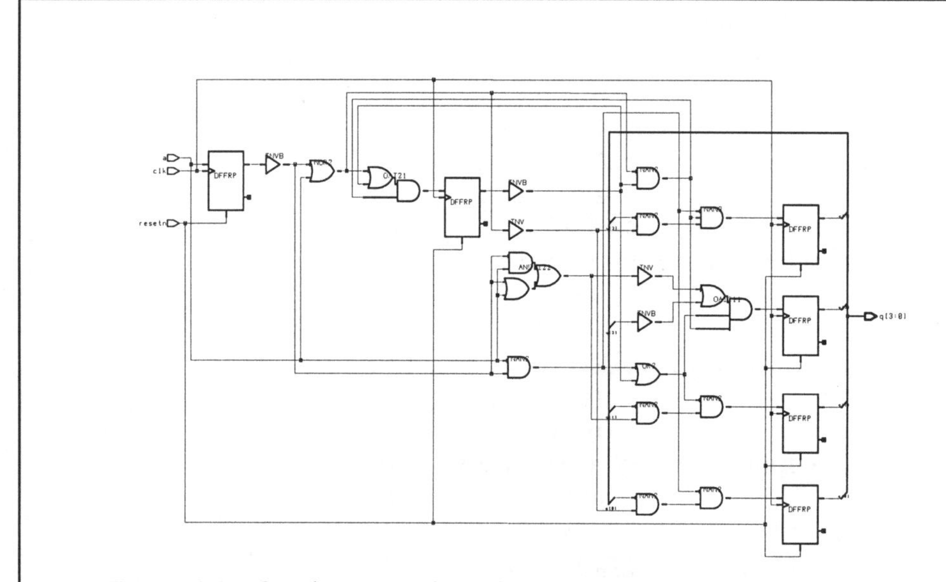

Figure 9.25: Synthesis result without defining the value of all the output signals in each clock cycle.

Three gates per output may not seem much, but in small designs or if a number of state machines have been used, these extra gates can result in a larger circuit being required to accommodate the logic. A few gates can also be "the straw that breaks the camel's back" in the case of large designs. Moreovers, the timing in the state machine will be worse if an unnecessary multiplexor is included in the data path. The capacitance of the outputs will also increase if they are fed back to the multiplexor. This capacitance means that the outputs will be slightly slower in terms of timing.

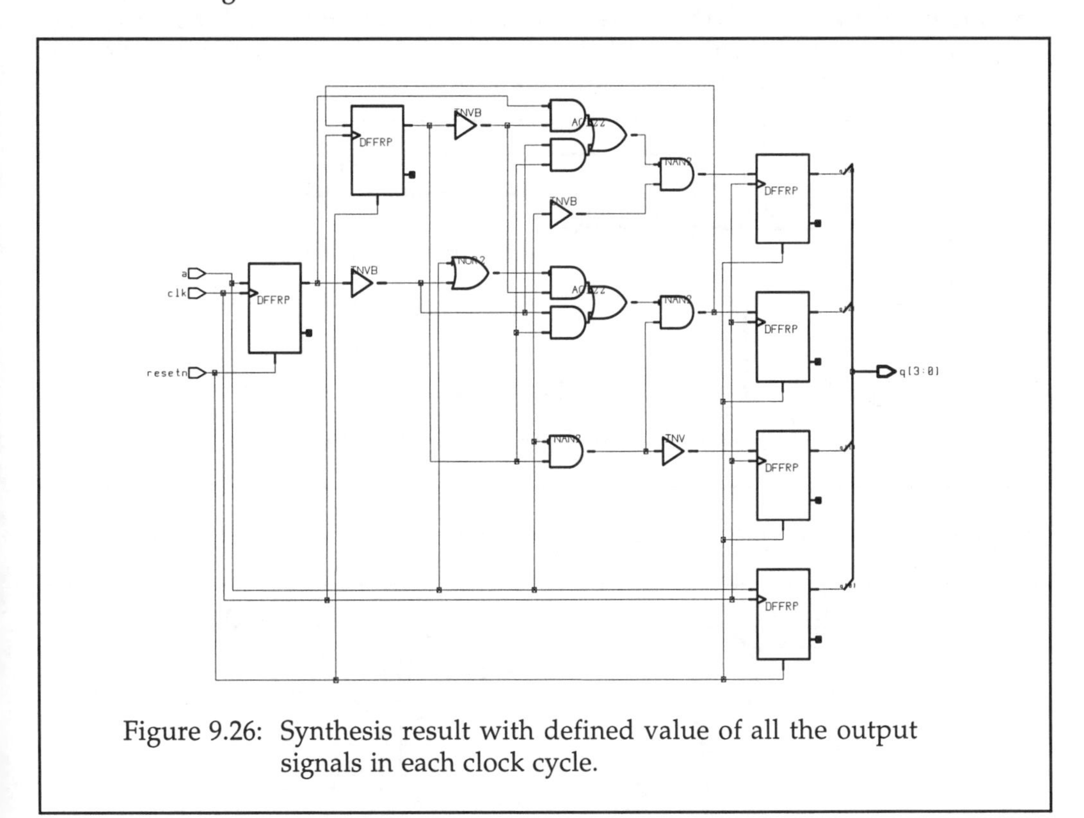

Figure 9.26: Synthesis result with defined value of all the output signals in each clock cycle.

9.10. Asynchronous state machines

The state machines discussed so far have all been clocked, i.e. synchronous. If, instead, the state vector is fed back without gates being used (see Figure 9.27), the state machine will be asynchronous. The advantage of asynchronous state machines is that usually they are faster than clocked ones. From the point of view of design methodology, asynchronous state machines are not the optimum solution (see Chapter 11, "Design methodology"). It is only if the performance requirements are so tough that they cannot be met in any other way that an asynchronous state machine can normally be justified.

When designing asynchronous state machines, the designer must make sure that what is known as "racing" cannot occur. Racing refers to the state machine changing several states at a time in an uncontrolled manner. The state coding is very important in an asynchronous state machine. The state vector may only change one bit at a time at a state change, which means that the state coding must be included in the VHDL code. Many synthesis tools do not support asynchronous state machines either. Those which do support them do not manage to optimize their timing and area effectively.

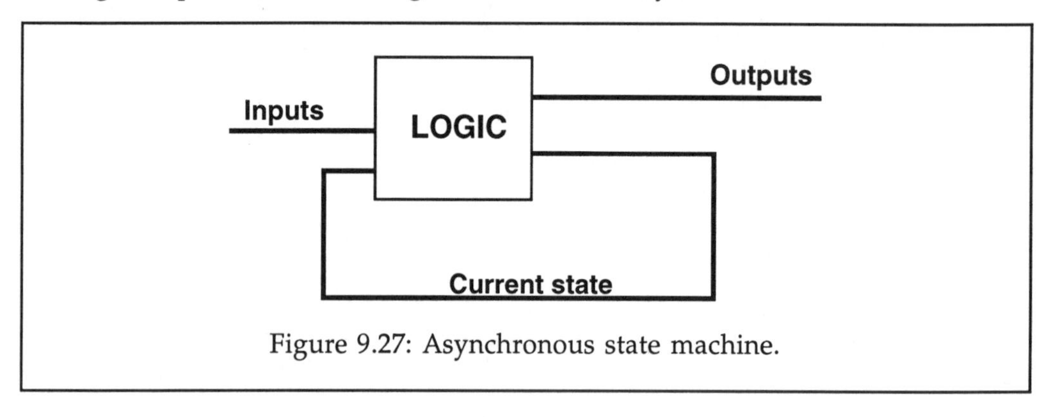

Figure 9.27: Asynchronous state machine.

Suppose that the state machine in Figure 9.27 has to be designed asynchronously.

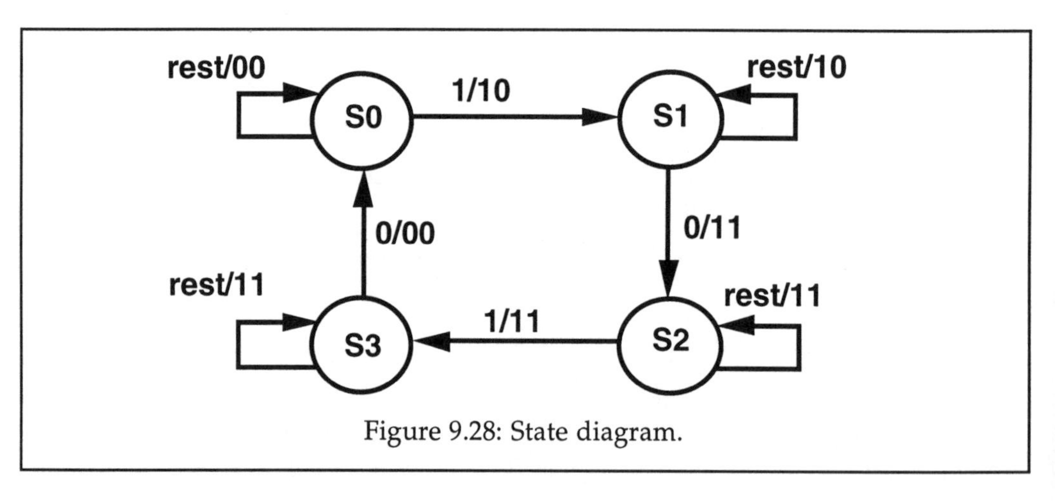

Figure 9.28: State diagram.

The VHDL code can then be written as follows:

```
Library ieee;
Use ieee.std_logic_1164.ALL;
```

```
Entity test is
 port( a,resetn:   in   std_logic;
       q:          out  std_logic_vector(1 downto 0));
end;

Architecture rtl of test is
type state_type is array (1 downto 0) of std_logic;
constant s0: state_type:="00";
constant s1: state_type:="01";
constant s2: state_type:="11";
constant s3: state_type:="10";
signal state: state_type;
begin
  process(resetn,state,a)
  begin
    if resetn='0' then
      state<=s1;
      q<="10";
    else
      case state is
        when s0 =>  if a='1' then
                       state<=s1;
                       q<="10";
                    else
                       state<=s0;
                       q<="00";
                    end if;
        when s1 =>  if a='0' then
                       state<=s2;
                       q<="11";
                    else
                       state<=s1;
                       q<="10";
                    end if;
        when s2 =>  if a='1' then
                       state<=s3;
                       q<="10";
                    else
                       state<=s2;
                       q<="11";
                    end if;
```

```
                  when others =>  if a='0' then
                                    state<=s0;
                                    q<="11";
                                  else
                                    state<=s3;
                                    q<="11";
                                  end if;
               end if;
             end if;
          end process;
        end;
```

The synthesis result is shown in Figure 9.29.

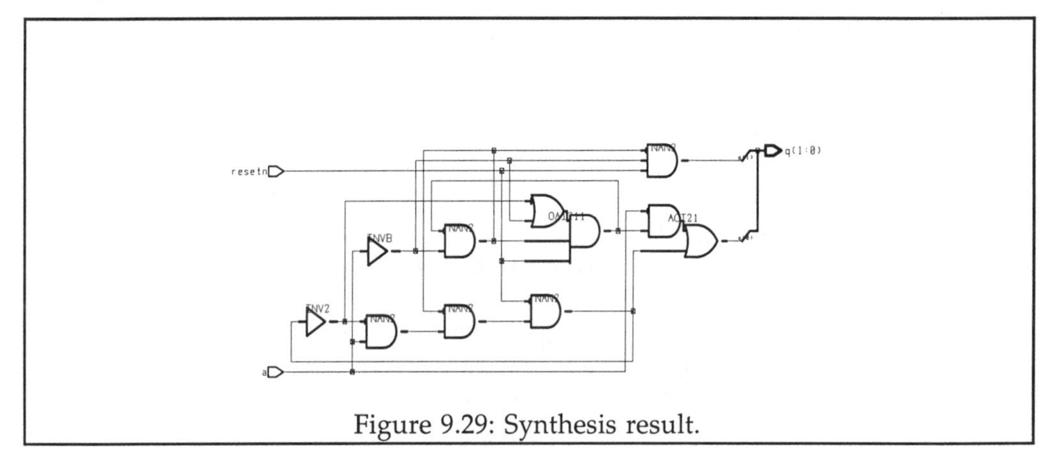

Figure 9.29: Synthesis result.

9.11. Exercises

1. What is the difference between a Mealy machine and a Moore machine?

2. (a) Are the outputs from a Mealy or Moore machine guaranteed spike-free?
(b) Which state machines have guaranteed spike-free outputs?
(c) Specify two situations in which spike-free outputs are required from a state machine.

3. Is it possible to combine different types of state machine in the same machine?

4. (a) What are residual states in a state machine?
(b) What alternative description methods are there with regard to residual states?
(c) Specify two ways in which a state machine can reach a residual state?

5. (a) Does the state coding affect the function of a Mealy/Moore machine?

(b) In which of the state machines described in the chapter is the function affected by the state coding?

(c) What is meant by Gray, sequential and one-hot coding of the state?

(d) Can different state coding be equally suitable for different technologies?

(e) If the state vectors are defined as follows:

```
type state_type is (s0, s1, s2, s3, s4);
signal state: state_type;
```

How many flip-flops does sequential/one-hot decoding require?

6. In which of the state machines do the outputs change over fastest once the inputs have changed over?

7. (a) What is required of the output signals if it is to be possible to use an output=state machine?

(b) Is there any way round the problem in question 7(a)?

8. What happens if an output signal is not assigned a value in each clock cycle in a Moore/Mealy machine with clocked outputs?

9. (a) The state machine should be an output=state machine. The figure below shows the state and the machine's block diagram. Write the VHDL code for the entity and architecture. At reset_n='0' the state machine should be initiated to state S0.

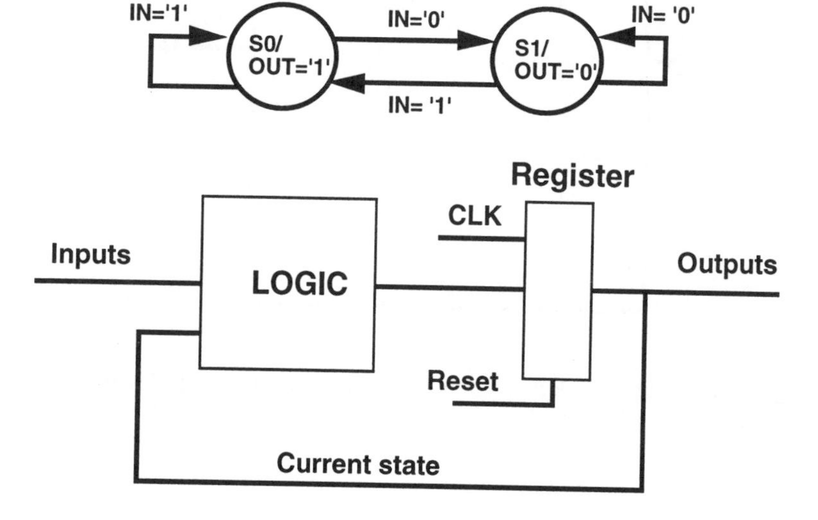

(b) Draw a finished block diagram and write the code for the following Moore machine (not spike-free).

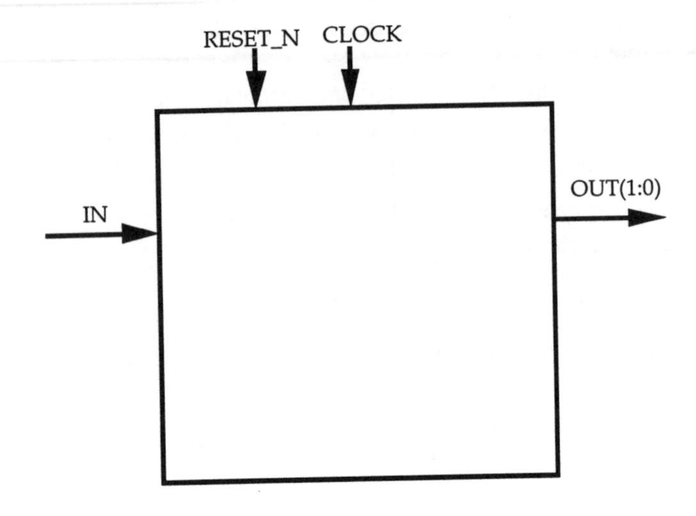

(c) How can exercise 9(b) be made spike-free if solution 9(a) is not used.

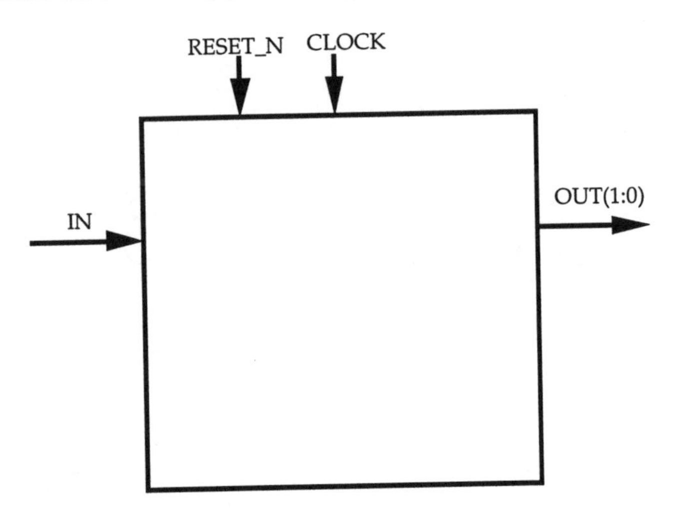

(d) Draw a finished block diagram and write the code for the following Mealy machine (not spike-free).

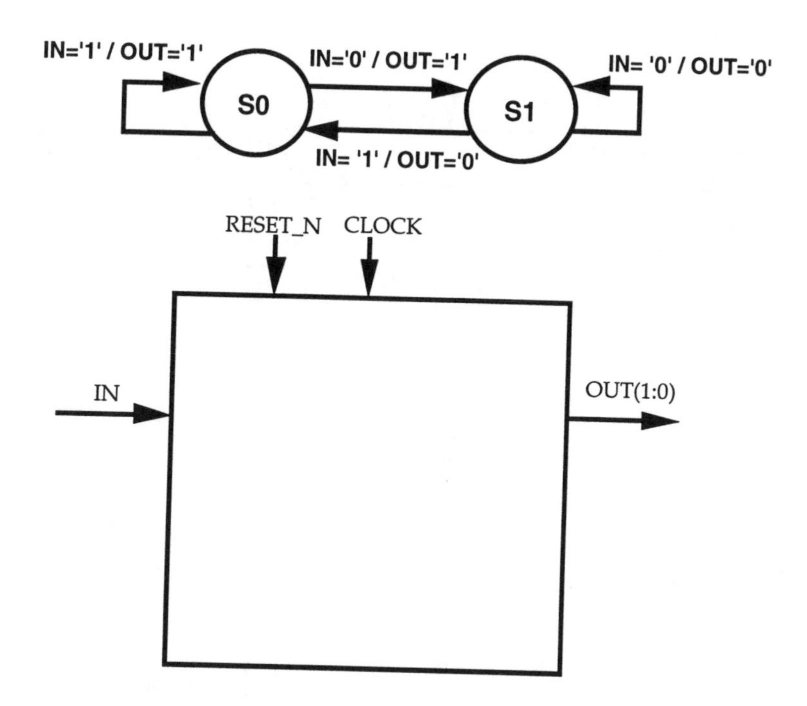

(e) Repeat exercise 9(d) but with spike-free outputs.

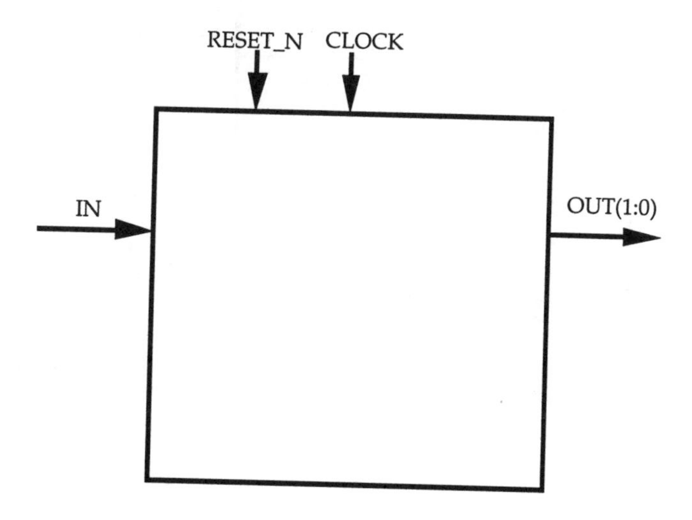

(f) Draw a finished block diagram and write the code for the following state machine (spike-free).

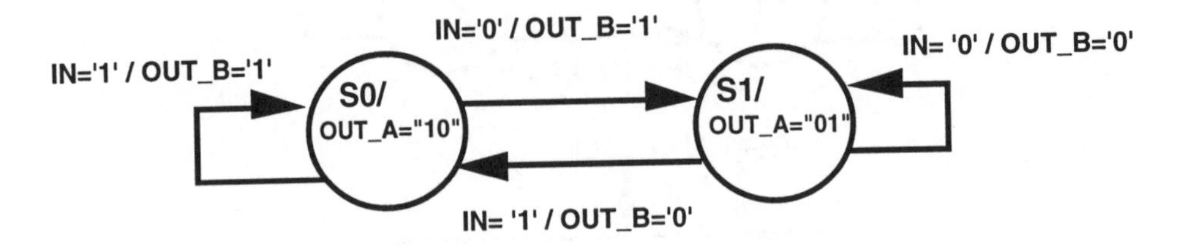

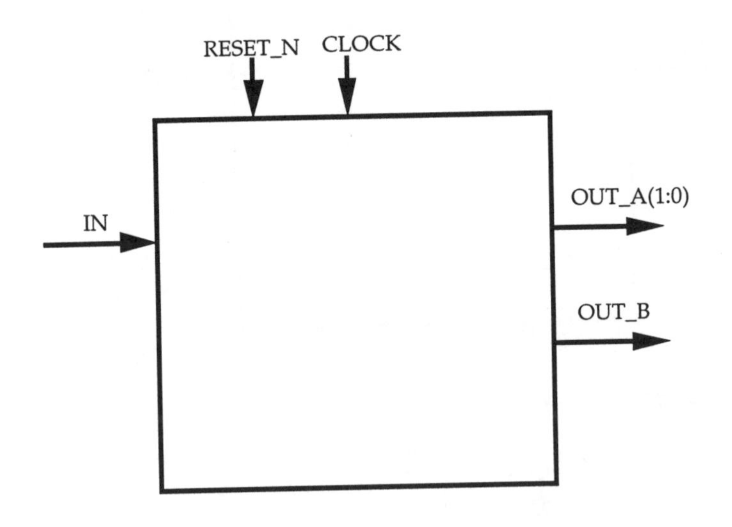

10
RTL synthesis

This chapter deals with synthesis from VHDL at RT level (see Figure 10.1). Synthesis from behavioural level is dealt with in Chapter 17, "Behavioural synthesis".

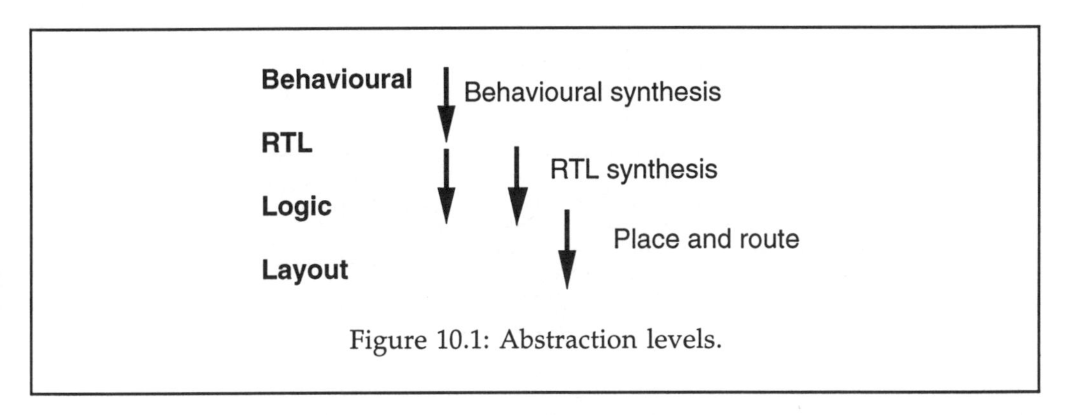

Figure 10.1: Abstraction levels.

VHDL was initially designed for documentation and simulation only. The idea was that it should be possible to describe and simulate systems and components in a high-level language. The late 1980s saw the first tools which were able to translate some parts of the VHDL standard into a netlist. In the early 1990s this development speeded up. More and more people began to see the great advantages of being able to use the same language for both simulation and design. If a design had been described in VHDL, it was a great boost to be able to take that VHDL code as a basis and translate it automatically into a netlist.

Nowadays synthesis to a netlist is done in a hardware description language for the majority of ASIC designs. A synthesis tool can accept languages other than VHDL. Some, for example, also accept Verilog. VHDL and Verilog are the two largest hardware description languages and have approximately 50 per cent of the market each. VHDL is growing faster than Verilog, however, and will therefore soon be the most widely used language for both simulation and synthesis.

It is not possible to synthesize the entire IEEE standard in VHDL: only a subset of the commands is supported by synthesis tools. There are also differences between what the various tools will support. Previous chapters have given examples of what can be synthesized and what can only be simulated. Work is currently being done to try to standardize what it should be possible to synthesize in VHDL. This standard will probably set a minimum level for what a synthesis tool should support. Usually, however, it is quite simple to change between the advanced synthesis tools on the market. Much more work is required, on the other hand, when going from an advanced to a simpler synthesis tool.

Synthesis is similar to a compiler for software. It translates a higher level of abstraction to a lower one. Various parameters, such as speed, area and special conditions for individual signals can be set.

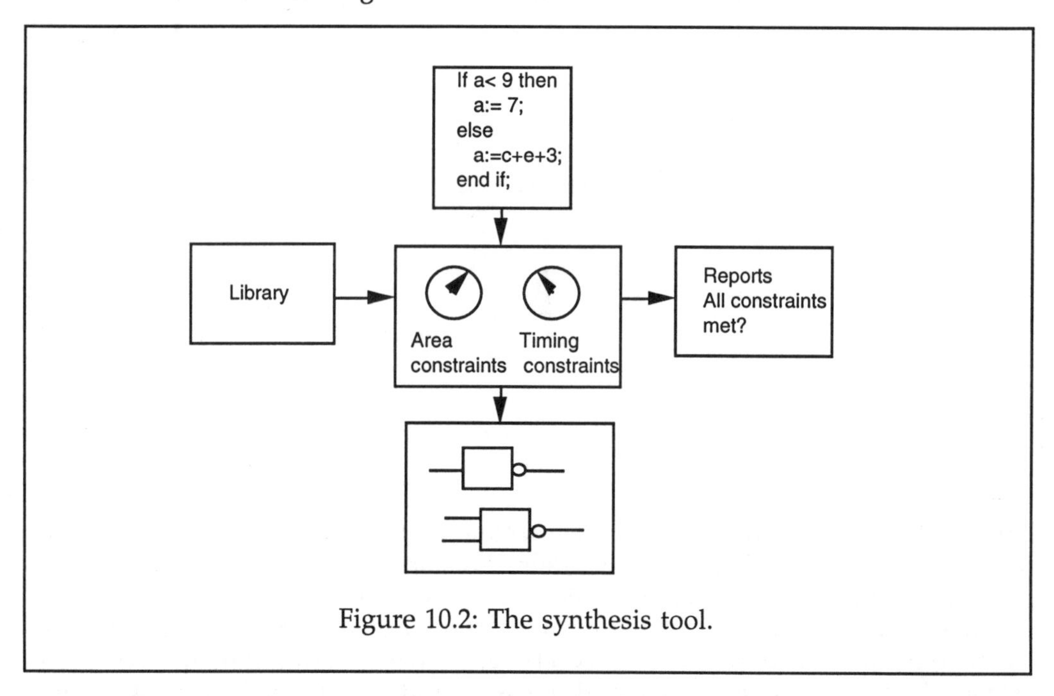

Figure 10.2: The synthesis tool.

The input data of the synthesis tool are VHDL, a technology file and constraints (Figure 10.2). The output data are a report file and a netlist. The netlist is a schematic with gates. The report file provides information on how the synthesis has been done and what the result is.

A typical flowchart for the synthesis and optimization process is shown in Figure 10.3.

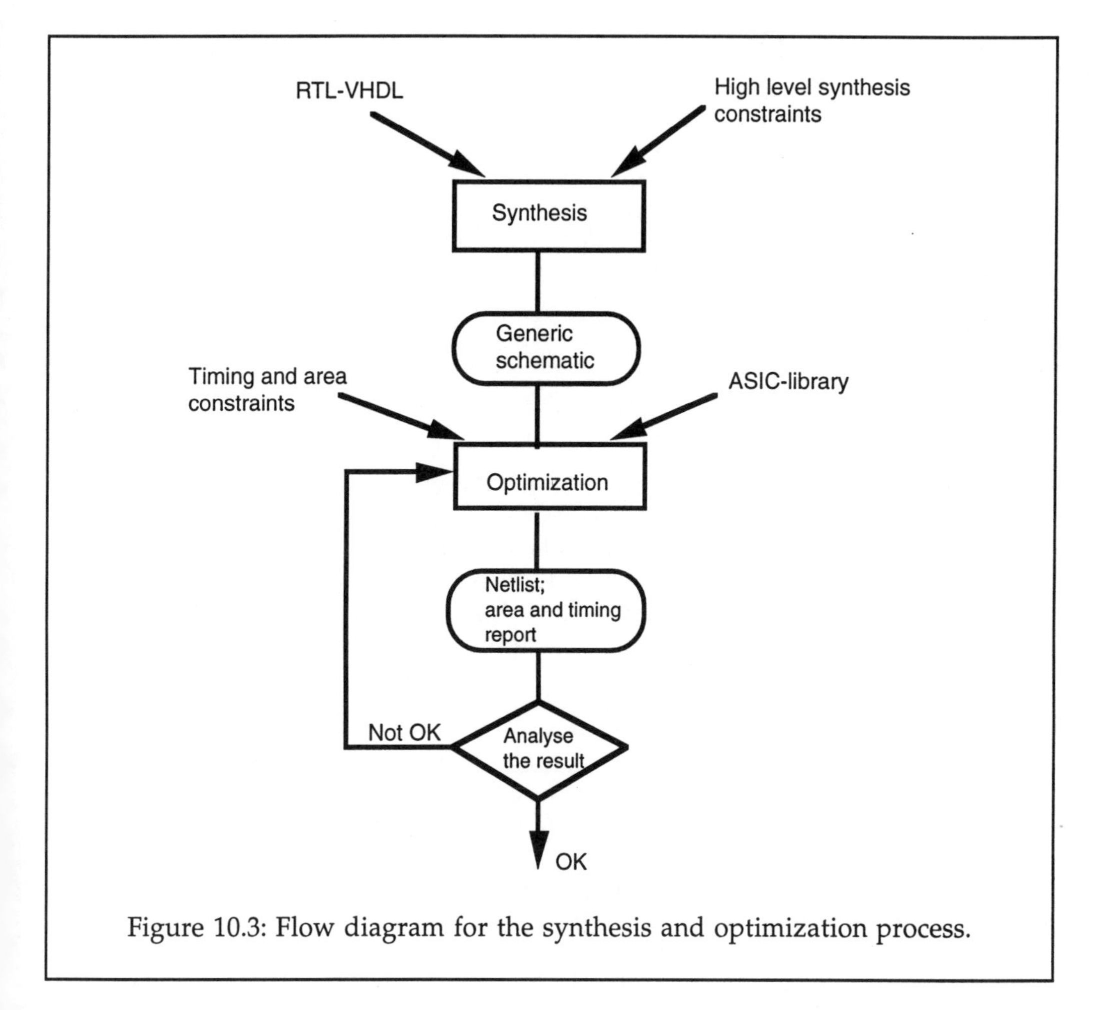

Figure 10.3: Flow diagram for the synthesis and optimization process.

Most advanced synthesis tools start by generating a generic (technology-independent) schematic on the basis of the VHDL code. When creating the schematic, the synthesis tool allows high-level optimization such as resource-sharing by adders. The size of the schematic can now also be specified. This part is called synthesis and the following part optimization. During optimization a technology-specific library has to be chosen. As the circuit should have been verified before synthesis was started, there should now be no functional errors left in the VHDL code. Problems which can arise now are timing and area problems, and specific design requirements such as rise and fall times on internal networks.

10.1. Optimization and mapping

When the synthesis tool translates the VHDL code into a generic schematic, it will identify adders, multipliers, multiplexors and registers, etc. All adders, multipliers and the like will receive special treatment during the optimization phase. These arithmetic functions are recognized from the VHDL code by the symbols "=", "+", "-" and "*", and they can be implemented in several ways. An adder, for example, can be implemented in the following ways:

- Ripple carry adder
- Carry look ahead adder
- Carry look forward adder

If we compare the adder area and timing for the various adders, the difference is considerable. The ripple carry adder is a very small adder and therefore slow too. The carry look forward adder, on the other hand, is fast but relatively large. As already mentioned, the emphasis should always be on the circuit's function and not on the choice of implementation. It is the synthesis tool that should decide how the adder is to be implemented. If we want to add two vectors, the VHDL code can be written as follows:

```
q<= a + b;
```

Based on the constraints set in the synthesis tool, the implementation which takes the least area while meeting the timing constraints will then be chosen.

Ordinary combinational logic will go through several of what are known as mapping optimizations. During mapping optimization the synthesis tool will look at a small window of the logic, e.g. five inputs and two outputs. The size of the window differs slightly from tool to tool. In this logic window the synthesis tool will then try to improve the gate choice by selecting other gates from the specified library (Figure 10.4). The synthesis tool will also select driving strength for the gates being included, all to achieve the right timing and to minimize the area. Thus the synthesis tool does not try to find the fastest implementation of this logic window, but just the implementation which satisfies the set timing constraints exactly. Then it tries to minimize the area. The mapping optimization carried out in this small logic window is then done on all the logic in the circuit by moving the window until the constraints are met. With some tools it is possible to set the level for how much the synthesis tool should exert itself when mapping. There are normally three levels to choose from: low, medium and high effort.

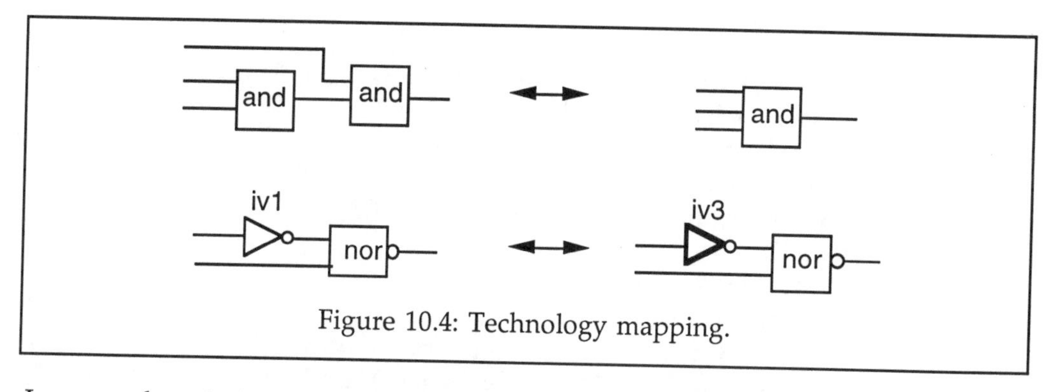

Figure 10.4: Technology mapping.

In normal optimization, the synthesis tool will optimize in relation to the set constraints. It is usual to talk about moving along a "banana curve" on the area and time axes. This means that the tougher the timing constraints, the larger the design will be, and vice versa (Figure 10.5).

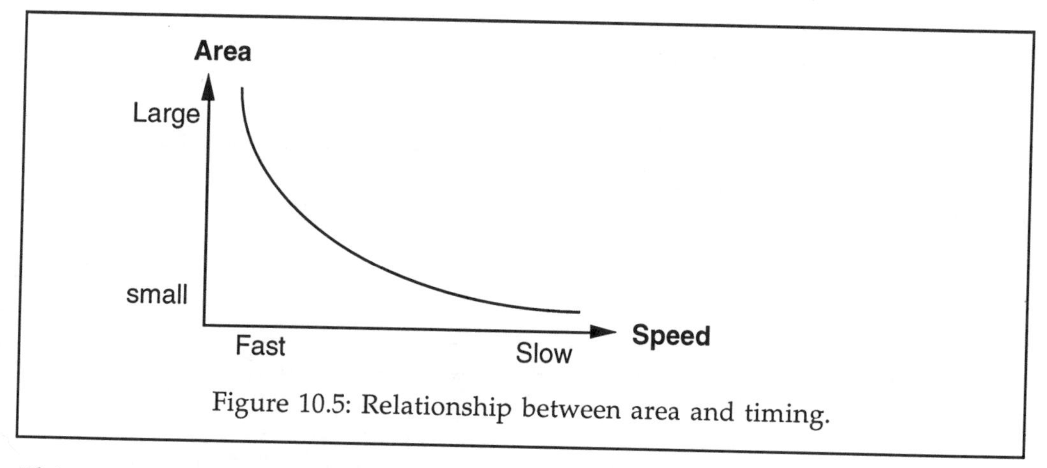

Figure 10.5: Relationship between area and timing.

If incorrect timing constraints have been specified for the synthesis tool, the wrong point on the banana curve will be reached. If we take the following VHDL code and try different optimization goals, the result will vary considerably.

```
Library ieee;
Use ieee.std_logic_1164.ALL;
Use ieee.std_logic_unsigned.ALL;

Entity syn is
  port( a,b:    in  std_logic_vector(4 downto 0);
        q1,q2:  out std_logic);
end;

Architecture rtl of syn is
signal i1,i2:std_logic;
```

```
begin
   i1<='1' when a+b<12 else '0';
   i2<='1' when a=10 else '0';
   q2<=i1 or i2;
   process(a,b)
   variable int:integer range 0 to 63;
   begin
     int:=conv_integer(a);
     case int is
        when 2 | 5        => q1<=a(0) and b(0);
        when 3 | 4        => q1<=a(1) and b(1);
        when 1 | 6        => q1<=a(2) and b(2);
        when 8 to 10      => q1<=a(3) and b(3);
        when 11 to 24 => q1<=a(4) and b(4);
        when others      => q1<=a(0) xor b(0);
     end case;
   end process;
end;
```

If we then take the same VHDL code and optimize for the smallest possible area, we obtain the result shown in Figure 10.6.

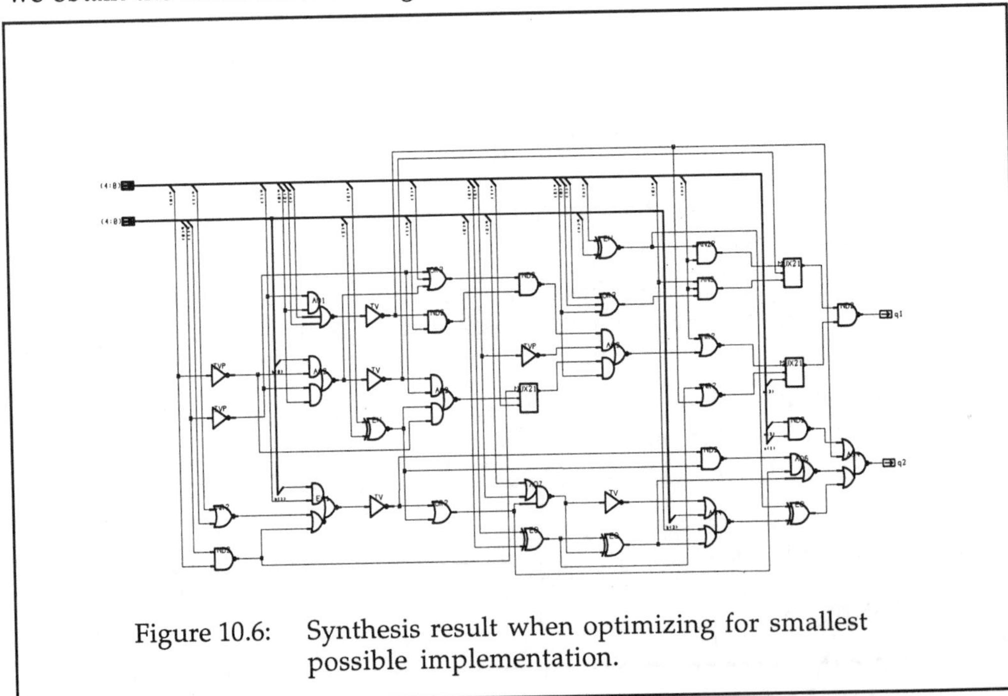

Figure 10.6: Synthesis result when optimizing for smallest possible implementation.

If we optimize the above VHDL code so as to obtain the fastest implementation possible, the schematic shown in Figure 10.7 is obtained.

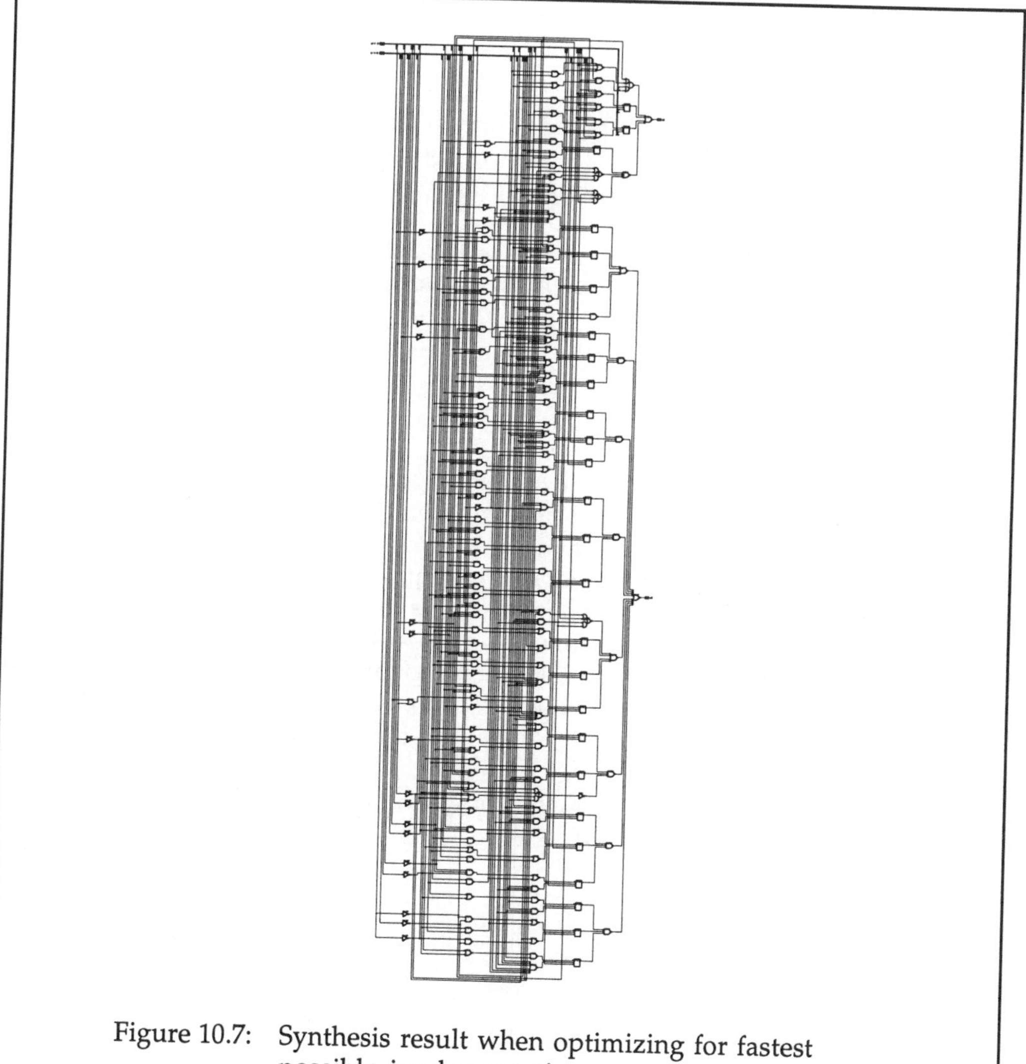

Figure 10.7: Synthesis result when optimizing for fastest possible implementation.

If we compare the results, we obtain the following table:

	Number of gates	**Speed (ns)**
Timing constraints	487	4.5
Area constraints	69	10.8

Clearly, the difference in result is considerable. This shows how important it is to set correct, realistic timing constraints for the synthesis tool.

10.2. Constraints

All advanced synthesis tools are based on optimization in relation to constraints. Below is an example of how constraints are set in Synopsys. The principles are largely the same in all advanced synthesis tools, although the syntax and options may vary.

Constraints can be divided into three main groups:

- Design rules
- Timing constraints
- Area constraints

No synthesis tool can perform magic. If the set constraints cannot be met at the same time, the synthesis tool will prioritize them in the above order. The order of prioritization can normally be changed if so desired. It is the norm, however, for timing constraints to have a higher priority than area constraints. If, for example, the constraint is 40 MHz, it is no use having a design which works at up to 38 MHz. If the area constraint cannot be met, on the other hand, a larger circuit can normally be chosen. This will result in an increase in component price, however. The alternative is to change technology (usually more expensive), redesign the circuit (time consuming) or reduce the performance requirement.

If the design is wholly synchronous, only the timing constraints have to be specified for the I/O pins at the top level.

The following always have to be specified:

- Clock inputs and their period
- Input delay relative to the clock
- Output delay relative to the clock
- Any pin-to-pin delay (point-to-point constraint)
  (combinational logic from input to output)
- Any false path

The following also has to be specified if the synthesis tool is to be able to calculate the timing in the circuit and carry out meaningful optimization:

- Supply voltage
- Temperature range
- Wire load model

If the circuit is large (>20,000 gates) the hierarchical blocks normally have to be optimized separately first. In this case the same timing constraints as above should be specified for all the inputs and outputs at block level.

10.2.1. Defining clock inputs

Defining the clock is very simple. In Synopsys, for example, the following command shown in Figure 10.8 can be used to define a clock.

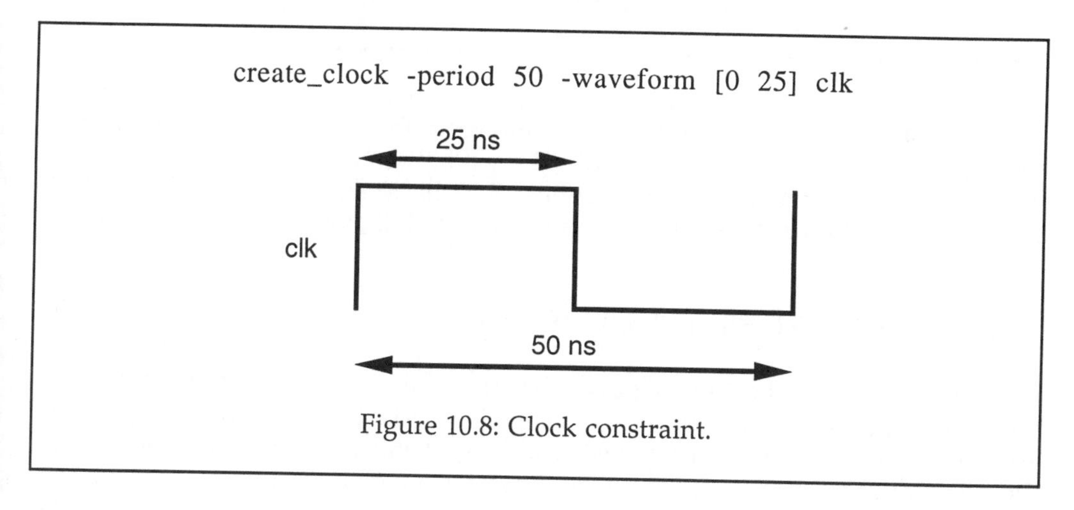

Figure 10.8: Clock constraint.

Alternatively, you can select the clock pin and then choose how the clock is to be defined from a pull-down menu. Other advanced synthesis tools work in the same way, although the syntax may different. If the system contains several clocks, they only have to be defined in the same way as above. The synthesis tool will then calculate the timing constraints between all the internal flip-flops automatically. If we defined the clock period as 50 ns, the synthesis tool will calculate that the combinational logic between all the flip-flops clocked with this clock can have a maximum delay of 50 ns less the delay through the flip-flop less the set-up time for the next flip-flop (Figure 10.9).

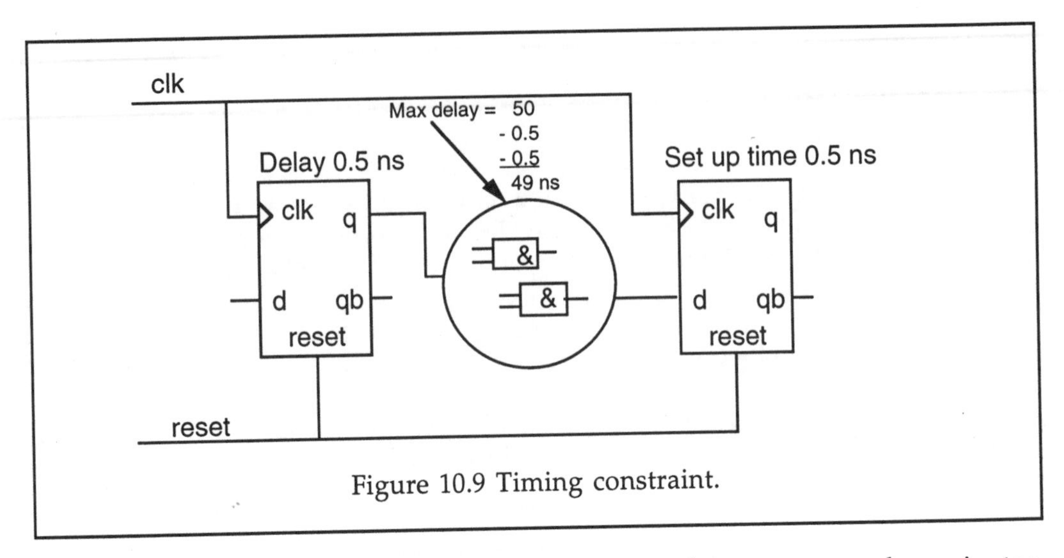

Figure 10.9 Timing constraint.

If the system contains several clocks, there must be a common denominator between them. The synthesis tool will automatically calculate the denominator and what demands on the combinational logic between the flip-flops it leads to. If, on the other hand, there are no connections between the two different clock systems, there does not have to be a common denominator between the clocks.

10.2.2. Defining input and output delay

The time at which the inputs have a stable value accessible at the I/O port relative to the clock (input delay constraints) must be specified for the synthesis tool. The synthesis tool must know how long at most any combinational logic from the input to the flip-flop can take. If we assume that the input delay is 20 ns relative to clk (50 ns period), we are requiring that it should take 50 - 20 = 30 ns minus any set-up time from the input to the flip-flop (Figure 10.10).

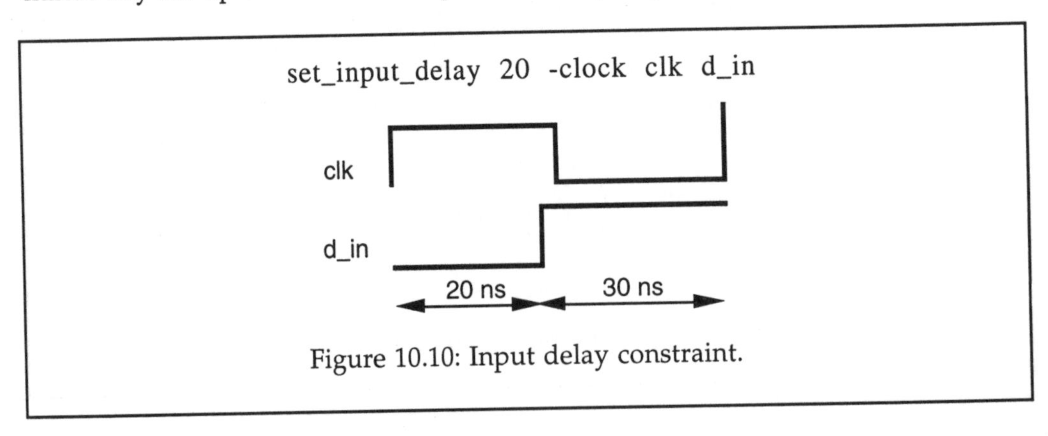

Figure 10.10: Input delay constraint.

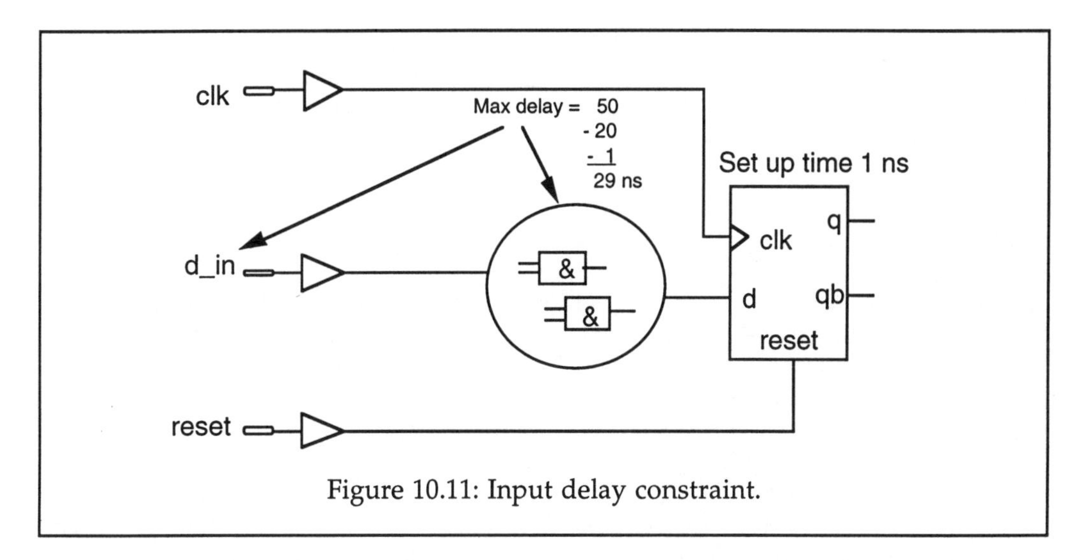

Figure 10.11: Input delay constraint.

The input delay should be set separately for each input in the circuit, as the timing conditions normally differ from input to input (Figure 10.11).

Just as with the inputs, constraints must be specified with regard to when the outputs should be accessible relative to the clock in the circuit (output delay constraints). If 28 ns is specified as the output delay, the time from the clock for the internal flip-flop to the output must be 50 - 28 = 22 ns maximum.

If there are purely combinational paths from inputs to outputs, the maximum delay can be specified by defining what is known as point-to-point timing (Figure 10.12).

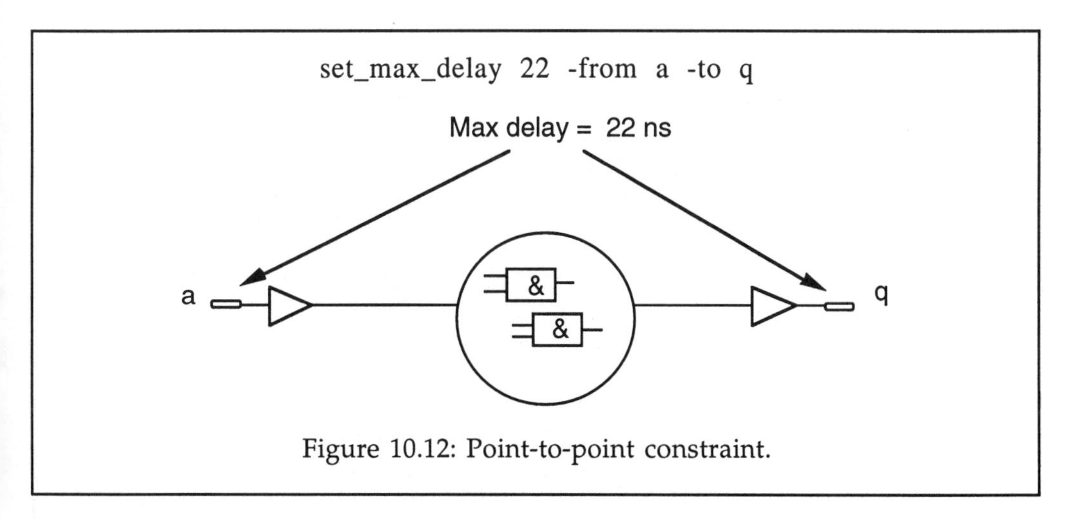

Figure 10.12: Point-to-point constraint.

Alternatively, we could have related inputs and outputs to a clock and specified the time constraint relative to the clock. This method is usually to be recommended, as it generally facilitates the constraints picture in the case of complex timing constraints.

10.2.3. False path

The next thing to be specified is whether or not there are any false data paths in the design. ASICs usually have test mode pins so that the circuit can be tested during manufacture. This will be connected to either 5 V or 0 V. All the timing paths which start from such a pin should therefore be defined as a false data path. This is because the synthesis tool will otherwise optimize these paths unnecessarily from a timing point of view. There are sometimes other data paths in the circuit which the designer knows will never occur in functional mode; these data paths should also be specified. All possible internal data paths are calculated by the synthesis tool. It is only those data paths which the designer knows to be false, based on the application or environment of the circuit, that should be specified, the rest will be taken care of by the synthesis tool.

Suppose that, based on external circumstances surrounding the circuit, we know that the data path from input *a* to output *b* will never occur during normal use of the circuit (see example below). The data path is then specified as false to avoid it being optimized unnecessarily with regard to time, as shown in Figure 10.13.

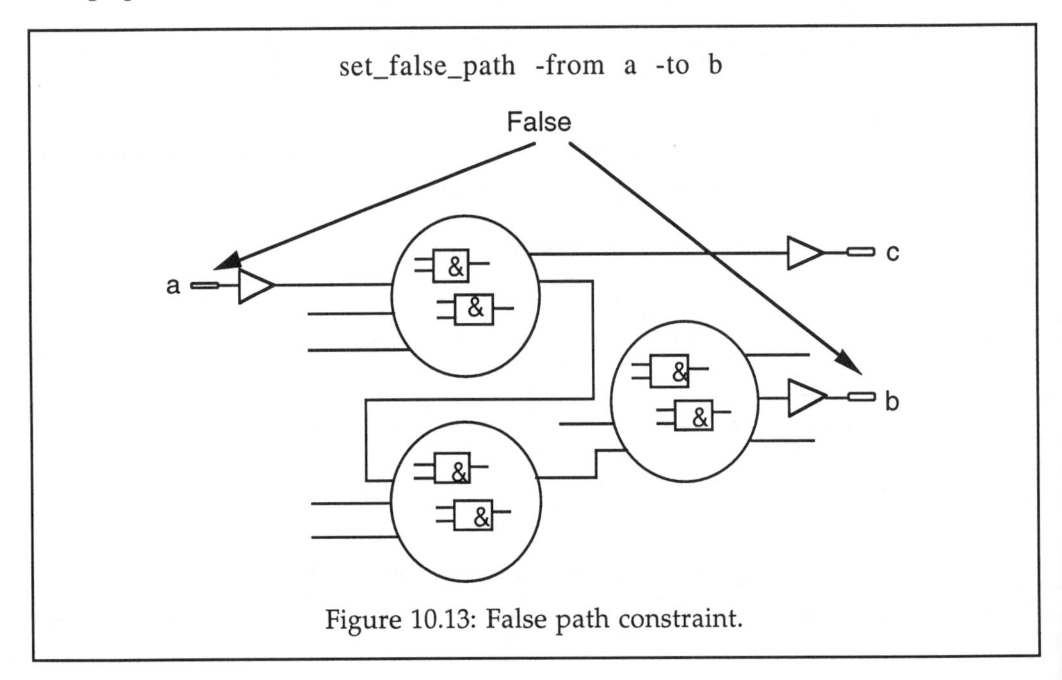

Figure 10.13: False path constraint.

It is particularly important to specify any false data paths in the case of FPGA design because the router needs this information to down-prioritize the data path when selecting a routing path. If a fast routing path is chosen for a false data path, it is normally at the expense of another data path being slower owing to limited routing resources in the FPGA circuit. The same principle applies to ASICs. The effects are not usually as serious in an ASIC, though, as the routing generally takes far less time as a percentage of the entire delay in an ASIC than is the case in many FPGA architectures. With very small line widths in ASIC technology (<0.6 µm), however, the routing delay starts to be of the same order of magnitude as the gate delays in the circuit.

10.2.4. Area constraints

Once all the timing constraints for the circuit have been set, the next step is to define the area constraint. This constraint is normally specified in NAND equivalents, i.e. gates.

Example:

```
set_max_area 20 000
```

If this constraint is not set, the synthesis tools will not make as much effort to optimize the area once all the timing constraints have been met.

10.2.5. Design constraints

The third constraint which should always be set for ASICs is the design constraints. Different ASIC manufacturers have different design constraints. If these constraints are not met, the ASIC manufacturer will return the netlist and not manufacture any circuits before the problem has been resolved. The usual design constraints set by ASIC manufacturers are rise and fall times, fan-out and capacitance of all internal networks. If these constraints are not specified for the synthesis tool so that they can be taken into account in optimization, major problems will normally result. It is difficult to fix the violations manually for large ASICs, if not impossible.

An example might be:

```
set_max_capacitance 10
set_max_transition 5
set_max_fanout 4
```

As all timing is dependent on supply voltage, temperature and process factor, they too must be specified. It is usual to specify, for example, that the circuit should function at 4.5-5.5 V supply voltage, 0-85°C junction temperature and a certain process factor. The process factor is normally set to worst case at max. timing and best case at min. timing. What is known as a wire load model should also be specified. The synthesis tool needs this to be able to estimate the capacitance for the internal signals. This table indicates what capacitance the internal networks will have depending on the fan-out. All that you normally have to do is to estimate the size of your final circuit and select the workload model which represents the nearest size.

Example:

```
set_operating_conditions  WCCOM        (worst case commercial
                                        0-70°C,  4.75-5.25V)

set_wire_load  20 000                  (gates)
```

10.3. Best case optimization

In ASIC design, min. timing also has to be taken into account. All the previous timing constraints can also be specified with min. timing instead of max. It is, for example, possible to require that the delay from input *a* to output *b* should be at least 10 ns and at most 30 ns. The greatest problem with best case time is usually hold, however. Suppose that we have a clock skew of 0.5 ns. Also assume that the delay through a flip-flop is 0.4 ns, the flip-flop has a hold constraint of 0.1 ns and the interconnection a routing delay of 0.1 ns. This means that the D-input on the flip-flop can change at the same time as the clock. As the hold constraint was 0.1 ns, this is not acceptable (see Figure 10.14).

The solution to this problem is to let the synthesis tool optimize in relation to best case timing as well. First the synthesis tool should be reset so that it works with best case process and maximum permitted supply voltage. The constraint for the synthesis tool can then be that the circuit must be able to cope with a clock skew of 0.6 ns, for example, between all flip-flops. It is also necessary to specify when the input signals will be available at the inputs at the earliest. This is because hold problems can also arise at the first input's flip-flop.

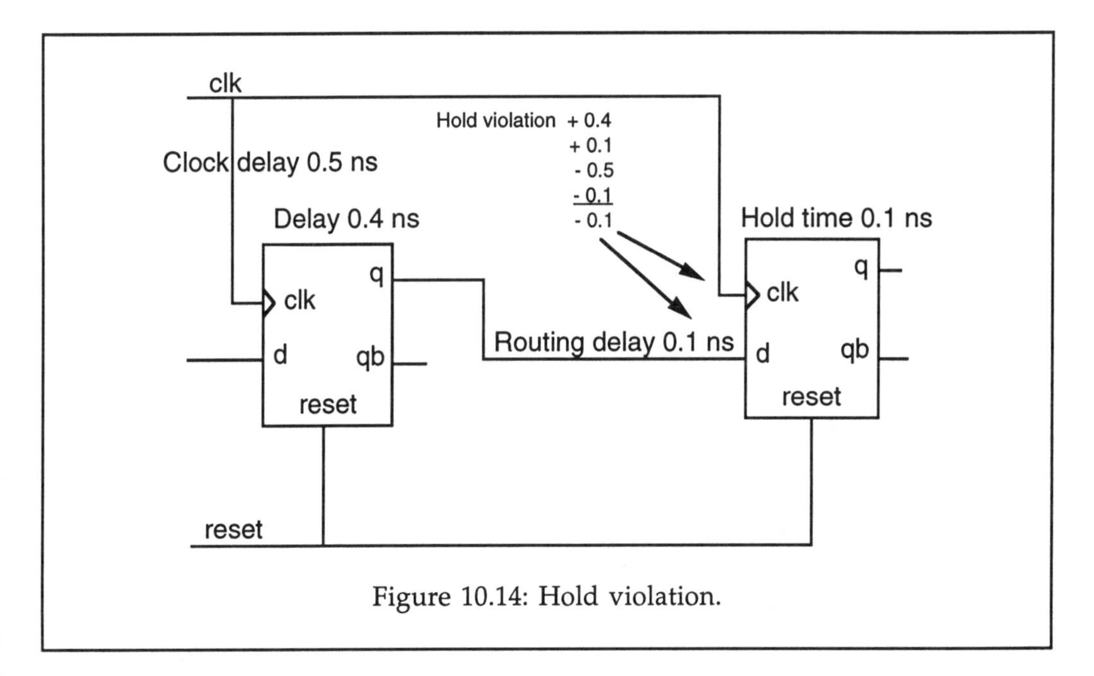

Figure 10.14: Hold violation.

Example:

set_operating_condition BCCOM (best case commercial)

set_clock_skew -uncertainty 0.6 clk

set_fix_hold clk

set_input_delay 2 -min -clock clk d_in

This optimization should also be done once all the maximum timing constraints have been met. With best case optimization there is a risk that the optimization will disrupt the worst case timing. This can be very difficult to get round if the synthesis tool cannot be started without touching any gates which do not have to be touched owing to hold violations. The Synopsys method for this, for example, is to start best case optimization in such a way that the tool is not allowed to touch data paths which do not violate the hold constraints.

In optimization of this type the synthesis tool will solve the hold problem by delaying the signal between the flip-flops with a buffer (Figure 10.15):

```
set_operating_condition BCCOM
set_max_clock_skew -uncertainty 0.6
set_fix_hold clk
compile -prioritize_min_path -only_design_rule
```

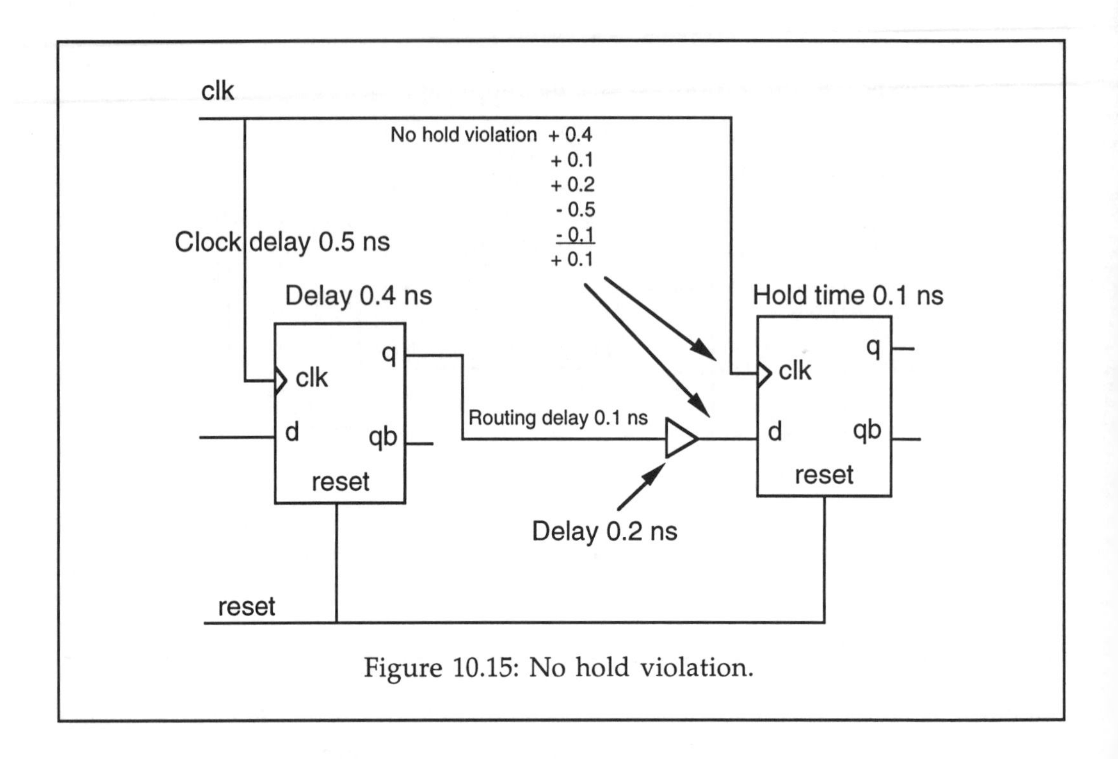

Figure 10.15: No hold violation.

10.4. What to do if the synthesis tool does not achieve the optimization goals

If problems arise with achieving the set optimization goals, the designer has the following options to choose from:

1. Try another optimization algorithm in the synthesis tool.
2. Change the state coding (for state machines only).
3. Rewrite the VHDL code.
4. Increase the minimum permitted supply voltage to the circuit.
5. Reduce the temperature range in which the circuit has to function.
6. Change technology.
7. Do not have as much error coverage (for area problems in particular).
8. Change to a better synthesis tool.
9. Optimize the logic manually.
10. Change the optimization goals.

It is recommended that the above are tried in the order in which they are listed.

Alternative 1: The advanced synthesis tools have various algorithms which can be chosen for optimization. At the beginning of the chapter the differences were shown which can be obtained by setting different area and time constraints, and by using different optimization algorithms. The area and timing for the two optimizations were very different. Of course, all synthesis tools have limits with regard to what they can optimize in a specific technology in terms of both area and timing, hence alternative 1 is not always viable.

Alternative 2: If the timing or area problem is in a state machine, the state coding can be changed using the state machine optimizer which is built into the advanced synthesis tools. This can have a relatively large effect on both area and timing (see Chapter 9, "State machines").

Alternative 3: If the VHDL code has to be rewritten, the designer has two alternatives:

- To change the VHDL code without changing it functionally
- To change the VHDL code functionally

It is often possible to rewrite the VHDL code without having to change its function. This is particularly true if you have a synthesis tool which is not among the best on the market. Such synthesis tools are usually affected to a large extent by the way in which the VHDL code is written. An advanced synthesis tool can normally reach the same synthesis result irrespective of the description method in the VHDL code for a small combinational block. If the combinational block is complex, the synthesis result can vary according to the way in which VHDL code is written for these tools as well. The synthesis result can, for example, be different if the **if-then-else** command or **case** command is used in the VHDL code. In general it can be said that the **case** command results in a faster but larger area than if the **if-then-else** command is used in the case of complex design.

As the VHDL code should be verified before the synthesis stage, the design will have to be verified again if the VHDL code is changed. Resimulating the design is normally time-consuming, so it is recommended that formal verification be used, if possible, to verify that the modified VHDL code has the same behaviour. Using formal verification (see Chapter 11, "Design methodology"), the old, synthesized VHDL code can be compared with the new code. If the result shows that they are identical from a functional point of view, the design does not have to be resimulated. This comparison only has to be made for the hierarchical block in which the VHDL code has been changed, and thus the method allows different VHDL description methods to be evaluated very quickly. The above reasoning assumes that the VHDL code does not need to be modified functionally. It should also be mentioned that tools are starting to appear which help the designer to write VHDL code effectively with a view to achieving set design goals.

If the optimization requirements still cannot be achieved, the VHDL code can be modified functionally. If, for example, the timing constraints are the problem, the designer should begin by analysing exactly what the timing problem is. This can easily be done by obtaining a timing report from the synthesis tool. One way of solving such problems is to pipeline the data paths.

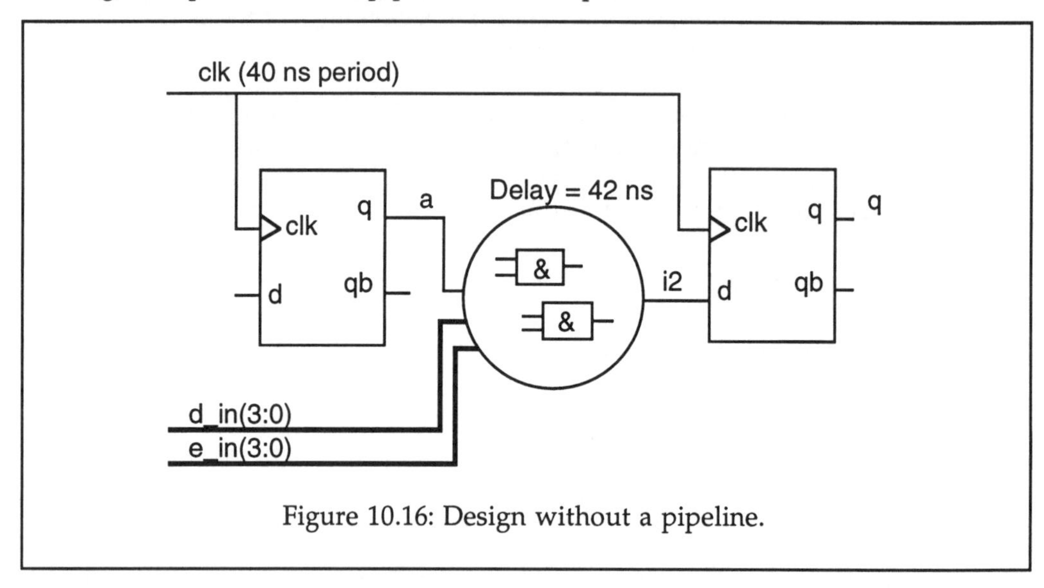

Figure 10.16: Design without a pipeline.

Suppose that Figure 10.16 represents the synthesis result from the following VHDL code:

```
Library ieee;
Use ieee.std_logic_1164.ALL;

Entity No_pipe is
  port( clk,d:        in   std_logic;
        d_in,e_in:    in   std_logic_vector(3 downto 0);
        q:            out  std_logic);
end;

Architecture rtl of No_pipe is
signal a,i1,i2:std_logic;
begin
  process(clk)
  begin
   if clk'event and clk='1' then
     a<=d;
     q<=i2;
```

```
  end if;
 end process;

 i1<='1' when d_in="1100" else '0';
 i2<='1' when i1='1' and e_in="0101" else '0';
end;
```

If we pipeline this design, the following VHDL code is obtained:

```
Architecture rtl of pipe is
signal a,b,i1,i2:std_logic;
begin
 process(clk)
 begin
  if clk'event and clk='1' then
    a<=d;
    b<=i1;
    q<=i2;
  end if;
 end process;

 i1<='1' when d_in="1100" else '0';
 i2<='1' when b='1' and e_in="0101" else '0';
end;
```

The disadvantage of doing this is that the data path is delayed by one clock cycle. The synthesis result is shown in Figure 10.17.

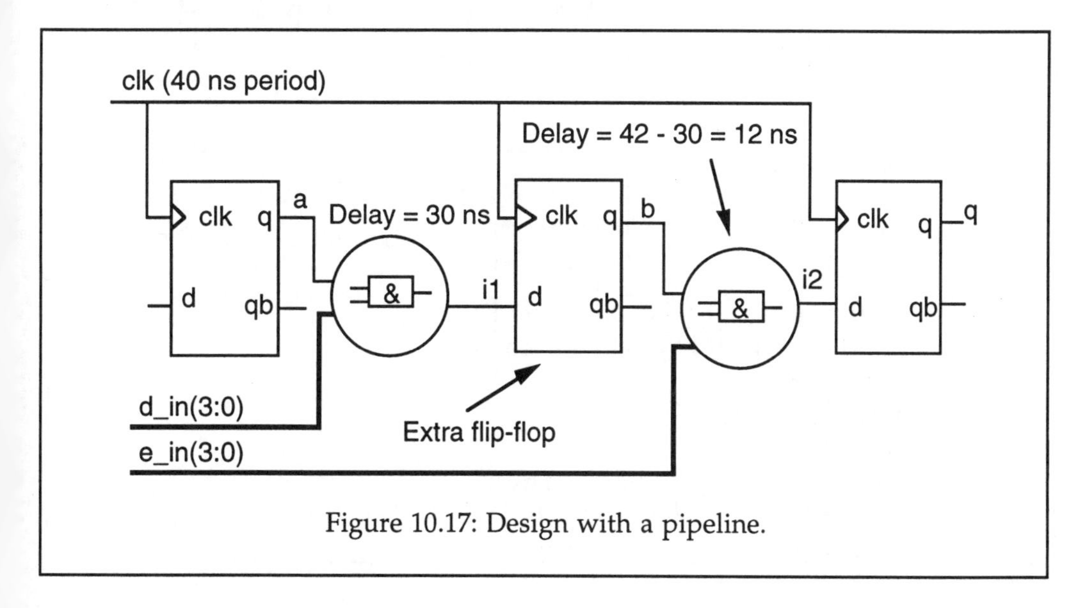

Figure 10.17: Design with a pipeline.

Another alternative is to carry out what is known as register balancing. This can be performed by some advanced synthesis tools at RT level. A design with and without register balancing is shown in Figures 10.18 and 10.19.

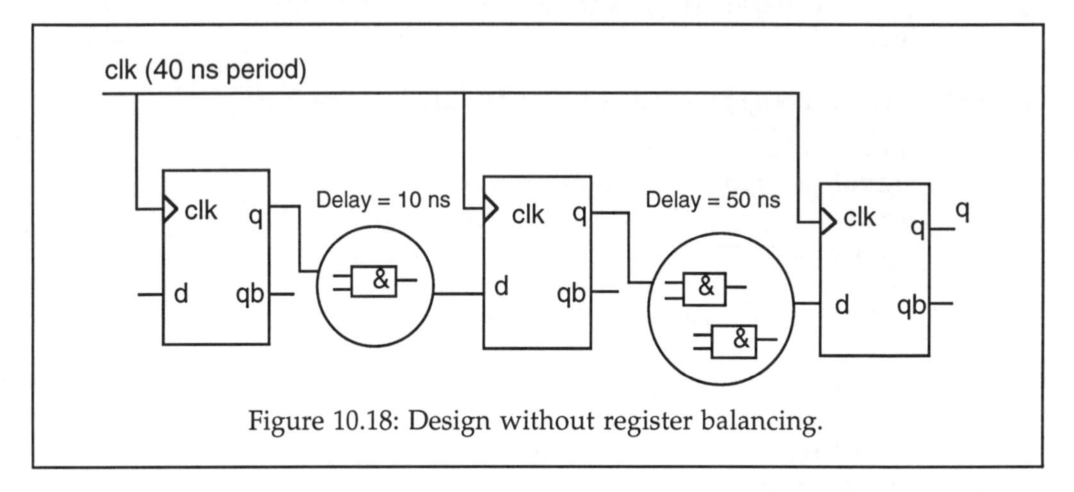

Figure 10.18: Design without register balancing.

The number of flip-flops in the design can be changed by such register balancing.

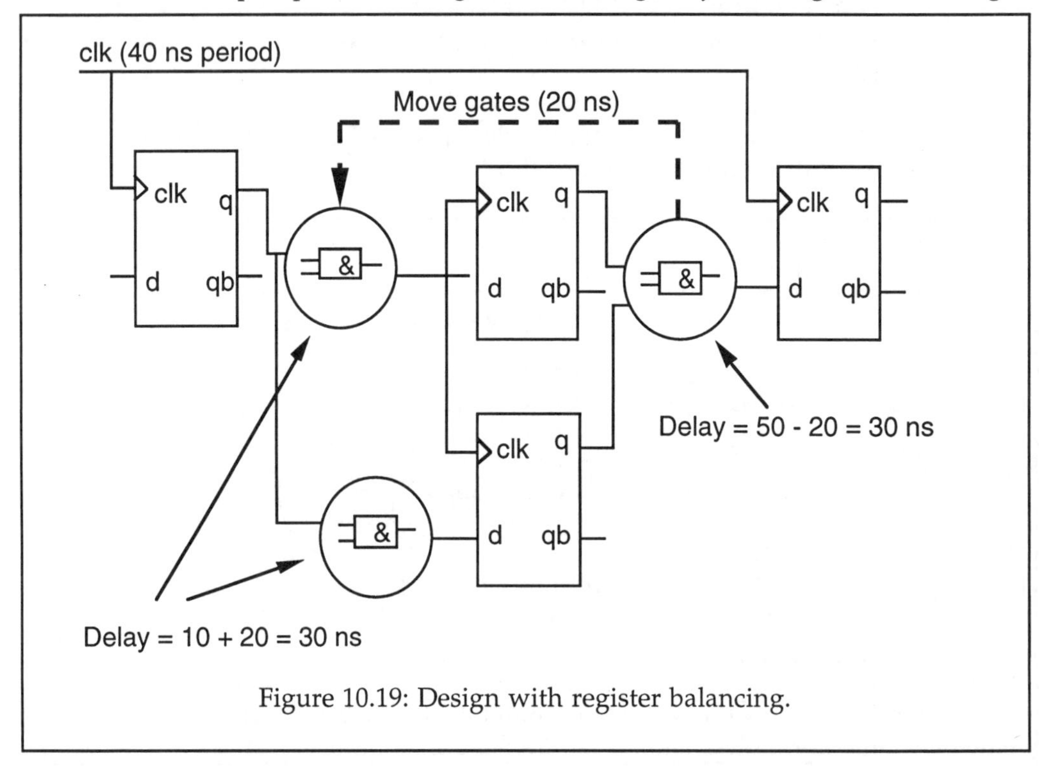

Figure 10.19: Design with register balancing.

The advantage of letting the synthesis tool do the register balancing is that the circuit does not have to be resimulated. On the other hand, formal verification can be used to prove that the circuit is the same functionally before and after balancing. This verification method is built into some advanced synthesis tools and is therefore very quick and easy to use. If register balancing is done, it must be documented carefully, as the number of flip-flops can be changed, as already mentioned. In the case of VHDL at RT level, the number of flip-flops is normally determined by the VHDL code. This does not apply to the part of the design for which register balancing has been done.

Alternative 4: The timing for ASIC/FPGA CMOS circuits is a function of supply voltage. If, for example, the supply voltage interval is 4.5-5.5 V, the timing in the circuit can be improved by 10-30 per cent for CMOS circuits by keeping the supply voltage within 4.75-5.25 V instead.

Alternative 5: The timing for ASIC/FPGA CMOS circuits is a function of the temperature of the silicon. The warmer the silicon, the slower the gates in the design. If we compare the speed at room temperature (25°C) with the speed at 70°C, for example, there is a difference by a factor of approximately 1.5-2.0. Better performance can therefore by achieved be lowering the maximum permitted temperature for the silicon.

Alternative 6: As the VHDL code is technology-independent, different technologies can be tested until the constraints are met.

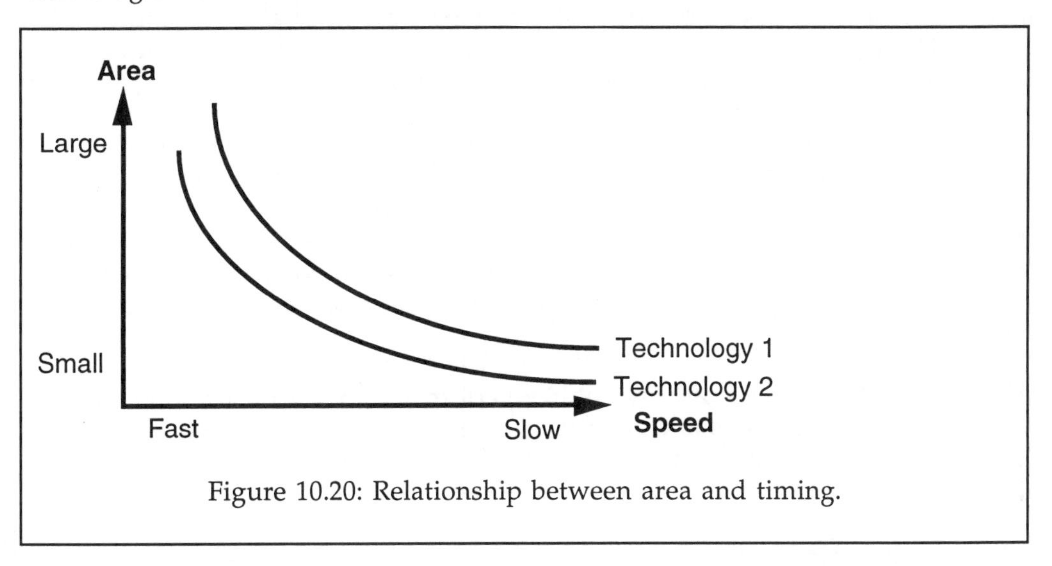

Figure 10.20: Relationship between area and timing.

By selecting a faster technology, for example, it is possible to achieve better performance while at the same time making the design smaller (Figure 10.20). However, this normally results in an increase in component price.

Alternative 7: If the area is the problem, choosing full scan should be considered. Full scan means that all the memory elements in the design are exchanged for a scan equivalent and included in a scan chain. These scan flip-flops take slightly more gates than an ordinary gate. The number of gates can be reduced by excluding flip-flops from the scan chains, but this is normally at the expense of lower error coverage. In some cases it is also possible to obtain a faster design by not using scan flip-flops in the critical data paths, with scan flip-flops normally being slightly slower than ordinary flip-flops. For a more detailed explanation see Chapter 12, "Test methodology".

Alternative 8: If the best synthesis tool was not used, changing to a more advanced tool is an option. The risk here is that the VHDL code will need to be rewritten slightly, as the syntax for what can be synthesized varies between some synthesis tools. This is a very expensive alternative as a new synthesis tool will have to be purchased. The cost of the tool can often be recouped quickly if ASICs/FPGAs are designed regularly, as it will lead to a smaller area and better performance. The difference between synthesis tools can be up to 30-40 per cent for both area and timing.

Alternative 9: If a poor synthesis tool has been used, the synthesis result can sometimes be optimized better manually, but this is very time-consuming. The recommendation is to change to a better synthesis tool if the synthesis result can be improved with a reasonable amount of work. If a good synthesis tool has been used, manual optimization is often the result of lack of knowledge of how the synthesis tool should be used effectively. It is possible to create manually all the netlists that a synthesis tool creates; the problem is that it is time-consuming and it is often difficult to achieve the same result as the synthesis tool.

The requirements are slightly different if a full custom ASIC is being designed. In this case, the designer can create special macros to improve the timing or area in certain critical data paths. The synthesis tool cannot create these macros, as it can only select them from the defined library. This method is also time-consuming but can sometimes solve the problem. The macros should then be saved in a library so that the synthesis tool can choose them when next optimizing this technology.

Alternative 10: As a final way out, the performance or area constraints for the circuit can be reduced. In some projects a loss of performance can be compensated for by increasing performance in other parts of the system.

10.5. Summary

It is important to know how the VHDL code should be written in order to obtain a good synthesis result. Mastery of the synthesis tool is just as important.

Mastery of VHDL, the synthesis tool and design with top-down methodology offers major advantages over conventional design. An advanced synthesis tool can facilitate and speed up the entire design process considerably. If an advanced synthesis/optimization program is used, all the timing, area and design constraints can be set as required in order to obtain a functioning circuit. A static timing analyser, which makes it possible to analyse the timing in the circuit, is often included as well. Report files can also be obtained for area and possibly design constraints which have not been met. If these reports show that all the design constraints have been met, it can be said with a high degree of certainty that the circuit will work, as the circuit should have been verified functionally during the VHDL simulation.

To summarize, it can be said that it is synthesis that makes top-down design possible with VHDL.

11

Design methodology

There are several different procedures when it comes to designing ASICs/FPGAs (Figure 11.1). The design methodology which most people now given prominence to and try to follow is top-down.

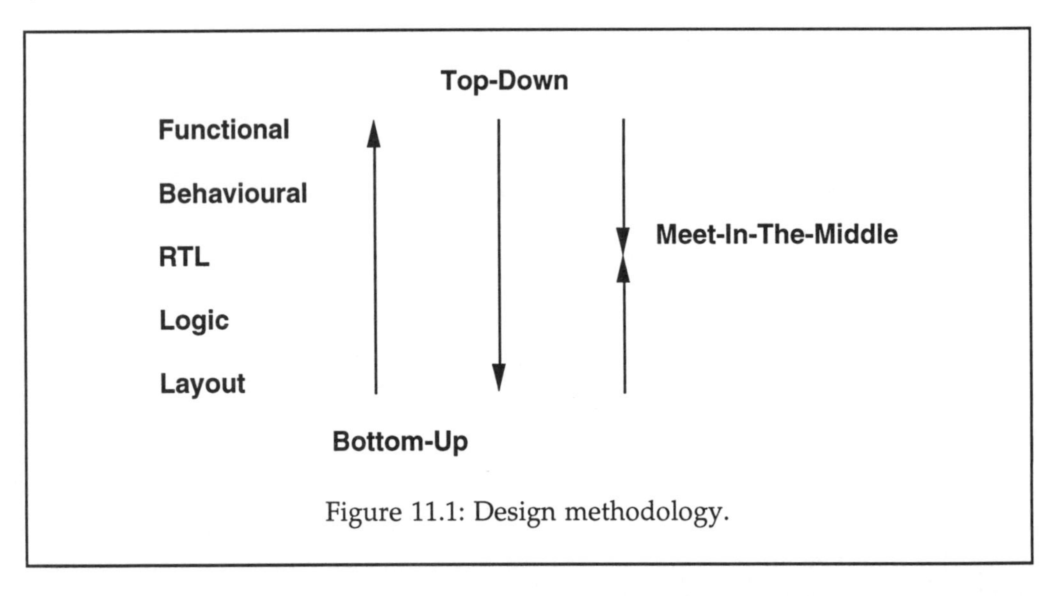

Figure 11.1: Design methodology.

The construction methodology most widely used before VHDL broke through was meet-in-the-middle. Most projects started with a requirement specification being written. Then design was started directly at gate level, i.e. the technology had already been chosen. The gates were positioned and connected manually in a schematic. The designer was also forced to think about matters which have nothing to do with the circuit's function at the same time as trying to design something which satisfied the requirement specification. In ASIC design the designer was obliged to choose a driving strength for all gates. The designer also needed to choose an architecture for the design at this early stage, i.e. the timing and area constraints controlled the implementation method of the design. The timing was calculated manually when the gates were being positioned on the schematic. While making these calculations, it was important not to exceed the number of gates that could be accommodated in the intended circuit (ASIC/FPGA), so that in the timing calculations the designer had to keep a manual check on the driving strength and capacitance of the gates and networks. If problems were then discovered in the timing simulation, or if there were found to be too many gates, this had to be corrected manually in the schematic. Sometimes entire hierarchical blocks had to be redesigned in order to achieve the design goals. If the project management then changed the requirements during the project, e.g. modified timing constraints or the internal bus size, the whole design often had to be consigned to the dustbin. It was just as bad if the chosen technology was changed suddenly as a result of a competitor quoting a cheaper price or the chosen supplier not being able to deliver. If this happened, the designer often had to start again from scratch. Thus the designer devoted just as much time to matters which did not affect the function of the circuit as to describing the circuit in a functionally correct manner.

The bottom-up methodology was even worse than meet-in-the-middle. With this methodology the designer started by doing the layout for the circuit board and then seeing what circuits could be positioned on the card. The next step was to investigate the function of the card. The project ended with the requirement specification being written on the basis of the simulation result. Virtually nobody used this design methodology.

As complexity increased in most designs, firms needed a new way in which to design their circuits. Even for those designs which did not increase in size, the design phase had to be greatly reduced in terms of time in order to meet time-to-market requirements and save money. If the designer could concentrate on the circuit's function instead of gate selection, capacitance, driving strength and the like, it should be possible to solve both these problems. For a great many firms the answer to these problems was a hardware description language, namely VHDL.

Designing with VHDL presupposes none of the design methodologies mentioned in Figure 11.1, with it being possible to choose the most suitable design methodology for the project. Top-down design has several advantages, however:

- Complex designs can be handled
- Design times are reduced
- Quality is increased
- Rapid prototyping with FPGA is possible
- Recyclability is increased

Below the same design is described first in the traditional way using schematic capture (Figure 11.2) and then using VHDL at RT level (Figure 11.3).

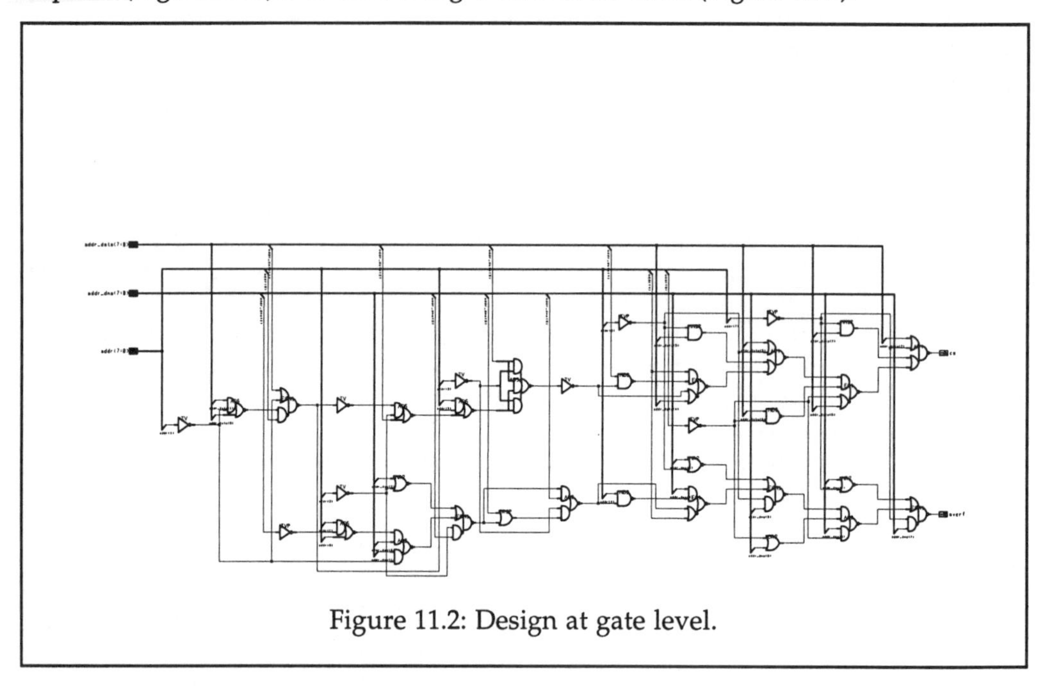

Figure 11.2: Design at gate level.

```
Architecture rtl of ex is
begin
    cs<='1' when addr > addr_data else '0';
    overf<='1' when addr_dma > addr else '0';
end;
```

Figure 11.3: VHDL design.

If we compare the old, traditional description method of schematic design and designing with VHDL, it is immediately obvious which is easier to design with, maintain and reuse. There are not many people who could look at Figure 11.2 and say what the function of the "tangle" of gates is in five or six seconds. Most people, on the other hand, can understand the function of the VHDL code. Of course, the difference is not always as great as it is here (though it often is), but it still illustrates the difference clearly. The entity is not shown above because the comparison should be as fair as possible. The equivalent of the entity in the schematic would have been a symbol. All the vectors above are eight bits in length. If we want to change their length, we only have to change the declaration in the entity in the VHDL example, whereas much larger manual changes would have to have been made in the schematic to change the bus size.

11.1. Top-down flow

In top-down design, you first concentrate on specifying and then on designing your circuit functionally (Figure 11.4). The degree of detail will increase the further down in the flow the project goes. The technology does not have to be chosen before the synthesis stage, although the circuit must have been verified functionally before synthesis.

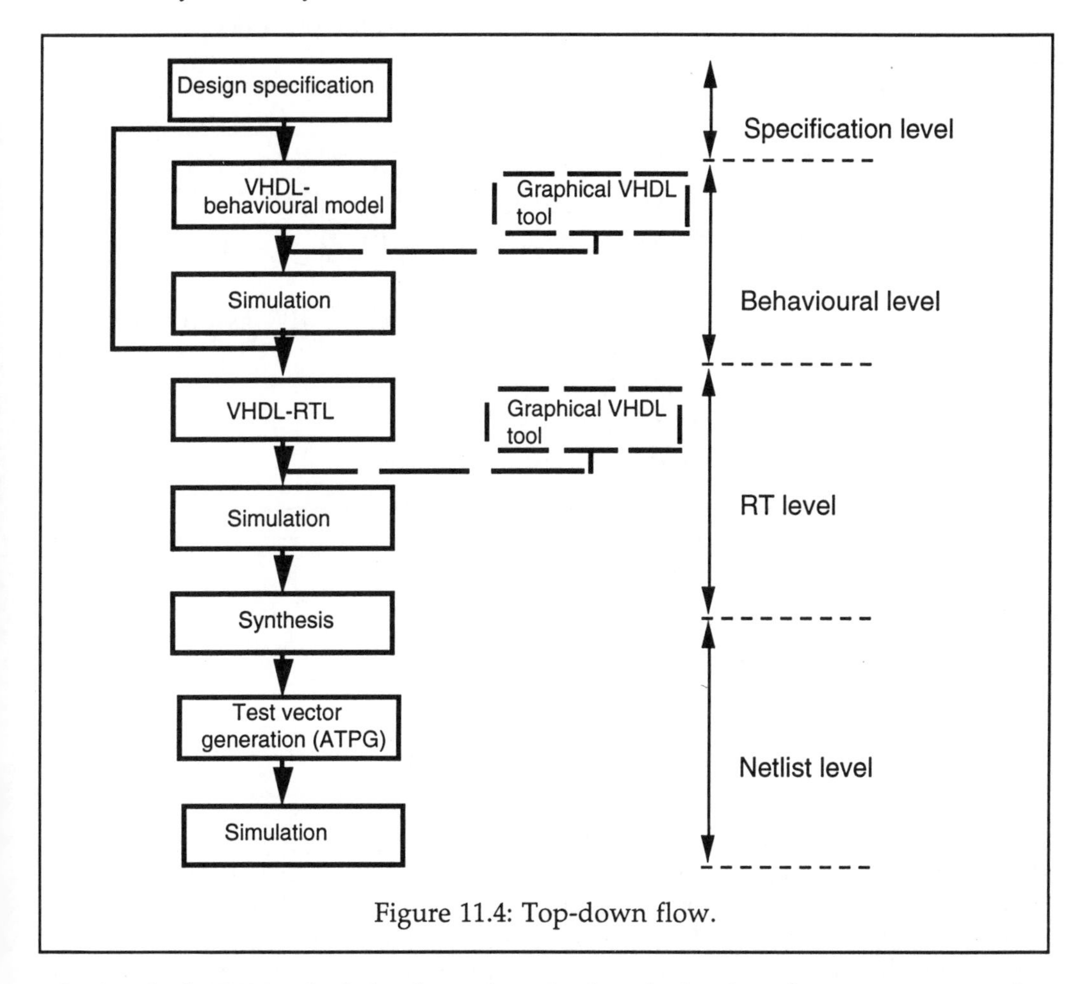

Figure 11.4: Top-down flow.

The level of VHDL which has been described in the book so far is register transfer level (RT level). This is the level which synthesis tools accept and synthesize to hardware. Behavioural VHDL uses the same VHDL commands as already described, but without a thought of hardware. The aim of the behavioural model is merely that it should have the correct behaviour and be simulatable. Abstract data types are usually used in a behavioural model. It is also usually more or less asynchronous and therefore does not contain a system clock in many cases. In addition, time delays can be described in the VHDL code using the **after** command instead of using a counter as required at RT level to create delays, for example.

An example of behavioural code for a watchdog might be:

```
Architecture behv of watchdog is
begin
   time_out<= '0' when (trig='1' or resetn='0') else
             '1' after 10000 ns;
end;
```

The same example at RT level would be:

```
Architecture rtl of watchdog is
signal count:std_logic_vector(6 downto 0);

begin
   process(clk,resetn)  -- 1 clock cycle = 100 ns (10 MHz)
   begin
      if resetn='0' then
         count<=(others=>'0');
         time_out<='0';

      elsif clk'event and clk='1' then
         if trigg='1' then
            count<=(others=>'0');
            time_out<='0';

         elsif count/=100 then
            count<=count+1;
            time_out<='0';

         else
            time_out<='1';
         end if;
      end if;
   end process;

end;
```

This behavioural model must not be confused with the behavioural code which can be used for behavioural synthesis. If the intention is to synthesize the behavioural model, account must be taken of the hardware in the VHDL code, i.e. the model must also contain clocks and registers at behavioural level. There is more on this in Chapter 17, "Behavioural synthesis".

With top-down design, the circuit can first be modelled by making a behavioural model. The advantage of a behavioural model (not synthesizable) is that the designer can simulate the circuit at an early stage and discover any system errors. This level can, however, be skipped in the top-down flow in many cases. It is only in the case of highly complex circuits (systems) or if a lot of circuits have to be designed at the same time for a system that this description level can normally be justified. This is because a simulatable model can also be produced quickly at RT level and the entire code usually has to be rewritten to obtain a description at RT level. If, on the other hand, the application lies within the area in which behavioural synthesis is effective, it is recommended that a behavioural model should always be used. This model should then be written in such a way that it is synthesizable using behavioural synthesis. In this case the VHDL code does not have to be rewritten for RT level, as synthesis is done from behavioural level.

One of the reasons why it is so important to master top-down design is that the design effort grows exponentially with the number of gates (Figure 11.5). This means that, if the size of the design is doubled, the design effort is normally more than twice as large.

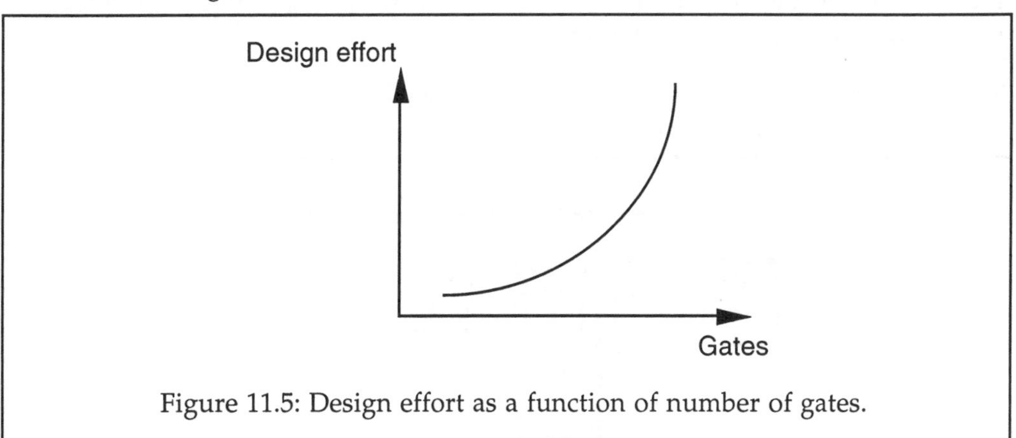

Figure 11.5: Design effort as a function of number of gates.

11.2. Verification

There are three verification steps in the top-down flow:

- The behavioural model
- The RTL VHDL model
- Gate level both before and after layout

The component can be verified as follows in all three of the above verification steps:

- Creation of test vectors in the simulator's internal language
- VHDL testbench
- System simulation

The traditional way of verifying the designed component, before the arrival of VHDL, was to create input stimuli for the models in the simulator being used. The disadvantage of this method is that the simulation becomes platform-dependent. There is also a great risk that, if the same designer does the test vectors, a possible logical error in the component can also appear in the test vectors and the error in the circuit will not be found. All that this method verifies, therefore, is that the designer designed what he thought he was designing, not that it will work when the circuit has been mounted on a circuit board.

The other method is to write a testbench. A testbench is a VHDL program which generates stimuli for the model and checks that it responds correctly. The degree of detail in the testbench can vary a great deal from case to case, from testing all the functions in the circuit which are described in the requirement specification, including timing, to testing only a few functions.

The testbench can also be used to test VHDL code at RT level and for final gate level simulation (Figure 11.6; see also Chapter 8, "Testbench").

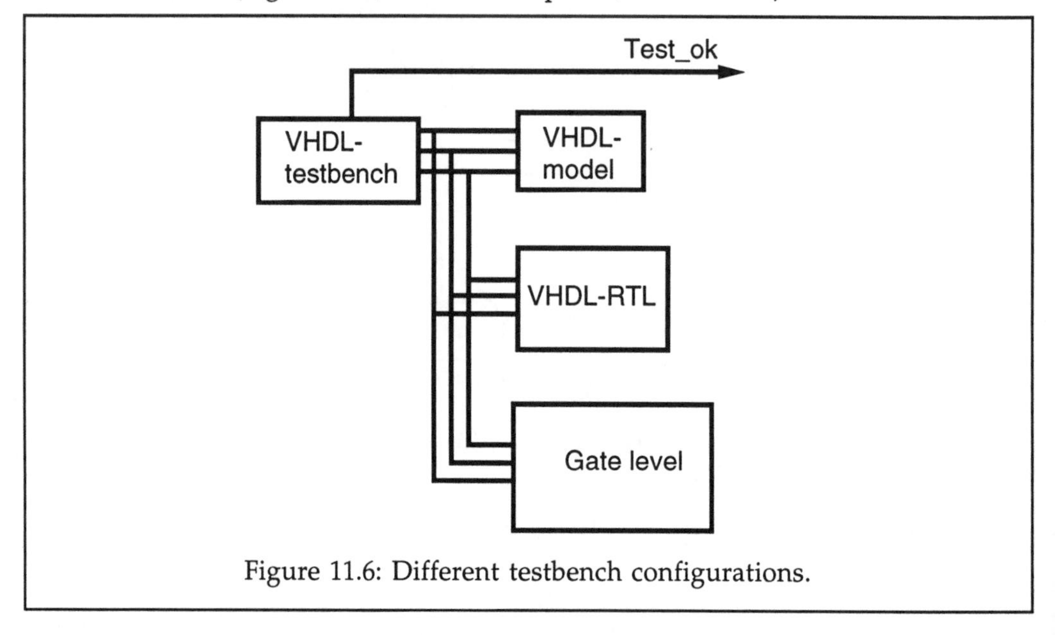

Figure 11.6: Different testbench configurations.

If the testbench is used in this way, verification is obtained that no mistakes were made when the design was broken down in detail. This is particularly important if different people are doing the behavioural model, the RTL code and the logic synthesis.

One of the advantages of testbenches in VHDL is that they are tool- and platform-independent. Here too there is a risk that the testbench will have the same logical errors as the design. One way of reducing this risk is to get someone else to write the testbench, e.g. the person who wrote the requirement specification. The testbench can then be included in the requirement specification.

The third method is to use models of the adjoining circuits for the model and do what is known as a system simulation. If VHDL models are not available for all the surrounding components, the designer must have access to a simulator which can cope with mix level (e.g. a backplane simulator), i.e. both VHDL and gate level. If, on the other hand, there are VHDL models for all the surrounding components, the usual VHDL simulator can be used for the system simulation. It is, therefore, access to VHDL models which determines whether a mix level simulator or a VHDL simulator should be used for the system simulation.

This verification method (system simulation) is the best when it comes to discovering possible design errors in the interface with the surroundings. The circuit will often interface with a microprocessor, EEPROM, RAM, control logic of the 74HCxx type, for example, other VHDL models, PLDs, FPGAs and other ASICs. It is now possible to buy a number of different microprocessor models. The PLD, FPGA and ASIC circuits can be either VHDL models or simulatable schematics, or the circuit can be placed in what is known as a hardware simulator which checks and simulates the circuit.

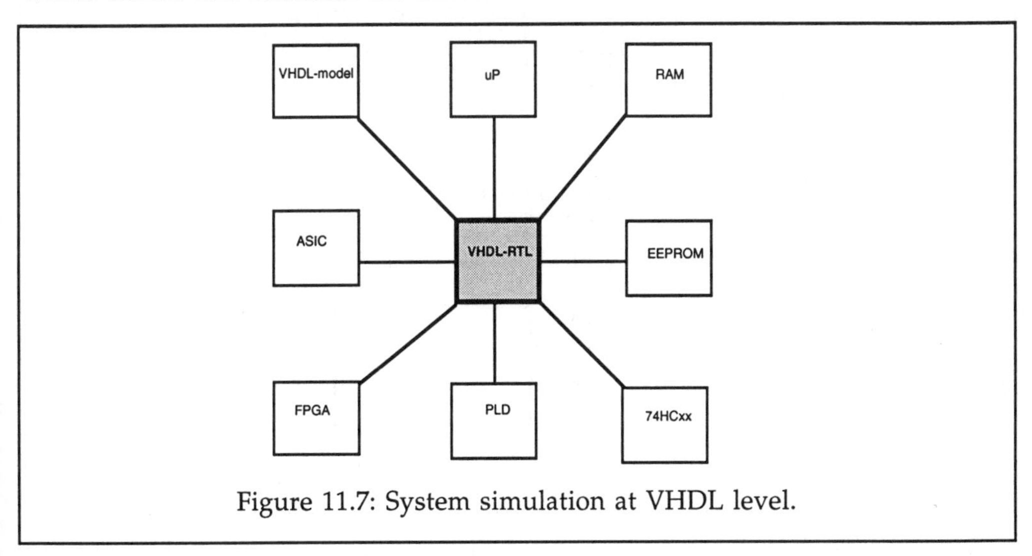

Figure 11.7: System simulation at VHDL level.

This system simulation should always been done on the VHDL model at RT level (Figure 11.7).

System simulation can also be done at behavioural level and gate level. Normally the circuit is only verified functionally at VHDL level. Timing verification is done in both the synthesis tool and gate level simulation. The timing constraints are set in the synthesis tool, which can generate reports concerning whether the timing constraints have been met or not after optimization. To verify that no mistakes were made when the timing constraints were set, the gate level model should also be verified in a system simulation (Figure 11.8).

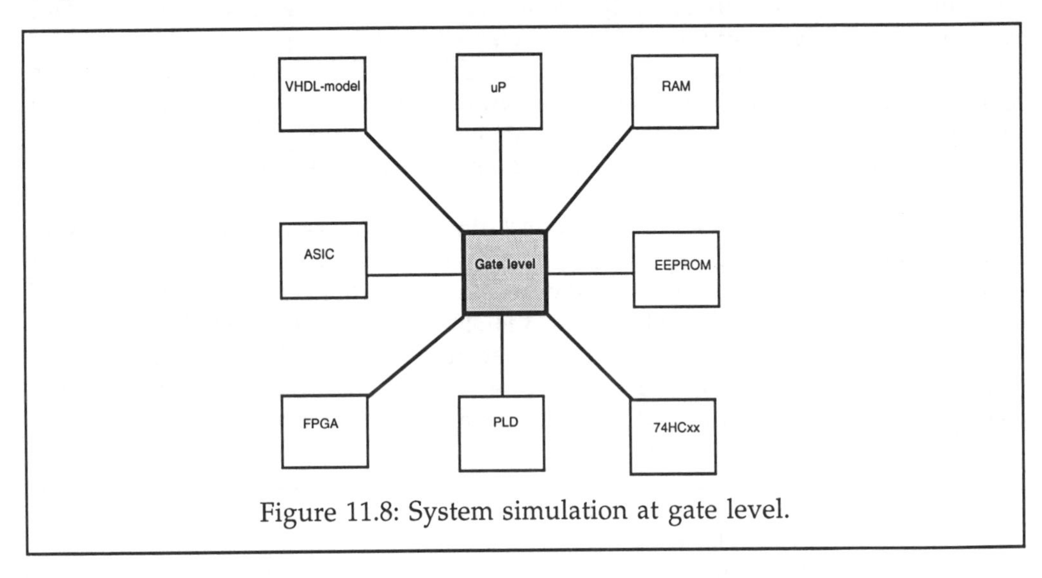

Figure 11.8: System simulation at gate level.

The disadvantage of system simulation compared with the other alternatives is that the simulation time increases.

After layout, the circuit should be verified to make sure that it still meets the timing constraints. The timing will always be different from the timing before layout because, before layout, for the internal networks the capacitances are only estimated.

Using the test vectors (see Chapter 12, "Test methodology") it is easy to decide whether any functional errors have occurred in the layout. The timing can be verified in two ways:

- Simulation
- Static timing analyser

Verifying the timing by means of simulation normally takes quite a long time, because it is difficult to check that all the data paths through the circuit meet the timing constraints. A much faster and more efficient method is to use a static timing analyser. With an analyser of this type the timing verification can be done in a few minutes with a high degree of reliability. A timing analyser which can be used for this purpose is also incorporated in the advanced synthesis tools.

11.2.1. Summary of the different simulation alternatives

The advantages and disadvantages of the simulation alternatives are summarized in Table 11.1.

	Simulation speed	Simulator- and platform-independent	Close to reality
Test vectors in the simulator's language	4	1	2
Testbench	4	4	3
System simulation	2	2	4

4 = Very good, 3 = Good, 2 = OK, 1 = Bad.

Table 11.1: Summary of the simulation alternatives.

The recommendation is to combine VHDL testbenches with system simulation. The individual VHDL blocks (in hierarchical design) can be verified with a simple testbench to start with in order to eliminate simple design errors. To speed up the simulation, it is also possible in many cases to do a system simulation at block level. Suppose, for example, that we have designed a bus protocol or a CPU interface. We can then simulate these design blocks in relation to their surroundings, i.e. other bus interface circuits or a model of the CPU in this example.

In the future the designer will probably be able to do the system simulation with an ordinary VHDL simulator. Some VHDL simulators can already be connected to hardware simulators and other "soft" models. Work is also being done on producing a standard, VHDL Initiate To ASIC Library (VITAL), with a view to being able to simulate gate level on a VHDL simulator as well. There are already some VHDL simulators and ASIC suppliers who support this standard. This means that the ASIC suppliers will also support what is known as sign-off from a VHDL simulator. Soon only a VHDL simulator will probably be required, and not backplane simulators, to verify a design at all levels.

Most time and effort should be spent on RTL VHDL simulation. Once the VHDL code has been simulated, there should be no functional errors left in the circuit. This is vital, as the top-down methodology is based on finding functional errors as early as possible. The only errors which should be possible now are timing and area problems. By their nature, these problems are directly dependent on which technology is used. Until now the design has been completely technology-independent, making it unnecessary to worry about possible problems of this type. The technology should be chosen in the synthesis tool.

11.2.2. Simulation speed

During the 1990s the performance of VHDL simulators has increased considerably. The size of the circuits and systems which are simulated in VHDL has also increased. This means that there is every reason to be conscious of how VHDL code should be written in order to speed up simulation. Unfortunately, the ways of writing VHDL code so that it is easy to read or quick to simulate sometimes conflict with one another. In such cases the designer must decide what is most important. There follow a few tips regarding what the designer should think about in the VHDL code in order to reduce the simulation time:

1. The fewer the processes, the faster the simulation.
2. Avoid converting values several times.
3. Do not include any unnecessary signals in a process sensitivity list.
4. Check which data type the chosen VHDL simulator prefers.
5. Make the code smart.
6. Keep the number of hierarchies down.
7. Avoid TextIO reading and writing.
8. Use variables instead of signals.

1. The fewer the processes, the faster the simulation

There are normally two types of process in a VHDL design: combinational and clocked. Clocked processes are normally triggered by the system clock and reset signal. It is common practice to use several clocked processes with an identical sensitivity list at several points in the same architecture. This facilitates readability and maintenance; on the other hand, it is not efficient from a simulation point of view with regard to speed. The simulation time can be greatly reduced if all the clocked processes are summarized in a single process instead of several. The same principle also applies to combinational processes.

An example of easy-to-read VHDL which is not optimum in terms of simulation performance might be:

```
Architecture rtl of ex is
begin
   p0:process(clk,resetn)
```

```
        begin
            ...
            q1<=..
        end process;

        ...

        p4:process(clk,resetn)
        begin
            ...
            q4<=..
        end process;
    end;
```

Using the same example but with only one clocked process, the following description method is not particularly good from a documentation point of view, but the VHDL code is much faster to simulate.

```
    Architecture rtl of ex is
    begin
      p0:process(clk,resetn)
      begin
            ...
            q1<=..
            ...
            q4<=..
      end process;
    end;
```

2. Avoid converting values several times

Not converting the same value several times in the VHDL code may seem obvious, but the mistake is easily made. If, for example, the following VHDL command is used in a VHDL testbench, it will result in the value being converted every time that the line is executed:

```
wait for 2 * period;
```

If the compiler is smart, conversion can be avoided in this instance. The safest way of solving this problem is to declare a constant as follows:

```
constant period2:time:= 2 * period;
```

The VHDL code can then be written as follows:

```
wait for period2;
```

In this way the value will never have to be calculated several times during simulation.

3. Do not include any unnecessary signals in a process sensitivity list
It is the sensitivity list that decides when a process is to start. If unnecessary signals are included, slower simulation will result. One of the commonest mistakes is to include signals other than the clock and reset in clocked processes.

Example:

```
process(clk,resetn,d)
begin
  if resetn='0' then
        q<='0';
  elsif clk'event and clk='1' then
        q<=d;
  end if;
end process;
```

The above VHDL code has signal *d* in the sensitivity list. As the process is clocked, output signal *q* can only change value at a clock edge or in the event of reset. The design will function from a simulation and synthesis point of view, but simulation will take longer than necessary as the process will be activated at every event for signal *d* in addition to clk and reset. The above process should be written as follows:

```
process(clk,resetn)
begin
    if resetn='0' then
        q<='0';
    elsif clk'event and clk='1' then
        q<=d;
    end if;
end process;
```

4. Check which data type the chosen VHDL simulator prefers
Some VHDL simulators are designed in such a way that std_logic and std_logic_vector, for example, are faster to simulate than std_ulogic and std_ulogic_vector. The designer should therefore ask the supplier of the VHDL simulator whether it has any such built-in preferences.

5. Make the code smart

The designer knows which of the conditions in the **if-then-else** command, for example, will most often be true. This condition should be listed first in the command, which will result in the simulation being slightly faster, as normally only one of the many conditions in the **if-then-else** command will have to be evaluated.

Example:

```
process(a,b);
begin
 if a>12 and b<11 then     -- The most common condition
    ...
 elsif a=13 and b=2 then -- The second commonest condition
    ...
    ...
 end if;
end process;
```

6. Keep the number of hierarchies down

The more hierarchies there are, the slower simulation will be. One reason for this is that more VHDL code is needed when hierarchies are used.

7. Avoid TextIO reading and writing

Using TextIO in the VHDL code normally increases the simulation time. Reading from and writing to external files should therefore be avoided if the simulation time is very important.

8. Use variables instead of signals

Variables are much faster to simulate than signals. They should therefore be used as much as possible if the simulation time has to be reduced. Variables are faster because they do not have an event queue like signals do when the signal has to change value. A variable can only have a now value, whereas a signal can have a now value and several values in the signal's event queue (see Chapter 4, "Sequential VHDL").

11.2.3. Formal verification

Formal verification is a relatively new area which is expected to be especially significant for verification technology in the future. There are currently two different principles for formal verification:

- Use of a "golden model"
- Requirements written in a text-based language

Formal verification can be used to verify that two designs are functionally identical. This is an efficient and quick way of verifying a design if the VHDL code needs to be rewritten because of timing or area problems, for example. With formal verification the correct design can then be used as a "golden model" for comparison with the new design. This means that the design does not have to be redone if the VHDL code is changed, which saves a great deal of time (see Chapter 10, "RTL synthesis").

The other way of using formal verification is to describe the requirements for a component, for example, in a language which the formal verifier understands. Based on this requirement specification, the tool can then verify the design to see if it satisfies the requirements. This type of verification can replace or supplement simulation to some extent.

11.2.4. Recommendations for verification

VHDL level
It is recommended that system simulation should always be done for VHDL simulation at behavioural and RT level. Before system simulation is commenced, the design should be verified with a VHDL testbench. This is because, as previously mentioned, the VHDL testbench has much faster turnaround times that the system simulation. Verification is therefore facilitated if the majority of errors are found using the VHDL testbench.

Gate level
The design does not actually have to be simulated at gate level. The timing should be verified using the synthesis tool's built-in timing analyser. It is recommended, however, that both the earlier VHDL testbench and a system simulation carried out at gate level should be used because the synthesis tool may contain a bug, resulting in an erroneous netlist. It is also advisable to verify that the set timing constraints are correct by doing some system simulations at gate level.

After layout
It is therefore recommended that the test vectors should be used (in ASIC design) to verify that the design is still correct after layout. The timing should be verified using a static timing analyser.

11.3. How to write RTL VHDL code for synthesis

If a good synthesis result is to be obtained, it is important to describe VHDL code in the right way. One of the commonest mistakes made by novices is to describe their designs in a large number of very small (< 500 gates) entities. There are two major disadvantages of having all these small hierarchical blocks. Firstly, much more structural VHDL code is needed to bind all the components together. Structural VHDL code is not particularly readable either. The increased amount of VHDL code also means an increased risk of error. Secondly, the synthesis result will be worse with small hierarchical levels because the synthesis tool cannot divide the logic between hierarchical levels. This normally leads to more gates and poorer timing. Below is a very small example which aims to illustrate this principle. First the VHDL code with extremely small hierarchical levels:

```
Library ieee;
Use ieee.std_logic_1164.ALL;

Entity mux is
   port( d0,d1,sel: in     std_logic;
         q:         out    std_logic);
end;

Architecture struct_mux of mux is

 component and_komp
  port( a,b: in   std_logic;
        c:   out  std_logic);
 end component;

 component or_comp
  port( a,b: in   std_logic;
        c:   out  std_logic);
 end component ;

 component inv_comp
  port( a: in     std_logic;
        b: out    std_logic);
 end component ;

 signal i1,i2,sel_n:std_logic;

 For U1 : inv_comp Use Entity work.inv_comp(inv_beh);
 For U2,U3 : and_comp Use Entity work.and_comp(and_beh);
 For U4 : or_comp Use Entity work.or_comp(or_beh);
```

```
begin
   U1 : inv_comp port map(sel,sel_n);
   U2 : and_comp port map(d0,sel,i1);
   U3 : and_comp port map(sel_n,d1,i2);
   U4 : or_comp port map(i1,i2,q);
end;
```

The synthesis result is shown in Figure 11.9. If we describe the entire design in a single component instead, we obtain the following VHDL code and, in Figure 11.10, synthesis result.

```
Library ieee;
Use ieee.std_logic_1164.ALL;

Entity good is
   port( d0,d1,sel:  in   std_logic;
         q:          out  std_logic);
end;

Architecture rtl of good is
begin
  q<=(d0 and sel) or (not sel and d1);
end;
```

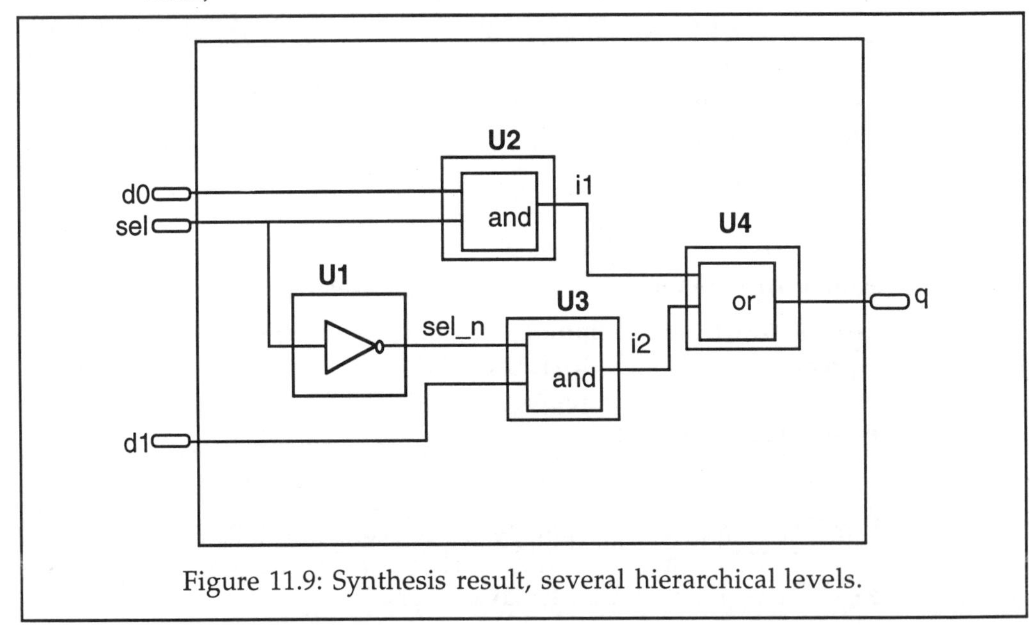

Figure 11.9: Synthesis result, several hierarchical levels.

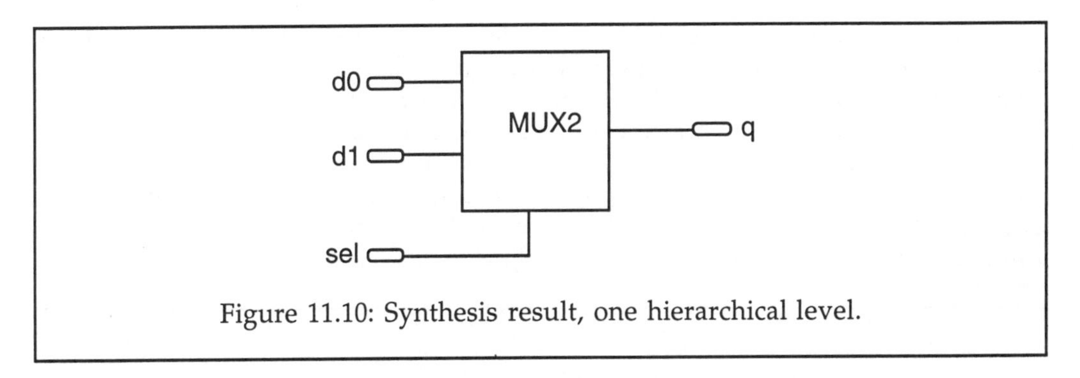

Figure 11.10: Synthesis result, one hierarchical level.

If we compare the area of the two synthesis results, we see that:

Lots of small hierarchical levels	7 gates
One hierarchical level	3 gates

Thus, in this example, the synthesis result for a lot of small hierarchical levels is more than twice as large. This example was extreme. If we compare more realistic component sizes, e.g. 100-150 gates and 500-6000 gates, the difference is usually 5-20 per cent for both time and area. This is because a hierarchical level locks up the logic and makes it difficult for the synthesis tool to optimize it. The above reasoning does, however, require the 500-6000 gates to belong together functionally. If the component consists of 150-gate blocks which have nothing to do with each other functionally, there will not usually be any difference between lots of small hierarchical levels and one large one. Hierarchical levels with more than 10,000 gates are not recommended either because they take far too long to synthesize and it is not good from a timing point of view for very large logic blocks to separate the gates, as they will probably be placed a long way away from each other on the silicon.

Recommendations

1. **Group logic which belongs together in the same architecture.**
2. **Separate logic which is to be optimized with different optimization strategies (see Chapter 10, "RTL Synthesis").**
3. **Separate logic which has different optimization goals (area/timing).**
4. **The size of the lowest hierarchical level should be 500-6000 gates.**

As 500-6000 gates represent quite a lot of code, it may be difficult to understand the functionality of each hierarchical block. An effective way of solving this is the **block** command in VHDL, which is a concurrent command that can be set around any other concurrent command in the VHDL code.

If the **block** command is used to mark the function of different parts of an architecture, it facilitates readability markedly. What is more, internal signals which are only used inside the block can be declared in the declaration part of the **block** command (Figure 11.11). This means that the signals can be declared in more detail where they are actually used in the code. If the blocks are then named in a smart, explanatory way, it is possible to achieve the same readability as for small hierarchical blocks. Nor is difficult-to-read structural VHDL code needed to bind the small hierarchical levels together. A better synthesis result is also obtained.

```
Architecture rtl of ex is
<Declarative part>  -- Visible in whole architecture
begin
...
<Block_name>:Block
<Declarative part>  -- Visible only within the block
begin
 Parallel VHDL commands
end block <Block_name>;

<Block_name>:Block
<Declarative part>  -- Visible only within the block
begin
 Parallel VHDL commands
end block <Block_name>;
...
end;
```

Figure 11.11: Block command.

During synthesis, a hierarchical level can also be obtained for each block command in the VHDL code, depending on which synthesis tool is used. As this leads to small hierarchical levels, it is not recommended.

Another advantage of the block command is that, during simulation, each block command can be seen as a hierarchical level in VHDL (Figure 11.12). This means that if, for example, you want to take all the signals which belong to a logic function, e.g. address decoding, you just have to click on the block name in the hierarchical menu and then select all the signals in that level. Not all simulators have this refinement, however, although it is available in Mentor Graphics Quick VHDL and V-system among others.

An example of how the block command should be used might be:

```
Entity ex is
  port( a:                    in   integer;
        clk,resetn,c:         in   std_logic;
        out1,out2,sel1,sel2:  out  std_logic),
end;

Architecture small_ex of ex is
signal b1:std_logic;
begin

  addr_dec:Block
  begin
    process(a)
      case a is
        when 123=> sel1<='1';
                   sel2<='0';
                   b1<='0';

        when 130=> sel1<='0';
                   sel2<='1';
                   b1<='0';

        when others=>sel1<='0';
                     sel2<='0';
                     b1<='1';
      end case;
    end process;
  end Block;

  State_M:Block
  type states is (s0,s1); -- Internal signal declaration for
                          -- block State_M
  signal state:states;
  begin
    process(clk,resetn)
    begin
      if resetn='0' then
        state<=s0;

      elsif clk'event and clk='1' then
        case state is
          when s0 =>  if b1='1' then
                        state<=s1;
```

```
                                  end if;
                when s1 =>        if c='1' then
                                     state<=s0;
                                  end if;

              end case;
            end if;
          end process;

          process(state)
          begin
            case state is

              when s0 =>    out1<='0';
                            out2<='0';

              when s1 =>    out1<='0';
                            out2<='1';
            end case;
          end process;

        end Block;
      end;
```

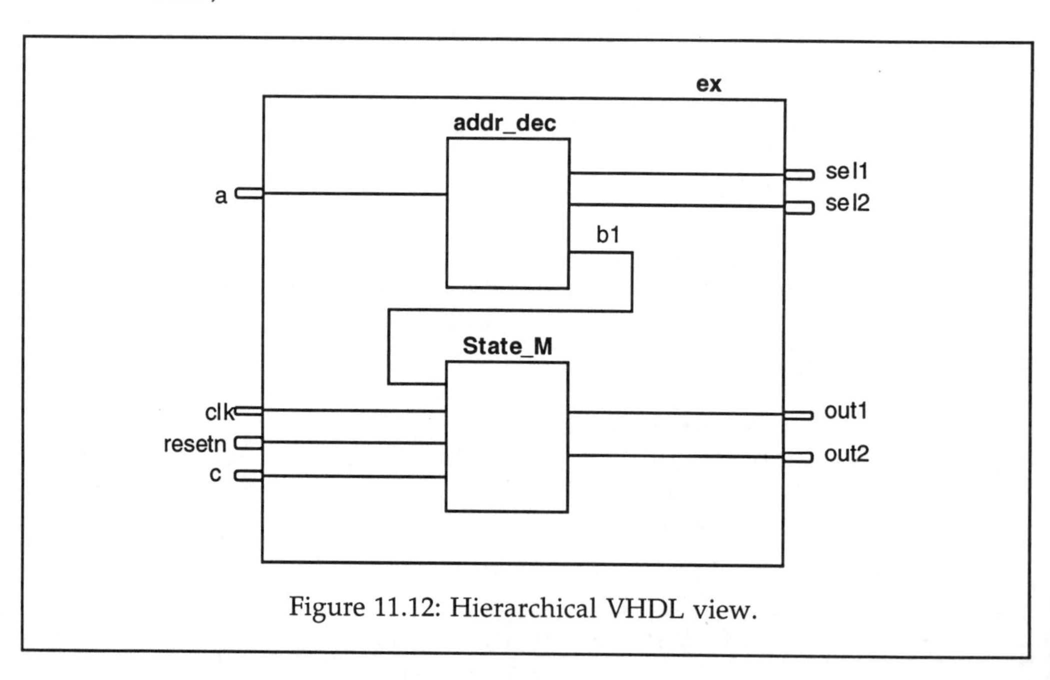

Figure 11.12: Hierarchical VHDL view.

Recommendations

1. **Always use the block command in large architectures in order to increase readability and facilitate debugging during simulation.**
2. **Use structural VHDL code to create hierarchies in the VHDL code and not the block command.**
3. **There should not be more than four hierarchical levels.**

The number of hierarchies should be kept down because, apart from the reasons mentioned above, the hierarchical signal names in the design will be much too long if more than four levels are used. Several ASIC manufacturers have restrictions on how long these names can be. Verification work is also made more difficult if the design contains many levels of hierarchy, as there is more to keep track of.

A graphical VHDL tool is sometimes useful when it comes to designing the behavioural model and RT level code. In these tools it is possible to describe some functions as follows as a supplement to the VHDL code: the state diagram, flowchart and truth table are shown in Figures 11.13-11.14 and Table 11.2 respectively.

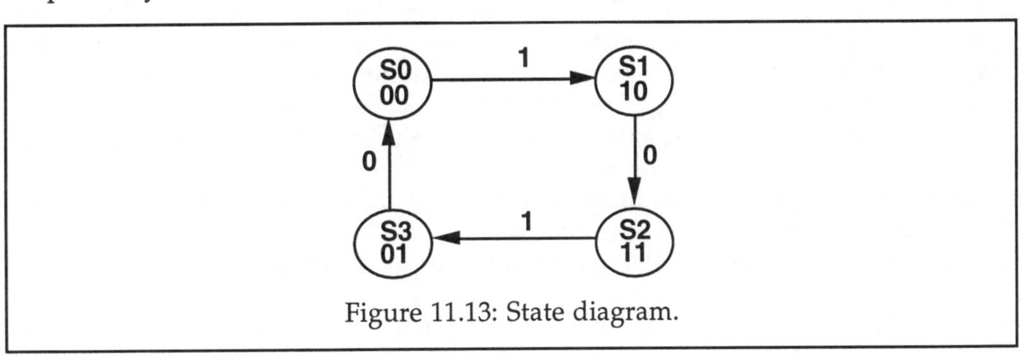

Figure 11.13: State diagram.

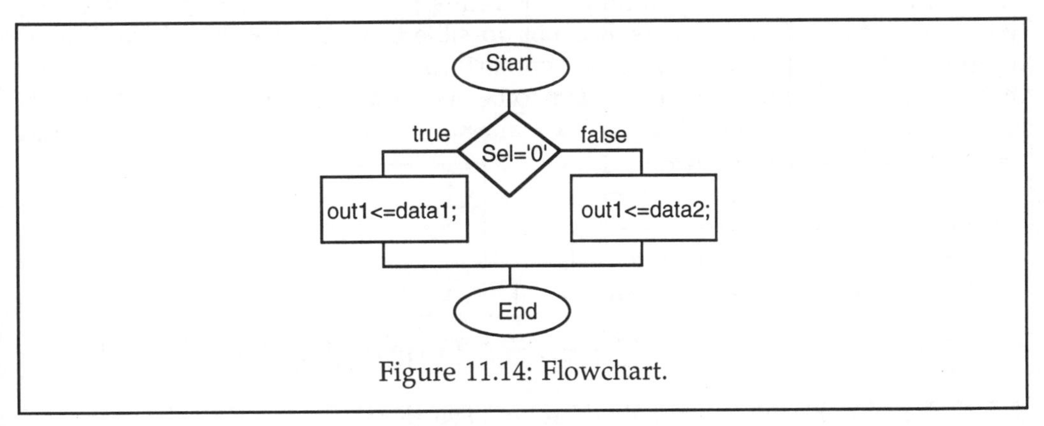

Figure 11.14: Flowchart.

a	b	c
0		0
	0	0
1	1	1

Table 11.2: Truth table.

C code, VHDL code for simulation or VHDL code (RTL) for synthesis can then be generated from these descriptions. If synthesizable VHDL code is chosen as output data (a must in hardware design), it is also normally possible to choose which synthesis tool the code should be optimized in relation to: this is important as the subset of VHDL which can be synthesized varies from synthesis tool to synthesis tool.

For people who are not used to designing with VHDL, this description method may be faster than writing directly in VHDL. It should be remembered, however, that most (all existing) graphical VHDL tools restrict the description method slightly and, in the case of state machines, for example, the choice of implementation (see Chapter 9, "State machines"). One of the greatest advantages of these tools is that the documentation for the circuit is neat and readable (the graphical picture), making it easier to understand and maintain. Their great disadvantage is that the design becomes platform- and tool-dependent again. If a design has to be modified after four years, for example, there is a great risk that the tool will have switched data format or been changed completely, as such tools are not based on an IEEE standard. It is true that there is the machine-generated VHDL code, which can be used as a basis and modified, but this code is normally much harder to read than VHDL code which has been written manually.

If the VHDL code is written incorrectly, the synthesis work can be made considerably harder. All the timing constraints have to be set in the synthesis tool. In the case of large designs, it is not possible to synthesize the whole design in one go. Each sub-block has to be synthesized separately instead. This means that the correct timing constraints have to be set for all the inputs and outputs of each hierarchical block. This work can be made easier if the following recommendations are followed.

Recommendations

1. **Design 100 per cent synchronously.**
2. **Use just one clock in each hierarchical block.**
3. **Each output from each hierarchical block should come straight from a flip-flop.**
4. **Each hierarchical block should be at least 500 gates.**

5. Avoid multicycle design.

If requirement 3 is to be realistic, the hierarchical levels used must not be too small, as this leads to unnecessary pipelining of the signals, subjecting them to a time delay. This results in requirement 4. Requirement 4 also agrees with previous requirements. Efforts should be made to have every output coming straight from a flip-flop because it makes it very easy to set the time constraints for the sub-blocks, provided the design is clocked. If, for example, we have a clock period of 25 ns and a flip-flop normally has a delay of 1 ns, the timing constraints of all the outputs should be set to the logic outside having 24 ns until the next clock edge. For all the inputs which come from another block, with the same clock, the constraint can be summarized as being that the logic inside the block must take a maximum of 24 ns. If, instead, we have outputs which come from large combinational blocks, the constraints will be difficult to assess before synthesis. It is even worse if several clocks are combined in the same hierarchical level. What is known as multicycle design also often leads to synthesis problems. Multicycle design means that a signal does not have to be stable after a clock cycle but can take several clock cycles, though the number is predetermined.

An example of multicycle design is:

```
q<=(a * b) * c;
```

Suppose that signals *a*, *b* and *c* come straight from a register and that output signal *q* is to be clocked in another register. If we have a clock period of 25 ns and the multiplications can take a maximum of 40 ns, we have a multicycle design of two clock cycles. The simplest way of getting round this problem is to store the result of *a* * *b* in a register and then multiply by *c*. However, this leads to the result being delayed by a clock cycle. Using a behavioural synthesis tool is often an effective way of handling such problems. The behavioural tool will take care of what is sometimes a difficult task by setting multicycle conditions and generating a correct synthesis result (see Chapter 17, "Behavioural synthesis").

The above requirements may seem unnecessarily strict, so many designers do not bother with them in their first design. If the circuit is small, this does not matter too much. If, on the other hand, the design is larger than 15,000 gates, the consequences can be serious and the entire project will be put at risk or seriously delayed. There are plenty of examples of designers being bitterly repentant after ignoring the above recommendations.

The disadvantages of designing asynchronously are many. The ten most important reasons for avoiding asynchronous design are as follows:

1. The design will often be technology- and layout-dependent.

2. The design violates the ATPG rules, reducing error coverage considerably (see Chapter 12, "Test methodology").
3. ASIC vendors' testers often do not support asynchronous designs.
4. There is very little tool support, e.g. not supported by behavioural synthesis tools.
5. It is harder to simulate and find problems.
6. The design is more difficult to maintain and reuse.
7. Synthesis tools cannot optimize or analyse the timing of the result.
8. In asynchronous designs the performance is often temperature-, supply-voltage- and process-dependent.
9. It is very difficult to use FPGAs for prototype verification of the system.
10. The total time from specification to the finished circuit normally increases by a factor of between 1.5 and 10.

The advantage of asynchronous designs, on the other hand, is that they are often faster and more frugal in their power requirements than clocked designs. The disadvantages far outweigh the advantages, however, and the recommendation is that every effort should be made to avoid asynchronous design. As asynchronous design leads to more difficulties and there is inadequate support from powerful tools, an ASIC project which tries its hand at asynchronous design normally takes between 1.5 and 10 times longer than if the design had been done synchronously. The picture may look different if tools which support asynchronous design are developed in the future. The trend in the 1990s, however, has been towards ever greater support for clocked designs. More and more firms have also given up and all but banned asynchronous design.

11.4. FPGA

The top-down flow we have described is intended for ASIC design, but it is not appreciably different in FPGA design (Figure 11.15). It is not until synthesis that the choice between ASIC and FPGA library has to be made. Another difference is that test vectors do not have to be generated for an FPGA, as they are normally tested by the manufacturer.

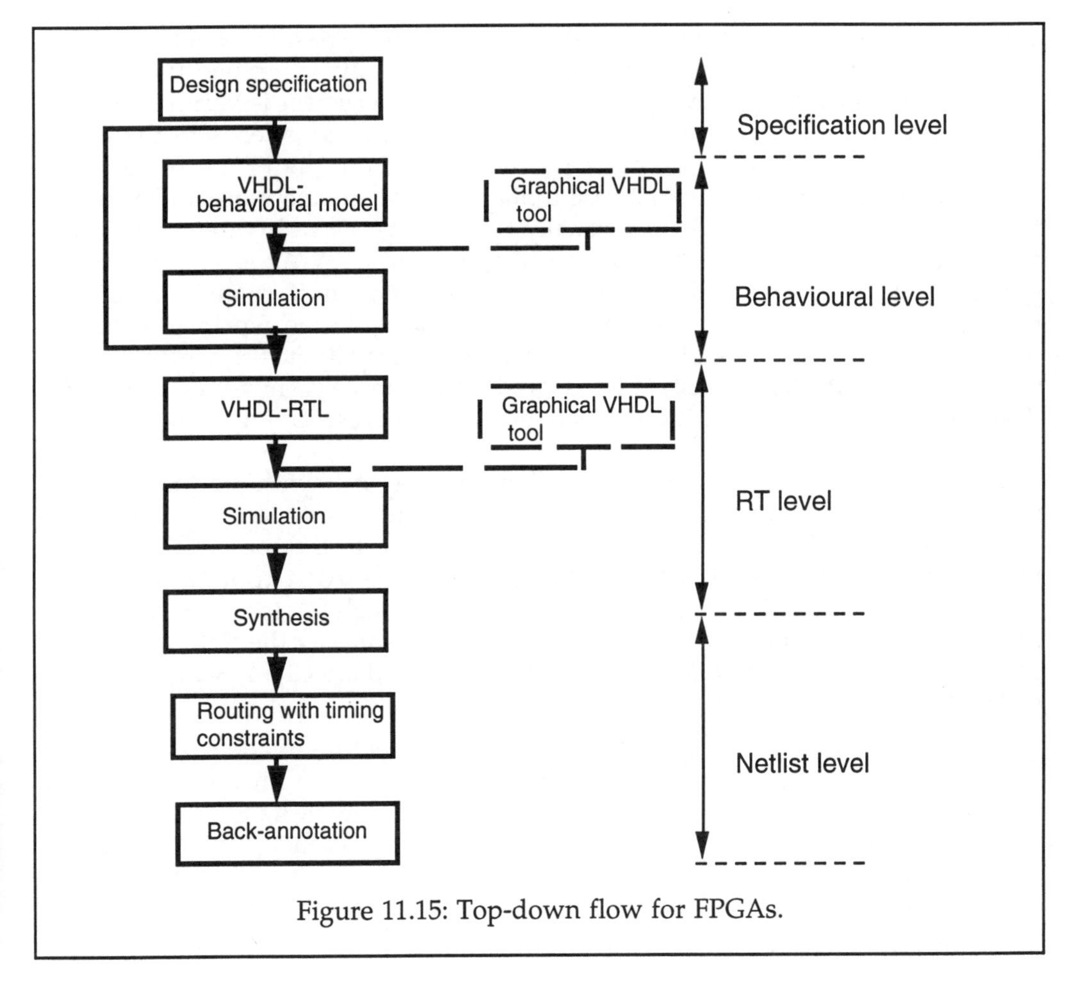

Figure 11.15: Top-down flow for FPGAs.

It is possible to use FPGAs as gate matrices in slightly simplified form. They are cheaper in small volumes, while gate matrices become cheaper at large volumes. Today the largest FPGAs are approximately 50,000 gates and ASICs approximately 3,000,000 gates. Some FPGAs contain a lot of flip-flops of the Xilinx type, while others have more logic of the Actel type.

As the FPGA flow is similar to that of ASIC design, it is possible to "map" the VHDL design, by synthesis, to an FPGA first and then use it as a prototype. Then the FPGA can be "remapped" to an ASIC in the synthesis tool when the system is verified using the FPGA prototype. An ASIC normally has a very high production cost. If the ASIC can first be verified in the system with one or more FPGAs, the risk of redesign and the major costs that would entail can largely be eliminated. FPGAs, on the other hand, are much slower than ASICs. This means that it is often impossible to verify the system at full speed with FPGAs. This need not be a problem, as the entire system can normally be verified functionally with a lower clock frequency. Normally, however, the system is verified at room temperature with a supply voltage of 5 V, for example. This leads to the FPGA being much faster than the timing report obtained from the FPGA router. This report is normally based on worst case timing. At room temperature and a supply voltage of 5.0-5.5 V, the FPGA is normally twice as fast as with worst case timing so that the system can still be run at full speed with FPGAs in some cases. A system with FPGAs cannot be delivered to a customer if room temperature timing is required for it to function as the system will stop working at higher temperatures. The purpose of FPGA prototypes is normally just to verify that the system is working before the ASIC is produced. The ASIC must, of course, be designed in such a way that it can function with worst case timing.

When FPGAs are designed, the connection between the synthesis tool and the router is very important. With a good synthesis and routing tool, all the timing constraints set in the synthesis tool can be transferred to the router. The routing tool can then route using the actual timing constraints. After routing the timing can be back-annotated to the synthesis tool's time analyser to see whether the timing was satisfied. This method of routing FPGAs is called timing-driven FPGA layout and is a must if FPGAs are to be designed effectively with top-down methodology.

12

Test methodology

When an ASIC is manufactured, verification is required to ensure that it has been made correctly. This test is normally performed by the ASIC supplier immediately after manufacture of the circuit. It must not be confused with the verification (simulation) which the designer carries out on the design prior to manufacture. The designer is trying to check whether the circuit has been designed in accordance with the requirement specification. The ASIC supplier, on the other hand, is only interested in whether the manufactured circuit is in accordance with the netlist which the designer created as the basis for manufacture. The ASIC supplier tests this by using the test vectors which the designer had to provide at the same time as the netlist for the circuit. Depending on whether the circuit passes the test vectors or not, it will either be supplied to the customer or scrapped. It is therefore important that the test vectors should be effective, as otherwise the ASIC supplier will end up delivering defective circuits. This chapter deals with how effective test vectors can be created automatically and what design rules must be followed in the VHDL code in order to make this possible. It also provides an introduction to various test methods, boundary scans and supplementary test vectors.

If we go back to the 1980s, test vectors were usually created manually. A subset of the test vectors used for verification (simulation) of the circuit was reused. Also, separate test vectors were written to achieve maximum fault coverage. Normally only 70-75 per cent fault coverage was achieved relatively easily. If the ASIC was 15,000 gates in size, it usually took four or five weeks to create test vectors which satisfied all the requirements of the specific circuit tester at the chosen ASIC supplier. Fault coverage seldom exceeded 85 per cent. If the ASIC supplier was then changed, the test vector format used for the previous ASIC supplier was not normally accepted.

In the 1990s this process also started to be automated. It became increasingly common to machine-generate test vectors to a large extent using an Automatic

Test Pattern Generator (ATPG). Using an ATPG too, fault coverage of more than 99 per cent can be achieved in one or two days' work. When comparing different fault coverage values, it is important to use the same definition of defects in the circuit. The defects which ATPG tools detect are what are known as "stuck-at one" and "stuck-at zero". "Stuck-at one" means that an internal node has been short-circuited to 5 V. "Stuck-at" defects normally represent a large proportion of the production defects in an ASIC. The production test generally has to be supplemented with additional test vectors above and beyond the ATPG test vectors. More about this later in the chapter.

Figure 12.1 shows a typical flow for ATPG.

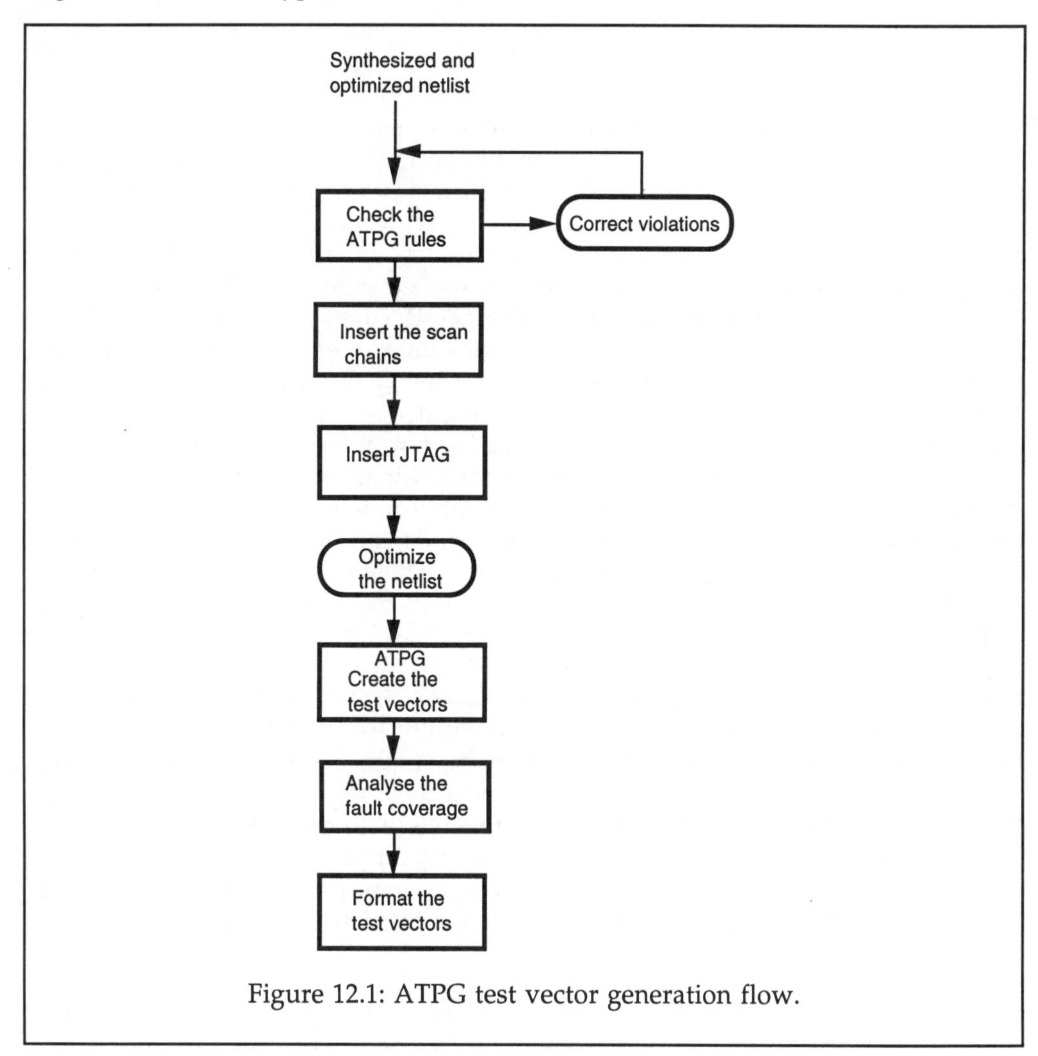

Figure 12.1: ATPG test vector generation flow.

12.1. Scan methodology

If ATPG is to be used, what are known as scan chains have to be inserted. These scan chains can be used to detect stuck-at defects, as already mentioned. Suppose that we have the very simple design shown in Figure 12.2, in which the internal input to the NAND gate is short-circuited to 5 V.

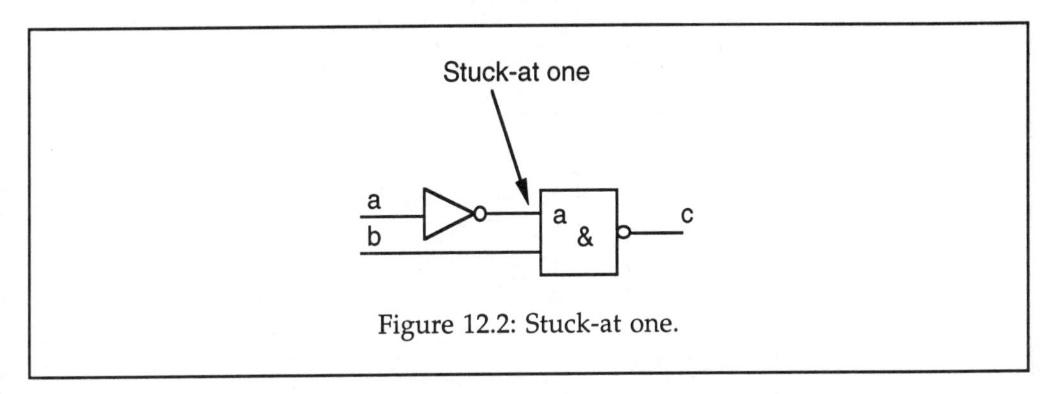

Figure 12.2: Stuck-at one.

This defect is easy to find if we apply the values a=1 and b=1 to the input signals. We then expect the output signal to be equal to 1. If the above circuit had a production defect, the output signal would be given the value 0. This means that the circuit can be labelled as defective.

If we are to be able to use this technology to check circuits which are more complex than the simple one above, all the outputs have to be controllable and all observable from the internal combinational blocks in the circuit. A normal design consists of flip-flops with combinational logic between the flip-flops (see Figure 12.3).

Assume that all the flip-flops could have been controlled to any value and that the content of the flip-flops could be seen from outside. It would then have been possible to test all the internal combinational logic in the circuit using the same principle as in the previous example. To make this test possible, therefore, the flip-flops have to be controllable and observable:

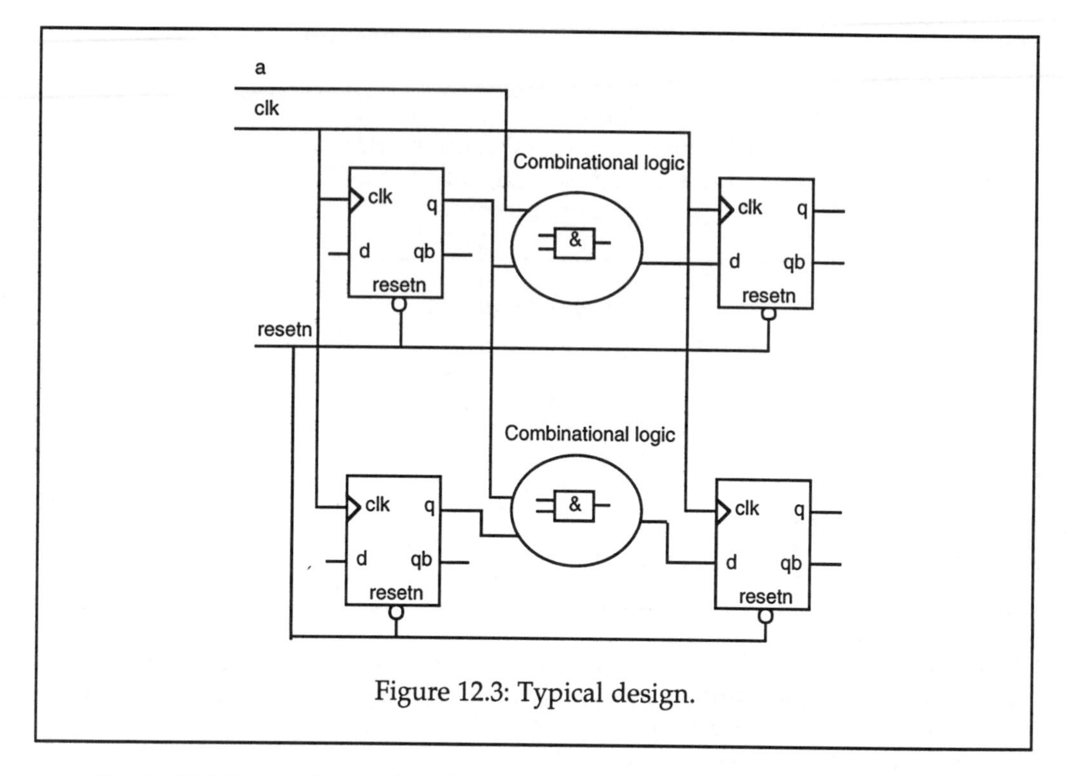

Figure 12.3: Typical design.

- **Controllable:** An internal node in the circuit is controllable if it can be driven to any value.

- **Observable:** An internal node in the circuit is observable if its value can be predicted and propagated to a circuit output for it to be checked.

This is precisely what the introduction of scan chains allows.

There are many different forms of scan method with ATPG, the commonest being:

- Multiplexed flip-flop
- Clocked scan
- LSSD

Multiplexed scan
The simplest and most widely used method is to swap the flip-flops in the circuit for what is known as a multiplexed scan flip-flop (Figure 12.4).

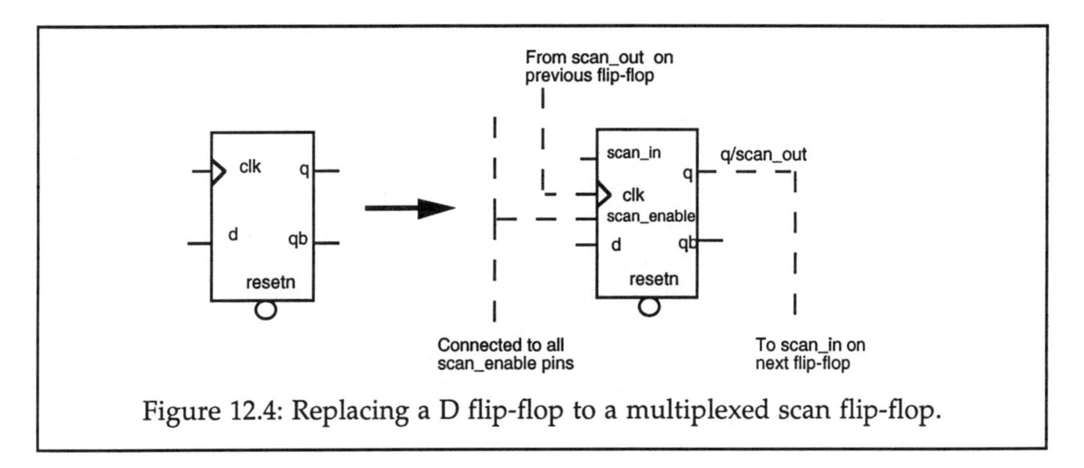

Figure 12.4: Replacing a D flip-flop to a multiplexed scan flip-flop.

In a multiplexed scan flip-flop it is possible using the scan_enable input to control which of the data inputs (*d* or *scan_in*) is to be clocked into the flip-flop. In functional mode, scan_enable is connected to 0 V. In the ATPG test, on the other hand, test mode is also used.

A multiplexed scan flip-flop requires the circuit to have at least one extra (scan enable) test pin. A multiplexed scan flip-flop is usually also three gates larger than a normal flip-flop.

Clocked scan

Clocked scan flip-flops have a separate clock input for *scan_in*. Depending on which clock is activated, the data will be clocked in from the usual *d*-input or *scan_in*. Clocked scan flip-flops therefore do not require a special *scan_enable* input. An example is shown in Figure 12.5.

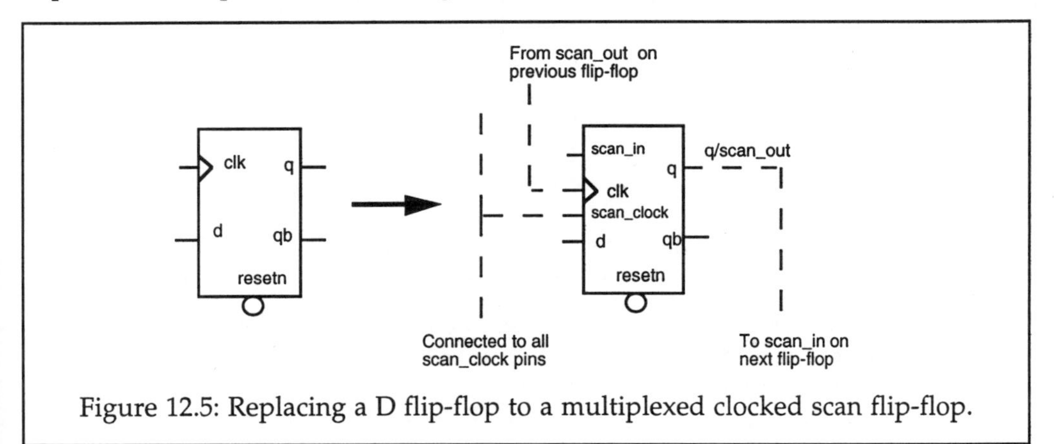

Figure 12.5: Replacing a D flip-flop to a multiplexed clocked scan flip-flop.

A clocked scan flip-flop requires the circuit to have at least one extra (scan clock) test pin. A clocked scan flip-flop is usually three gates larger than a normal flip-flop.

LSSD

There are different forms of LSSD (Level Sensitive Scan Design). LSSD can be used whether the design consists of flip-flops or latches. The difference between flip-flops and latches only exists in functional mode; in test mode the scan flip-flop and latch will behave the same. A complete LSSD scan flip-flop has two separate clock inputs (*a* and *b*) for clocking in the value from the *scan_in* input. Inside, the flip-flop consists of a master latch and a slave latch. Clocks *a* and *b* are what is known as a two-phase clock and should not normally overlap, i.e. be active simultaneously. Activating the *b* clock loads the value from scan_in to the master latch. If the *a* clock is then activated, the value is transferred from the master latch to the slave latch and therefore also to the *q/scan_out* output. The *a* and *b* clocks are not used in functional mode, where they both have to be inactive. In functional mode the flip-flop works like a normal flip-flop with an ordinary clock input and d-input. The advantage of this scan method is that the design is not clock skew sensitive in test mode, as it is based on a two-phase clock. The disadvantage is that the flip-flop is between one and three gates larger and normally slightly slower than the multiplexed scan flip-flop. Another disadvantage is that the circuit requires two extra clock inputs (*a* and *b*). An example is shown in Figure 12.6.

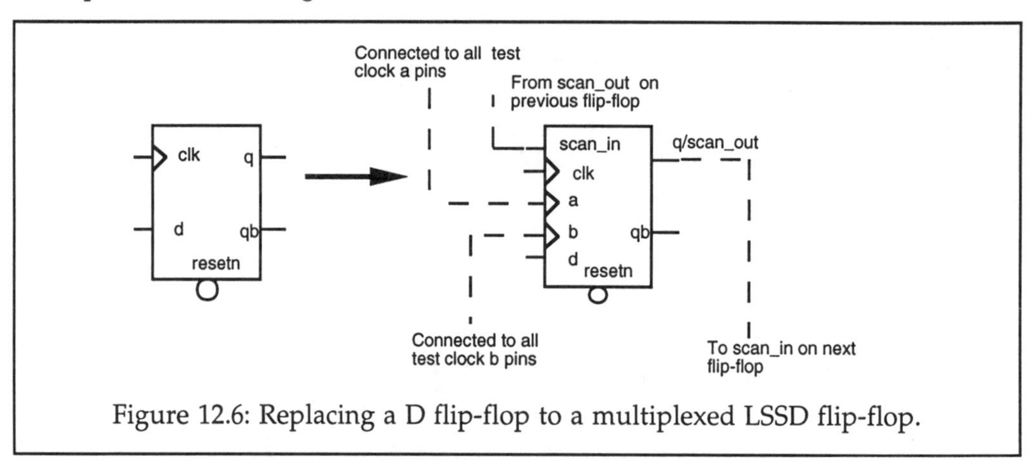

Figure 12.6: Replacing a D flip-flop to a multiplexed LSSD flip-flop.

For small circuits (<20,000 gates) and relatively slow technology (more than 0.8 μm) a normal multiplexed scan flip-flop is recommended (Figure 12.7). This is because a multiplexed scan flip-flop only requires one extra pin and has a relatively small area overhead. If the circuit is larger or if technology faster than 0.8 μm line width is used, some form of two-phase clock has to be used in test mode (e.g. LSSD) because the clock skew on the one-phase clock often causes hold violations for the flip-flops in test mode. This is caused by the scan flip-flops being connected in a long shift register, making them highly clock skew sensitive. See Chapter 11, "Design methodology".

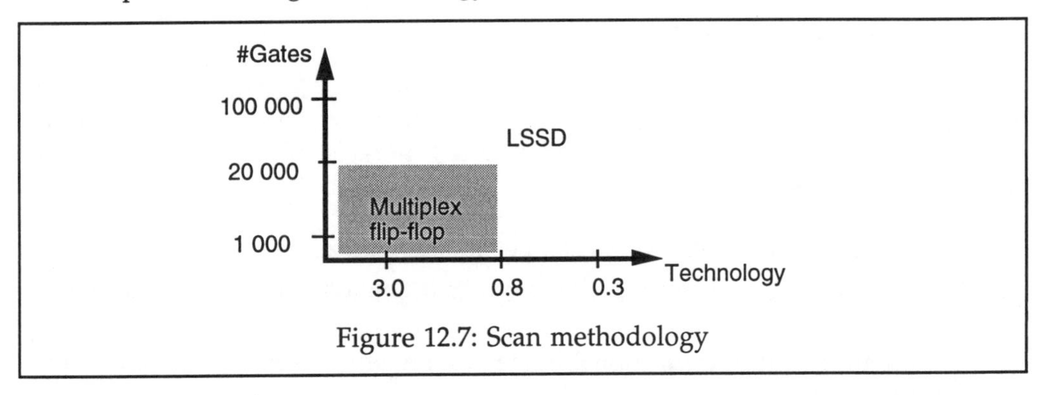

Figure 12.7: Scan methodology

If the flip-flops in the circuit are swapped for scan flip-flops, the flip-flops can be both controlled and observed. All the scan flip-flops function exactly like the flip-flop being replaced in functional mode. In test mode, on the other hand, the flip-flop changes function. If the flip-flops are joined in long scan chains, it is possible to shift known values into the flip-flops by activating *scan_enable* (multiplexed scan flip-flop). The values can then be sampled into the flip-flops from all the combinational blocks by switching to functional mode. If *scan_enable* is then activated again, the contents of the scan flip-flops can be shifted out for the values to be checked. This allows the combinational blocks inside the circuit to be tested even if the internal inputs or outputs in the combinational logic are not directly accessible through the circuit inputs and outputs. Figure 12.8 shows an example of a scan design.

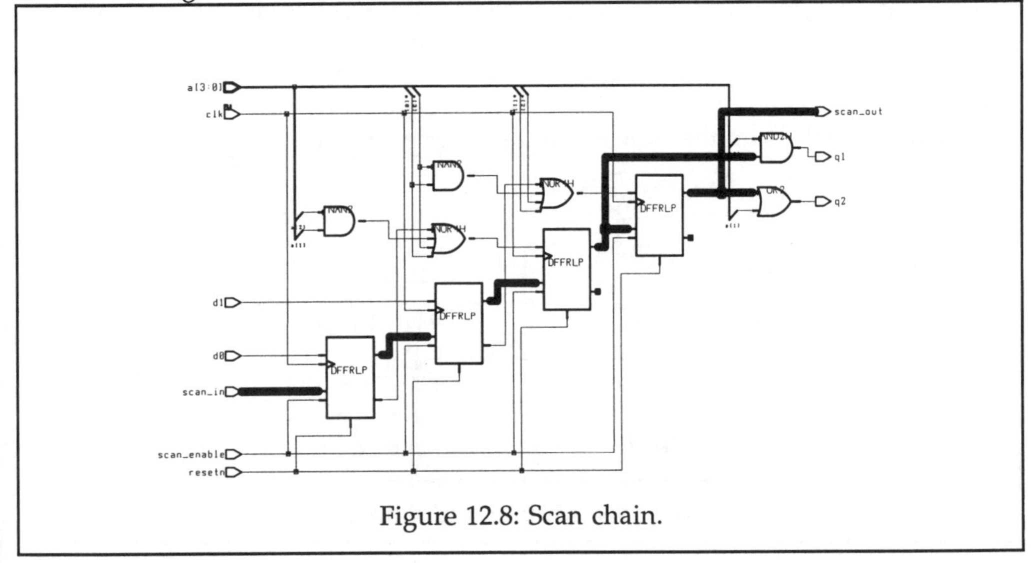

Figure 12.8: Scan chain.

An ATPG tool generates test vectors which can detect stuck-at defects. A test vector consists of the following parts:

- A collection of "1"s and "0"s which are applied to the circuit's inputs.
- The values expected at the outputs.

It is the ATPG tool which calculates what the expected values are on the basis of the values applied to the inputs. The expected values are then compared with the values which the newly manufactured circuit generates. If all the values agree, the circuit has passed the test.

The principles of how ATPG works can be summarized as five phases:

1. The circuit is put into scan_enable mode and new data are shifted into the scan chains.
2. The circuit is put into functional mode and known values are applied to all inputs.
3. The system clock is activated once to sample the result from all combinational blocks.
4. The values of the output signals from the circuit are checked to see if they have the expected value.
5. Scan_enable mode is activated again. The value is shifted out of all flip-flops to the *scan_out* outputs. The values at the *scan_out* outputs are checked to see if they have the expected value.

Normally the five phases overlap, i.e. new values are shifted into the scan chain as the old values are shifted out.

The first flip-flop's scan_in normally comes from a separate test input on the circuit (*scan_in*). The last flip-flop's output also goes directly out of the circuit to a special test pin (*scan_out*). In the event of a pin shortage, the *scan_out* output can be multiplexed with an ordinary output, in which case the multiplexor should be controlled by the *scan_enable* signal (Figure 12.9).

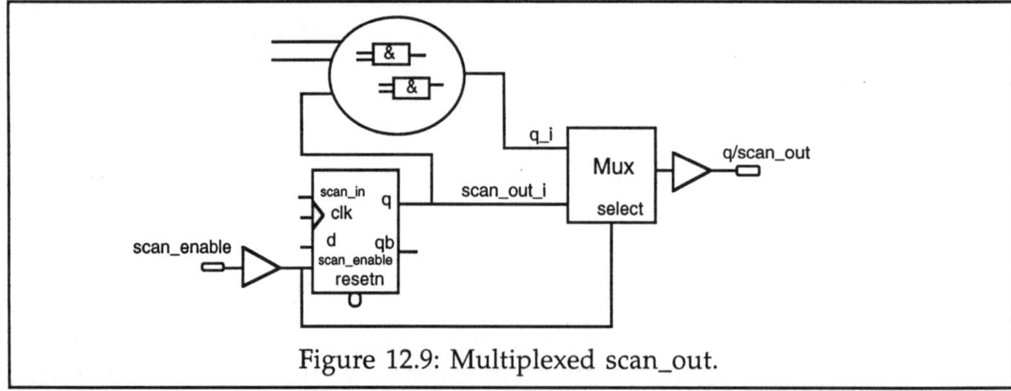

Figure 12.9: Multiplexed scan_out.

The *scan_in* input can also be shared with an ordinary input to save pins on the circuit (Figure 12.10).

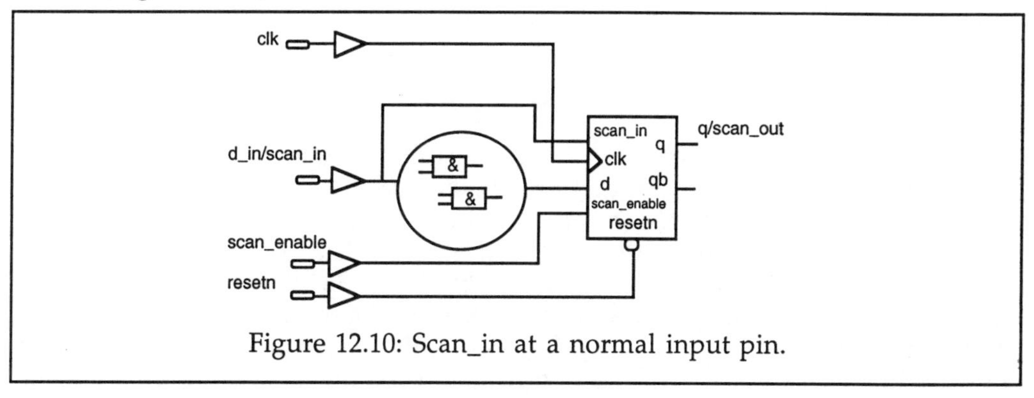

Figure 12.10: Scan_in at a normal input pin.

Swapping and connecting all the new scan flip-flops in the design is a big job. It is therefore possible to buy an add-on for today's synthesis tools, namely a test insertion tool. This software will insert the scan chains automatically and connect them correctly. As the number of test vectors is proportional to the length of the longest scan chain, several parallel chains are always used in large designs. This is a necessity, as ASIC suppliers normally accept only a limited number of test vectors. In the more advanced test insertion tools it is therefore possible to define the number of scan chains and/or their maximum length when inserting the scan chains. Suppose that an ATPG tool has generated N test vectors for the design in Figure 12.3. If the test insertion tool is given the command to insert two parallel scan chains instead of one, the number of test vectors will, in principle, be halved to N/2. This is because the ATPG tool will shift data in and out of the scan chains in parallel. Normal scan chains are usually 100-500 flip-flops long, depending on the size of the circuit and how many test vectors the ASIC manufacturer will accept.

The test insertion tool also allows the preferred scan methodology to be chosen by means of a simple command or menu selection. Swapping the normal flip-flops and connecting scan flip-flops is a quick and simple process with a tool of this type. For a circuit with 10,000 gates, for example, it just takes a few minutes. Generating the test vectors with ATPG takes about 10 minutes for an equivalent circuit. In the case of large circuits (>100,000 gates), on the other hand, the time needed to generate test vectors with ATPG can be quite long (more than 10 hours).

12.2. Full scan and partial scan

There are two types of ATPG algorithm. One is based on all the flip-flops in the circuit being connected to a scan chain, known as full scan. In the other, partial scan, only a subset of the flip-flops is connected in the scan chains. The per centage of gates in the overhead required for the two scan methods varies from design to design but a rough estimate would be:

Full-scan	**10-20 per cent extra gates**
Partial-scan	**5-15 per cent extra gates**

With these extra gates and an ATPG tool, the fault coverage is around 99 per cent for full scan, and slightly lower for partial scan - because, as in partial scan not all the flip-flops have been connected to a scan chain, they cannot be checked. The algorithm for partial scan is adjusted for this, but still does not reach the same fault coverage as that normally achieved by full scan. Partial scan is chosen instead of full scan if the area or, in some cases, timing constraints for the design are particulary stringent. A scan flip-flop is usually slightly slower than a normal flip-flop. The choice should be based on whatever has the greatest priority: area or fault coverage.

12.3. ATPG design rules

If top fault coverage levels are to be achieved, certain design rules have to be followed in addition to using scan chains. You have to be aware of these even when structuring your VHDL code. If you break the rules, fault coverage will drop considerably. The rules may vary slightly depending on the type of scan chain and ATPG tool being used, but the basic rules are the same. The most important of them are examined below.

Rule 1: Non-scan equivalent
For it to be possible to connect the scan chain in the circuit, the flip-flops have to have an equivalent scan flip-flop.

An example of when there is no equivalent scan flip-flop is shown in Figure 12.11.

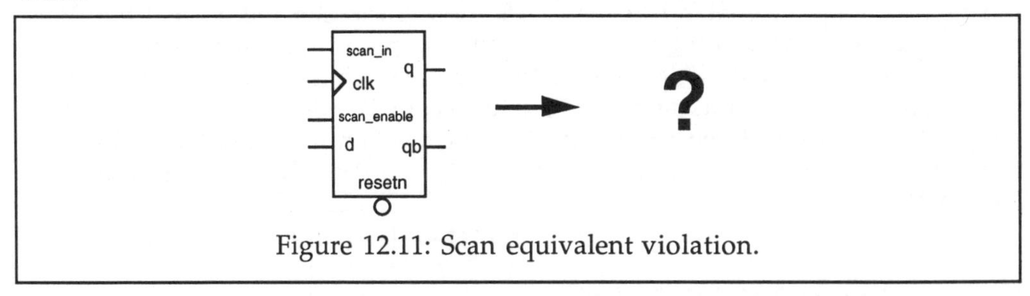

Figure 12.11: Scan equivalent violation.

Rule 2: Combinational feedback loop
Combinational logic may not have a feedback loop unless the feedback loop goes via a flip-flop. This feedback loop means that the design is asynchronous.

An example of a combinational feedback loop is shown in Figure 12.12.

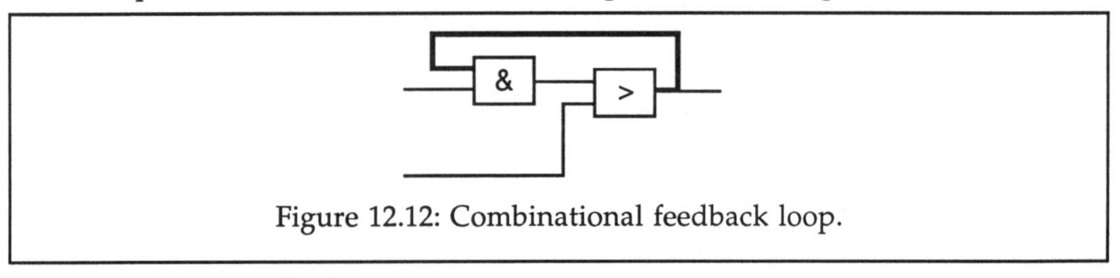

Figure 12.12: Combinational feedback loop.

Rule 3: Non-three-state driver
If a signal is driven by more than one driver, the drivers must be of the three-state type.

An example of illegal three-state driving is shown in Figure 12.13.

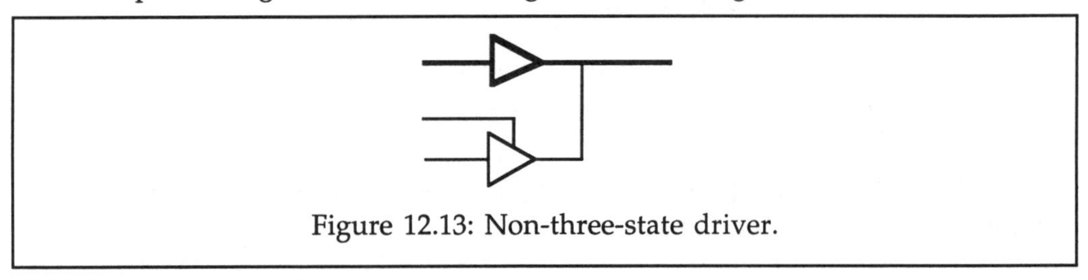

Figure 12.13: Non-three-state driver.

Rule 4: Non-controllable clocks
It must be possible to control the clock for all the internal flip-flops from an input on the circuit.

Two examples of when the clock is not controllable for an internal flip-flop are shown in Figures 12.14 and 12.15.

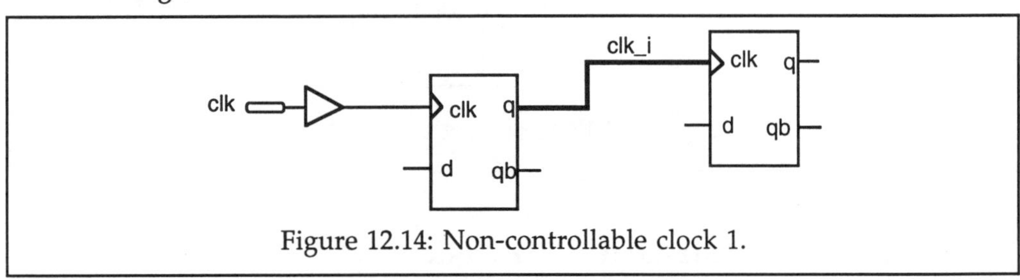

Figure 12.14: Non-controllable clock 1.

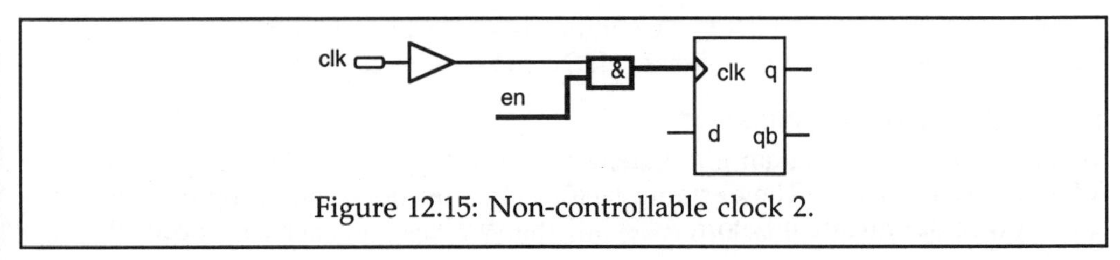

Figure 12.15: Non-controllable clock 2.

Rule 5: Non-controllable asynchronous reset
It must be possible to control the asynchronous reset for all the internal flip-flops from an input on the circuit.

An example of a design in which the asynchronous reset is not controllable is shown in Figure 12.16.

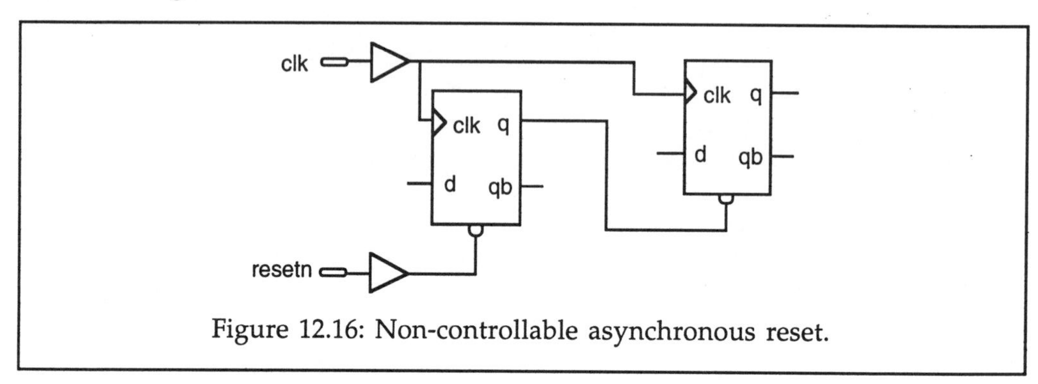

Figure 12.16: Non-controllable asynchronous reset.

Rule 6: Clock at data input
The clock may not be connected, either directly or via combinational logic, to the d-input of the internal flip-flops.

An example of a clock at data input is shown in Figure 12.17.

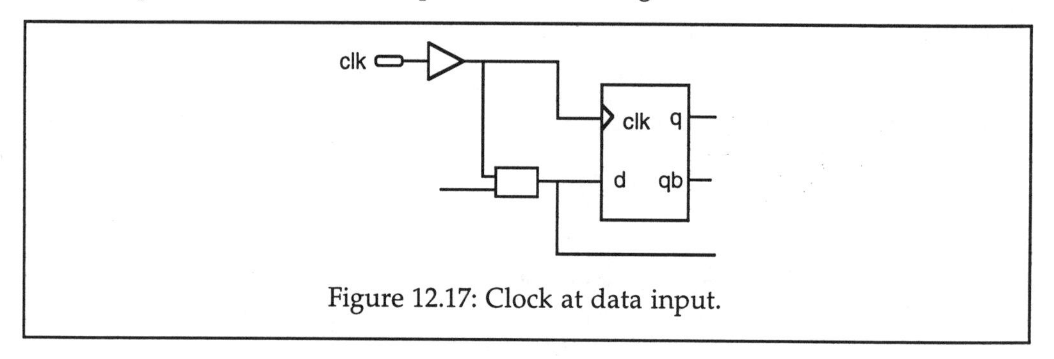

Figure 12.17: Clock at data input.

12.3.1. How to write testable VHDL code

Below we look at how VHDL code should be written to avoid breaking the ATPG rules. Examples of how not to write VHDL code are also given.

Rule 1: Non-scan equivalent
Whether or not there is an equivalent scan flip-flop is not decided by how the VHDL code is written. The deciding factor is the flip-flops chosen by the synthesis tool during optimization before ATPG. If the synthesis tool selects a scan flip-flop

during this optimization, the test insertion tool will not be able to swap it for an equivalent scan flip-flop. The same problem arises if the synthesis tool selects a flip-flop which does not have an equivalent scan flip-flop. The solution is to permit the synthesis tool to select only flip-flops which have an equivalent scan flip-flop. If some normal flip-flops have an equivalent scan flip-flop in a certain scan method and not in another, the synthesis tool has to know what type of scan method is to be used as early as the first optimization. This can be achieved in Synopsys, for example, by giving the following command:

```
set_scan_style LSSD
```

This command must be executed before the first optimization.

Rule 2: Combinational feedback loops

ATPG does not normally support asynchronous design. This means that combinational feedback loops are not permissible. Suppose that we synthesize the following VHDL code:

```
q<=q + a;
```

As signal *q* is on both sides of the assignment symbol (<=), a combinational feedback loop is obtained. This produces the synthesis result shown in Figure 12.18.

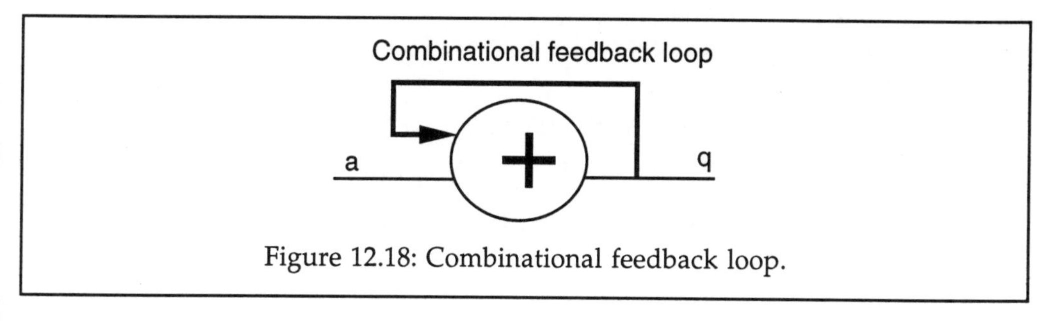

Figure 12.18: Combinational feedback loop.

The solution is to break the combinational feedback loop with a flip-flop. The VHDL code can then be written as follows:

```
Process(clk,resetn)
begin
  if resetn='0' then
    q_loop<=(others=>'0');
  elsif clk'event and clk='1' then
    q_loop<=q;
  end if;
end process;

q<=q _loop + a;
```

The synthesis result is shown in Figure 12.19.

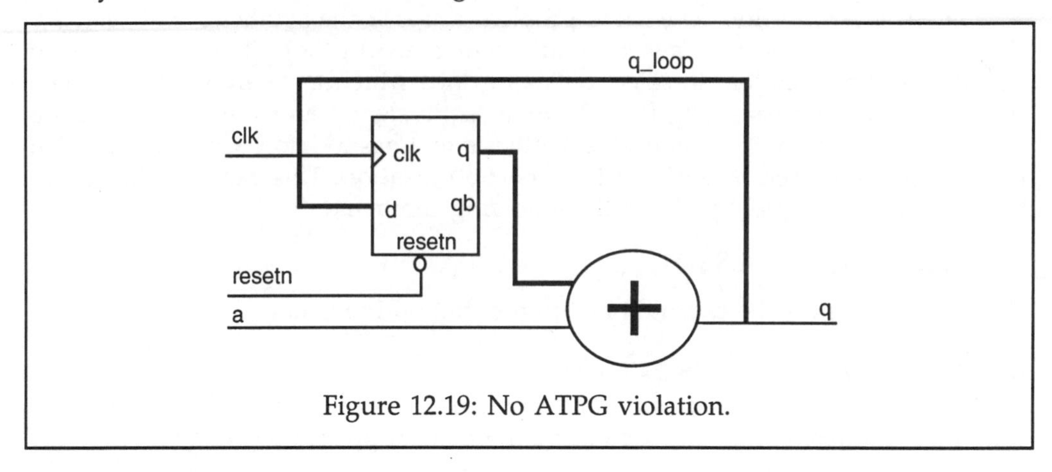

Figure 12.19: No ATPG violation.

Rule 3: Non-three-state driver
This error can occur if the VHDL code is written as follows:

```
q<=a;
q<=b when en2='1' else 'Z';
```

Signal *q* is driven by two VHDL commands, but only the last command results in three-state. This breaks the ATPG rules. The VHDL code should be written as follows instead:

```
q<=    a when en1='1' else 'Z';
q<=    b when en2='1' else 'Z';
```

This description method results in the error-free design shown in Figure 12.20.

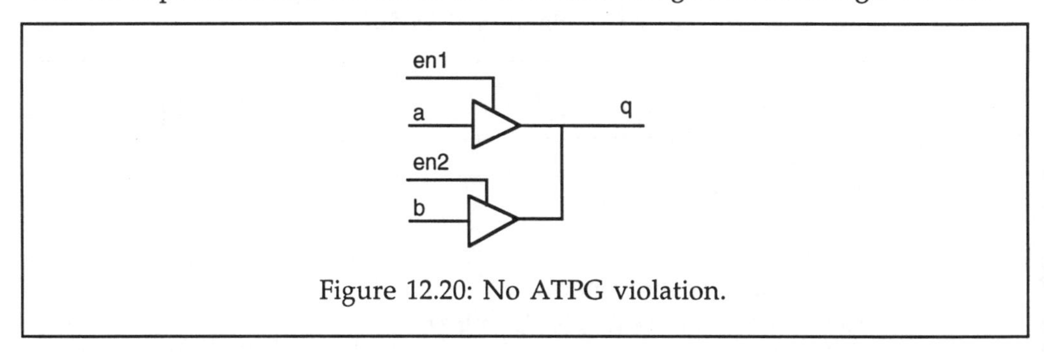

Figure 12.20: No ATPG violation.

Rule 4: Non-controllable clocks
The clock is often gated as shown in Figure 12.15 because it is required only to update the value in the flip-flop when en='1'. The VHDL code for gating the clock like this is as follows:

```
clk_g<=en and clk;

process(clk_g)
begin
  if clk_g'event and clk_g='1' then
    q<=a;
  end if;
end process;
```

Gating the clock in this way breaks the ATPG rules. There are two alternative solutions to the problem:

Alternative 1: The VHDL code can be written as follows:

```
process(clk)
begin
  if clk'event and clk='1' then
    if en='1' then
      q<=a;
    end if;
  end if;
end process;
```

The synthesis result of this VHDL code is shown in Figure 12.21.

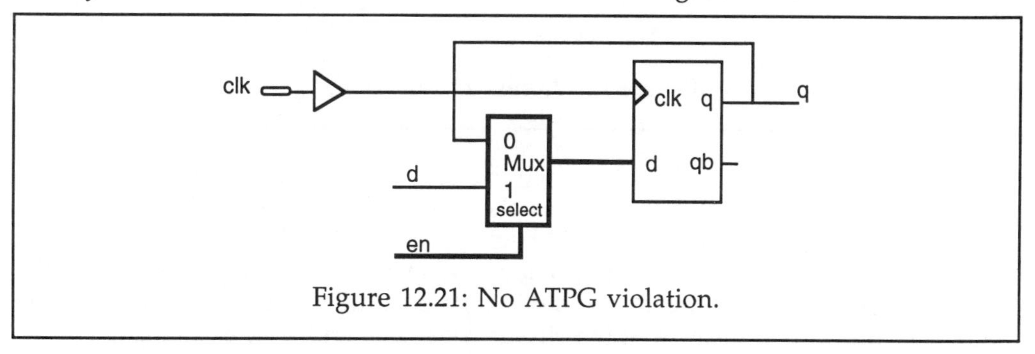

Figure 12.21: No ATPG violation.

Figure 12.21 shows that the output signal from the flip-flop is now fed back to a multiplexor. The multiplexor is controlled by signal en. If *en*='1', the flip-flop is given signal *a*'s value at a positive clock edge for *clk*. This is functionally identical to the preceding example. The difference is that the description method does not break the ATPG rules. This method is to be preferred, as it makes it possible to use the same ungated clock throughout the design.

Alternative 2: If the clock has to be gated for some reason, it must be possible to control this from the clock input on the circuit. If signal en is an input to the circuit, this is no problem, as in this case it is controllable. If, on the other hand, the signal is an internal signal, a special test signal, e.g. *test_hold*, has to be introduced. This test signal should ensure that the gated clock follows signal *clk* irrespective of signal *en*'s value in the ATPG test. The VHDL code should be written as follows:

```
clk_g<=(en or test_hold) and clk;

process(clk_g)
begin
  if clk_g'event and clk_g='1' then
    q<=a;
  end if;
end process;
```

The *test_hold* signal must be an input on the circuit, and it must be kept high during the ATPG test and low in normal mode. The synthesis result of the above VHDL code is shown in Figure 12.22.

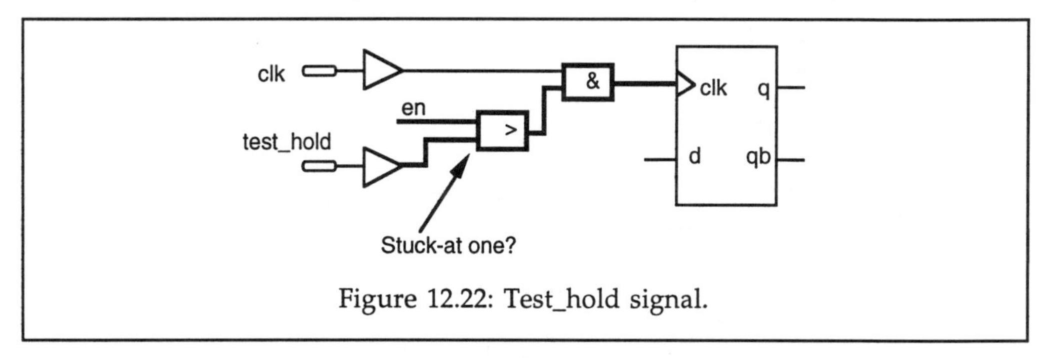

Figure 12.22: Test_hold signal.

One disadvantage is that the circuit requires an extra input. Another is that the ATPG test vectors cannot detect a possible "stuck-at one" at one of the OR gate's inputs (the test_hold input). This is because the *test_hold* signal is kept at '1' throughout the ATPG test, making it impossible to test this input. Some ATPG tools, however, accept the *scan_enable* signal being used as *test_hold* in the above example. This means that the circuit does not require an extra input.

The commonest reason for having to gate the clock is to save power. The clock can then be shut off for certain time periods by disabling signal *en*.

The above solution is not completely without its problems. Clock skew problems can arise in both functional and ATPG mode, as the clock goes via logic. Clock skew problems must therefore be given special consideration after layout if the clock has been gated.

Creating the clock for an internal flip-flop inside the circuit also breaks the ATPG rules (see Figure 12.14). The solution is either to write the VHDL code so that the external clock is used or to introduce a multiplexor. This multiplexor should be controlled in such a way that the internal clock is used in functional mode and the external clock in ATPG mode. The VHDL code can be written as follows to generate the multiplexor:

```
clk_i2<=clk_i when test_hold='0' else clk;

process(clk_i2)
begin
  if clk_i2'event and clk_i2='1' then
    q<=a;
  end if;
end process;
```

The synthesis result is shown in Figure 12.23.

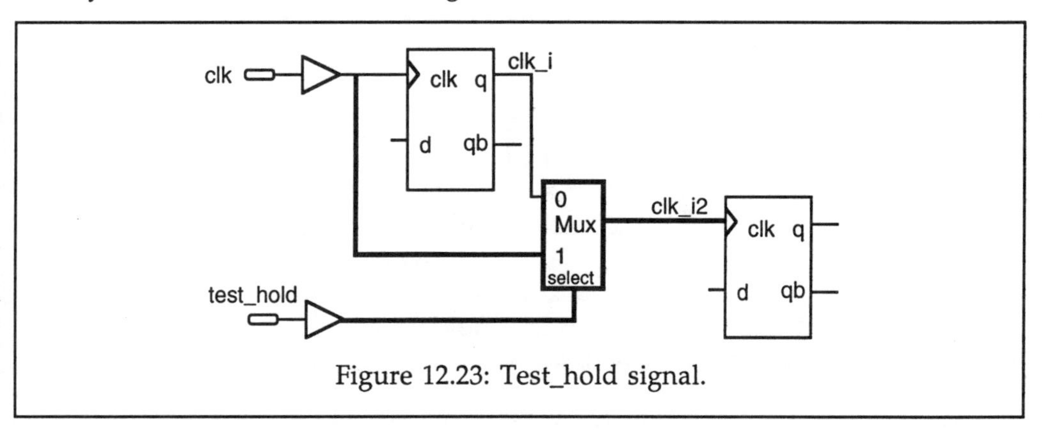

Figure 12.23: Test_hold signal.

Rule 5: Non-controllable asynchronous reset
The simplest way of solving this problem is to use the same asynchronous reset for all the flip-flops. This can be done by including the same reset signal in the sensitivity list of all the clocked processes, for example:

```
process(clk,resetn)
```

Signal *resetn* should be an input on the circuit. If an asynchronous reset is to be controlled from an internal signal, it must be controllable in ATPG mode from an input on the circuit. This means that a test signal, e.g. *test_hold*, has to be introduced. If signal *test_hold* has been introduced because of the clock being gated (see "Non-controllable clocks" above), the same signal can be used. The VHDL code for when the ATPG rules are violated is shown first:

```
process(clk,resetn_i)          -- resetn_i is a internal signal
begin
  if resetn_i ='1' then
    ...
end process;
```

The VHDL code should be written as follows to avoid violating the ATPG rules:

```
-- resetn external input
resetn_i2<=resetn_i when test_hold='0' else resetn;

process(clk,resetn_i2)
begin
  if resetn_i2='0' then
    ...
end process;
```

The synthesis result is shown in Figure 12.24.

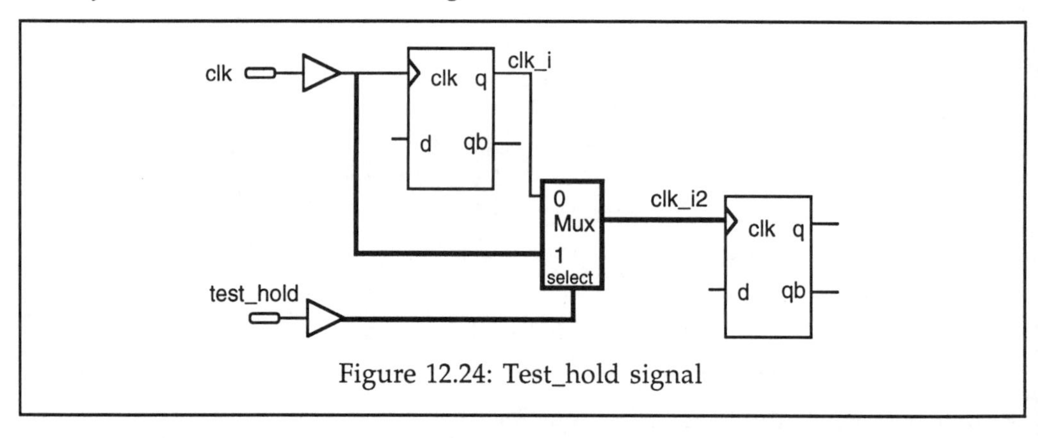

Figure 12.24: Test_hold signal

The internal asynchronous reset is now controlled by *test_hold*. In the ATPG test, test_hold='1'. This means that the internal reset can be controlled directly by input signal *resetn*. In function mode (test_hold='0'), the internal reset is equal to *resetn_i*, just as before.

Test_hold	reset_signal
'0'	resetn_i
'1'	resetn

Rule 6: Clock at data input

The VHDL code must be rewritten so that the clock is not used as a data signal. Such rewriting is totally dependent on what function is wanted.

To summarize, it is possible to generate ATPG test vectors even if some of the above rules have been broken. In this case the fault coverage around where the rules have been broken will be greatly reduced, sometimes right down to zero.

12.4. Boundary scan

Boundary scan (subset of JTAG IEEE-1149) is normally used to verify that the circuit has been mounted on the circuit board correctly. This test method is based on connecting all the I/O cells (inside the circuit) in a long shift register (see Figure 12.25). In the production test the outputs can then be set to either '1' or '0' by shifting a known pattern into the shift register. The value can be sampled at the inputs at the same time. The result can then be read off by shifting the result out of the shift register.

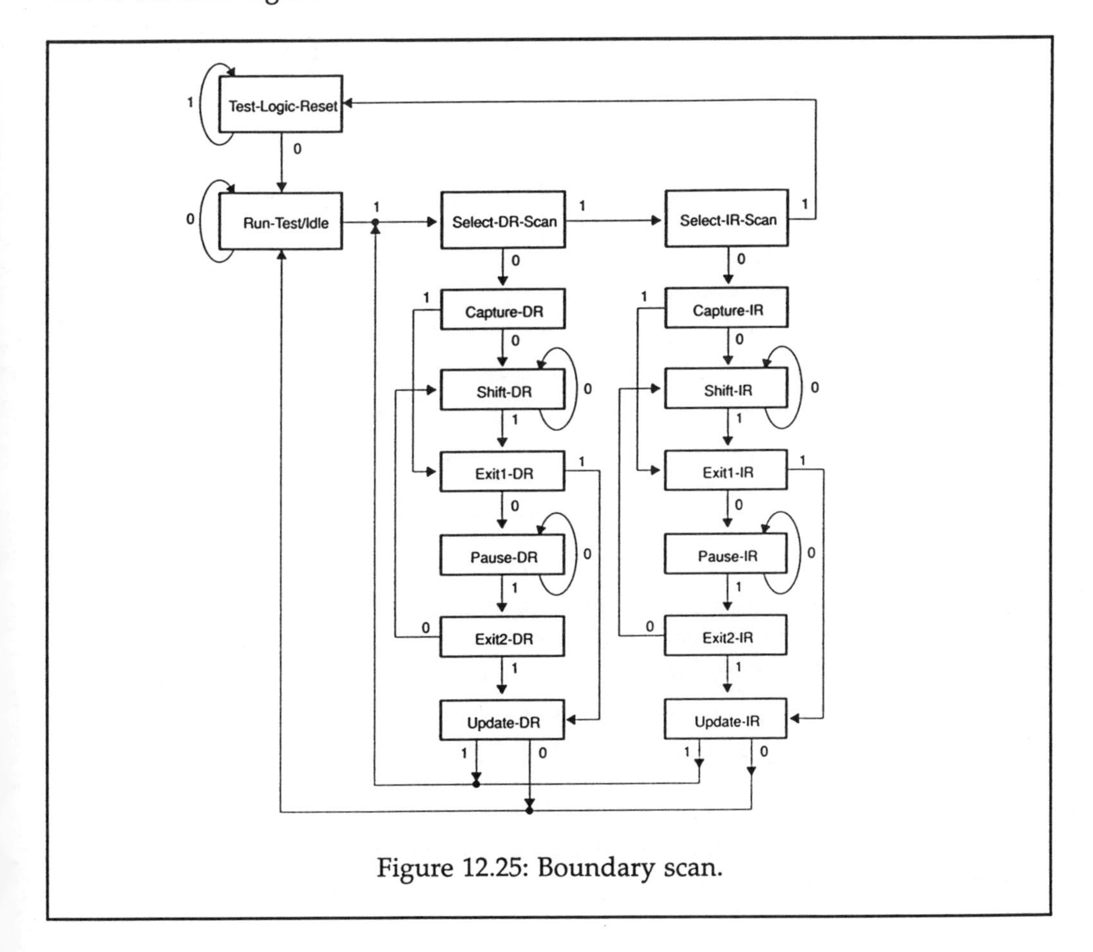

Figure 12.25: Boundary scan.

In order to control the shift register in the circuit, a state machine and an instruction register have to be included in the design in addition to the shift register. The state machine is called a TAP controller (Figure 12.26).

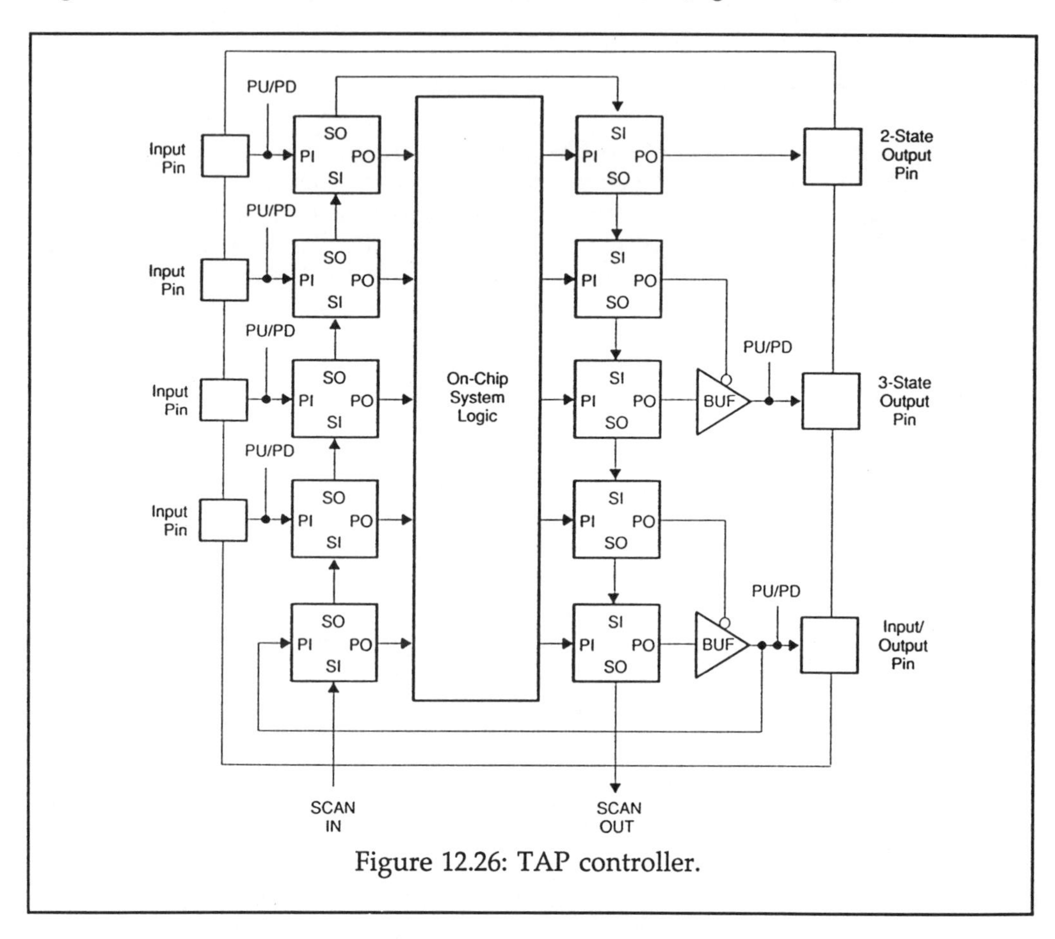

Figure 12.26: TAP controller.

In the late 1980s and early 1990s, the designer often had to spend one or two weeks designing boundary scan manually. Now there are tools to automate this process. There are two procedures:

- The synthesis tool can include boundary scan automatically.
- The test tool can generate synthesizable VHDL code for boundary scan automatically.

In some circuits, predominantly FPGAs and PLDs, boundary scan is always inserted on the silicon.

In the IEEE standard there are four obligatory I/O pins and one optional one which have to be included in the circuit:

- Test Data In (TDI)
- Test Data Out (TDO)
- Test Mode Select (TMS)
- Test Clock (TCK)
- Test Reset (TRST) (optional)

In addition to the shift registers and the TAP controller, the circuit has to contain an instruction register for boundary scan. The contents of this register decide what sort of test is to be performed by the boundary scan. Three instructions which must always be included, according to the IEEE standard, are:

- Bypass
- Sample/preload
- Extest

There are also a number of non-obligatory instructions: Intest, Runbist, IDcode and Usercode. There follows a brief outline description of the various obligatory instructions. For more detailed information, see the IEEE-1149.1 standard.

Bypass means that the circuit only sends the values which have been shifted in through to the boundary scan output direct, i.e. from TDI, via the bypass register, to TDO. This makes it possible to avoid shifting the values through the entire shift register in the circuit if several circuits are connected in a boundary scan.

Sample/preload means that the value of the inputs is sampled and can then be shifted out for checking.

Extest means that known values are shifted in through the boundary scan shift register. The I/O outputs are then updated to the values which have been shifted in. This makes it possible to check the value of all the outputs from the circuit and decide whether there is contact between the circuit board and the circuit's I/O pads.

Summary
If boundary scan is to be used, it is not particularly difficult to include this logic in the circuit, as there are tools which can automate the entire process. The procedure for including boundary scan is dependent on which tool is available and what type of circuit is to be designed.

12.5. Supplementary test vectors

The ATPG test vectors normally have to be supplemented before the ASIC can be signed off. Normally the designer has to supply the following test vectors for the circuit at sign-off:

- ATPG test vectors
- Parametric test vectors
- IDDQ test vectors
- Boundary scan test vectors
- Supplementary manual test vectors
- RAM test vectors

There follows a brief description of these test vectors. In ASIC design it is recommended that the designer should discuss the test vector requirements in detail with the ASIC supplier.

ATPG test vectors
The ATPG tool creates these automatically.

Parametric test vectors
In this test the supplier checks the circuit's I/O cells. They measure changeover levels and the like on all the inputs and outputs. These test vectors must set the outputs to '0', '1' and three-state (if possible). The inputs should be driven to both '1' and '0' to allow verification of the I/O cells. The test can include both DC and AC testing of the circuit.

IDDQ test vectors
The IDDQ test vectors are used to measure whether there is any variable power consumption when the circuit is at rest. If this is the case, it means that there is a short circuit in the circuit. Many ASIC suppliers accept just one test vector for testing this.

Boundary scan test vectors
The majority of ATPG tools cannot generate test vectors with acceptable fault coverage for the boundary scan logic. Some manual test vectors for this logic have usually, therefore, to be created.

Supplementary manual test vectors
These test vectors are not normally needed. The ATPG test vectors normally have very high fault coverage. In some cases it may be desirable to supplement them with some manual test vectors. These are usually generated from some of the simulations performed on the circuit.

RAM test vectors
If the circuit contains RAM, special test vectors are required, as the majority of ATPG tools do not generate test vectors for RAM.

Creating all these test vectors may seem like a lot of work. The time consumption is normally a couple of days, as the ATPG tool will have generated more than 90 per cent of all the test vectors. The number of other test vectors is normally very small compared with the number of ATPG test vectors.

It should also be pointed out that it is in the customer's (i.e. the designer's) interests for the test vectors to be as good as possible. If the circuit passes the test vectors without defects being found, it will be supplied to the customer. If the customer finds a defect in the circuit which has not been covered by the test vectors, the customer must normally bear the responsibility. What is more, ASIC suppliers do not accept an unlimited number of test vectors because testing takes time and is therefore expensive for the ASIC manufacturer.

13

Rapid prototyping

13.1. Introduction

In order to keep up with the competition, many companies have to reduce the development time for new products. This also means that the development time for prototyping has to be reduced.

One way of reducing the development time is to describe the system in a hardware description language such as VHDL, synthesize the description automatically and implement the result in an FPGA. If the product subsequently becomes a volume seller, the VHDL code can be synthesized to a more cost-effective solution, e.g. cell design.

This chapter describes how a small real-time kernel and a simple CPU can be implemented in a few FPGAs. During development VHDL and synthesis were used for transformation from RT level to gate level for the chosen technology. Verification included downloading a test program into main memory, which the CPU then executed during simulation (HW/SW Codesign). The finished circuits worked first time.

13.1.1. Rapid prototyping

Reducing the development time in the development phase is one of the most important goals for designers and project leaders. One way of achieving this goal is to use what is known as **rapid prototyping**. Today, rapid prototyping means using VHDL as a description language and verifying behaviour. In implementation technology, FPGAs are used to validate the concept or to emulate the physical prototype and software. Using FPGAs instead of ASICs (Application Specific Integrated Circuits) reduces the implementation time from a few weeks to a few hours. A change can be made in the design within a few hours or days.

13.2. Real-time kernel - a brief description

In most control systems there is a real-time kernel which manages a CPU. This means that the real-time kernel determines which program is to be executed and tells the programs whether they should wait for an external signal or a certain amount of time, for example.

In the following example the design consists of a real-time kernel, register files and a CPU. This system is called FASTCHART (a FAST and time deterministic CPU and HArdware-based Real-Time kernel). This is a first prototype for verifying that the new concept is good. Figure 13.1 shows the different parts.

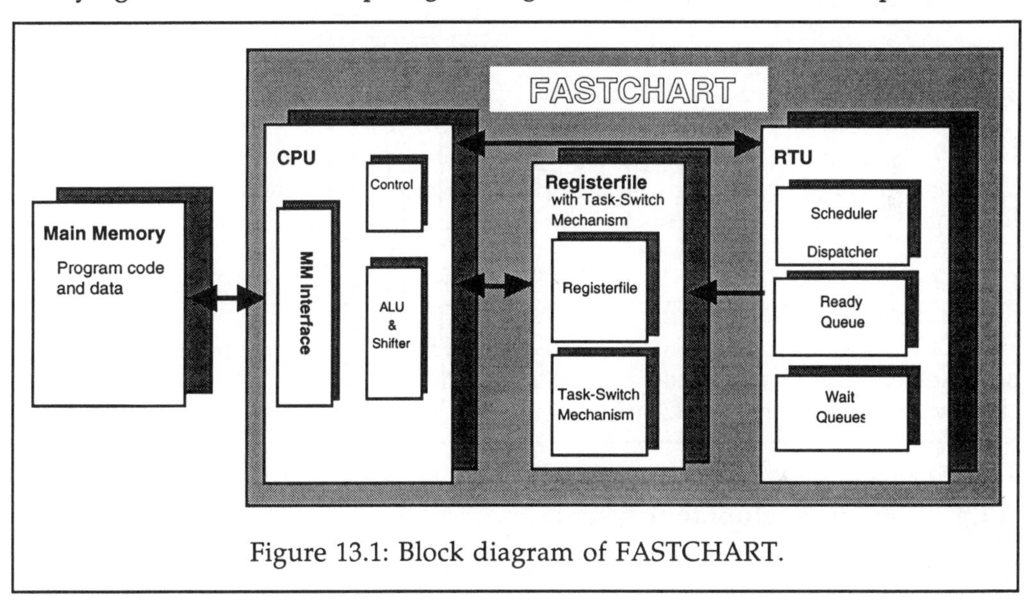

Figure 13.1: Block diagram of FASTCHART.

The CPU works with eight bits and has a few CPU instructions and real-time service calling. The register file contains the CPU register, instruction address register, etc., for the respective programs.

The RTU contains the most common functions in a real-time kernel such as scheduler, various types of queue, etc. The RTU can manage eight programs with two priorities. The main memory unit consists of standard EPROM circuits and is only modelled in the simulator (not synthesizable). If the CPU is integrated with the real-time kernel, the performance achieved for real-time functions is much higher than if the real-time kernel had been implemented in machine code. Thus it takes "no" time to change program. This means that it is possible to change program between every instruction without performance being reduced. The performance increases as far as calling the real-time kernel is concerned, which can occur during a CPU cycle.

All the parts are described in VHDL except for the main memory, which was taken from a library for the simulator. Figure 13.2 illustrates the complexity of the design. For a more detailed description, see the further reading at the end of the chapter.

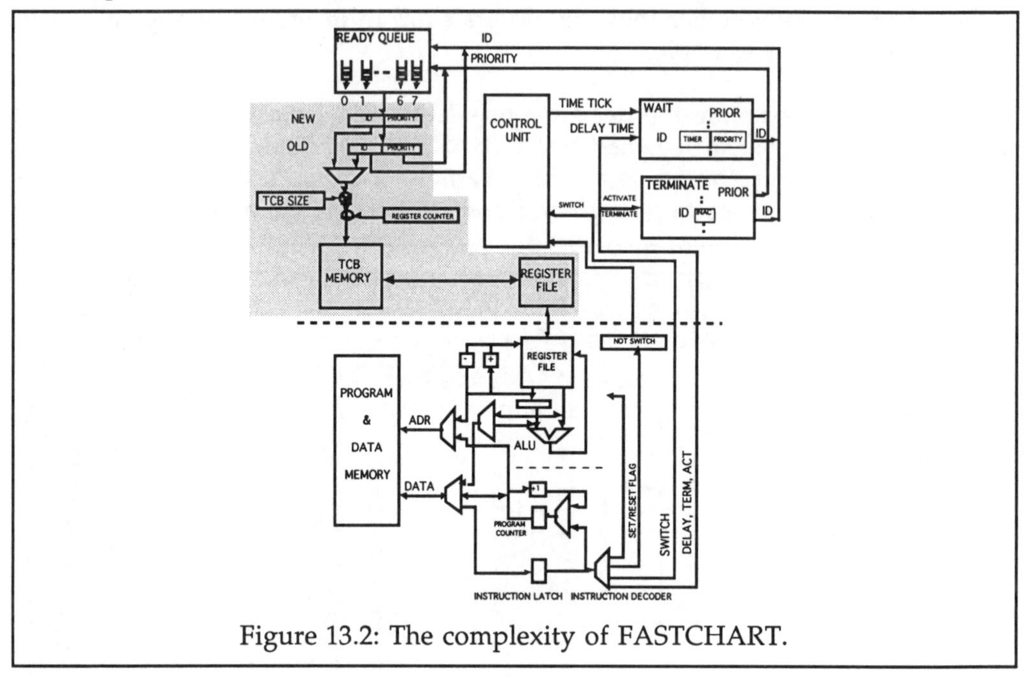

Figure 13.2: The complexity of FASTCHART.

13.3. The development system

In the early 1990s there were only a few development systems with synthesis to FPGAs. We chose a PC-based system called ViewLogic and ACTEL as the target technology.

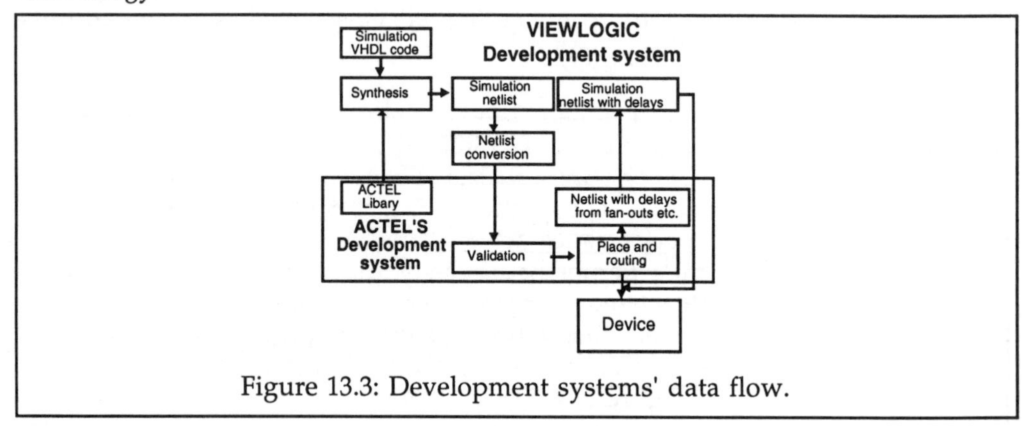

Figure 13.3: Development systems' data flow.

Figure 13.3 shows how the contact between two different development systems (ACTEL and ViewLogic) works. VHDL simulation and synthesis were done in ViewLogic. Routing, back annotation to the simulator (data on delay owing to wiring and fan-out) and programming of the ACTEL circuits were done in ACTEL's software. The interface between these pieces of software consisted of conversion programs.

13.4. Development phases

When working with untested ideas you go backwards and forwards through the steps in the development process between steps 1 and 5. Thus top-down methodology is used in principle.

The six steps in the development process are:

Step 1: Specification and system design
Step 2: Functional description of subsystems
Step 3: Integration and simulation of components
Step 4: Technology mapping
Step 5: Optimization, routing and precision time analysis
Step 6: Design and validation of prototype

In **step 1** the functionality of the prototype was specified. Some test programs (machine code) were also written for defining what the prototype should do when finished.

After specification the system was divided into subsystems (components). This step is called architectural design. This division was very simple because we had simulated the entire system in C during a feasibility study and the division was just like a "software system". While the division of the system was taking place, the interfaces between the subsystems were also defined. When the system has been divided up, you obtain a hierarchy (Figure 13.4). This hierarchy is just for mastering complexities. In the design, four levels with approximately thirty components were created. The division was done on the computer by using a graphics tool.

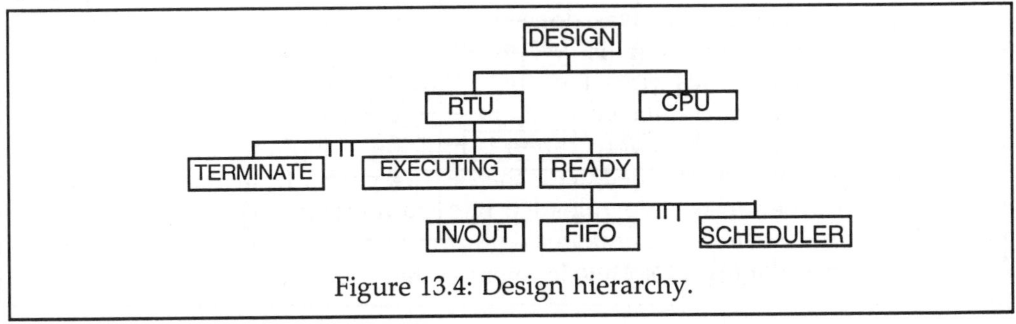

Figure 13.4: Design hierarchy.

In **step 2** the VHDL code was written at RT level for the leaf components. Leaf components are components which do not consist of other components. In Figure 15.4, FIFO is a leaf component, for example, while RTU is a component made up of several other components (Ready, Terminate, Executing, etc.). The straightest possible code was written and tested for synthesizability. In several instances the synthesis program did not manage to transform the code, so the code had to be rewritten in order to adapt it to the synthesis software. After a few days' work we had built up knowledge with regard to what could be synthesized.

The next example is not complete and was adjusted on the basis of the possibilities offered by the synthesis tool in 1991 (ViewLogic):

```
process
begin
  wait until clk = '1';
  if push_task_id = '1' then
     temp := addum(last_p,one);
      last_p(2 downto 0) <= temp(2 downto 0);
  elsif pop_task_id = '1' then
      temp := addum(next_p,one);
  end if;
   next_p(2 downto 0) <= temp(2 downto 0);
end process;

process(last_pointer, next_pointer)
begin
 if last_pointer = next_pointer then
    empty <= '1';
 else
    empty <= '0';
 end if;
end process;
```

Step 3: Integration and simulation of the components begin with verification of each component. Verification was done using input stimuli to the components and checks on the output signals. When all the components had been tested separately, the components were connected together to form new components (integration) which were then verified, too. This process continued until the main level was reached. Thus verification is basically bottom-up and design top-down. The top level was verified with the test programs from the specification phase. The test program was downloaded into main memory (RAM/ROM), after which the entire system was simulated with the test programs as input stimuli. The CPU executed the test program:

```
-- Description of the test program (assembler) --
--------------------------------
begin INIT_TASK
    ACTIVATE(TASK_1,PR_0);
    STARTADDRESS(20H);
    ACTIVATE(TASK_2,PR_1);
    STARTADDRESS(40H);
    ACTIVATE(TASK_3,PR_0);
    STARTADDRESS(60H);
    TERMINATE;
end    INIT_TASK;
--------------------------------
begin TASK_1
start: DELAY(16 cpu_clocks);
    JMP start;
end    TASK_1;
--------------------------------
begin TASK_2
start: DELAY(16 cpu_clocks);
    JMP start;
end TASK_2;
--------------------------------
begin TASK_3
start: DELAY(16 cpu_clocks);
    JMP start;
end    TASK_3;
--------------------------------
```

The above test program contains four separate programs. INIT_TASK is started after reset to the CPU, then TASK_1 to 3 are activated by INIT_TASK. Finally INIT_TASK is terminated. TASK_1, TASK_2 and TASK_3 call RTU (the real-time kernel) and request a delay of 16 clock cycles.

By executing the above test program, verification was obtained that certain functions in the RTU and CPU were working.

The first three steps are technology-independent. Everything was verified with VHDL code. If optimum solutions in terms of area are to be achieved, it is advisable to test different FPGA technologies during synthesis. This is because

Xilinx, for example, contains a lot of flip-flops, while an ACTEL circuit contains relatively few.

Step 4 is where the technology is chosen. The synthesis program needs a technology file in order to translate and optimize to the chosen technology (ACTEL 1A020). Now a new model consisting of gates has been generated.

Each gate has a unit delay of 10 ns. A new simulation and a first rough time analysis are carried out.

In **step 5,** packages and sizes for FPGAs are chosen. In our case, ACTEL 1020 and 1010 were chosen. The netlist from synthesis is routed to the chosen package, then a file containing all the delays from gates and interconnections is generated and new precision time verification is carried out. An alternative would be to write in time constraints for the synthesis tool. In the FASTCHART design we chose verification with the simulator instead. The optimum solution is otherwise to set up all the time constraints in the synthesis tool so that the tool can optimize to those constraints. It is then also possible to use the synthesis tool's internal timing analyser to analyse the timing both before and after routing. In order for this to work, the synthesis tool must both have an advanced timing analyser and be able to optimize the chosen technology with regard to timing.

Step 6 in the development process took the least time. Six FPGAs were programmed and a circuit board wired up. Then the test programs were programmed in the EPROM. Finally the program was executed. The prototype worked at the first attempt.

A rough timetable for the various activities is given below:

Steps 1-3	5 weeks
Steps 4 and 5	2 weeks
Step 6	1 week
Total	8 weeks

It is apparent that the initial stages of the development process take just as long as usual, while the final stages are faster because they are largely performed automatically.

Rapid prototyping does not produce an optimum prototype, but the goal was to be able to test and emulate the system faster. When the prototype was ready, it was possible to load new programs and execute them much faster than if the same behaviour had been modelled on the computer (at least 1000 times faster with the prototype).

13.5. Further Reading

The following references describe FASTCHART:

Lindh, L. and F. Stanischewski (1991) 'FASTCHART - a fast time deterministic CPU and hardware based real-time-kernel', *IEEE Real-Time Workshop, Paris,* 12-14 June.

Lindh, L. and F. Stanischewski (1991) 'A design of a real-time unit in hardware', *Svenska Nationella Arbetsgruppen i Realtid - SNART, Uppsala,* 19-20 August.

Lindh, L. and F. Stanischewski (1991) 'FASTCHART - idea and implementation', *IEEE International Conference on Computer Design (ICCD), Cambridge, Mass.,* 14-16 October.

More information on rapid prototyping can be found in the following article:

Lindh, L., K.D. Müller-Glaser and H. Rauch (1993) 'Rapid prototyping with VHDL and FPGAs', *Lecture Notes in Computer Science* no. 705, Springer-Verlag.

14 Common design errors in VHDL and how to avoid them

Perhaps the commonest error made by newcomers to VHDL is to try to use sequential commands such as **if** and **case** in the concurrent part of VHDL. The chapters on "Sequential VHDL" and "Concurrent VHDL" explain which parts of VHDL the various commands should be used in. All available VHDL commands are also listed in Appendix A, which also indicates whether the commands are used in sequential or concurrent VHDL.

14.1. Signals and variables

A common error when starting to design in VHDL is not distinguishing between when signals and variables have to be used. As shown in Tables 4.1-4.2, the behaviour of signals and variables can be completely different. Moreover, a signal can never correspond to a hardware signal line. Variables can only be used to store data temporarily in a process or subprogram, i.e. variable assignment is a sequential command. Therefore, a variable can only be declared in the sequential part of VHDL, whereas signal assignment can be both sequential and concurrent. However, signals can only be declared in the concurrent part of VHDL. As demonstrated in the example below, variables are needed because they are given their value immediately, whereas a signal is given its value after a delta delay. This means that VHDL has to be written differently depending on whether signals or variables are being used. The recommended solution is completely dependent on the design. The following example illustrates this distinction.

Suppose that the following Boolean function has to be described in a VHDL process:

```
int<=a and b and c
q<=int or d;
```

These concurrent signal assignments could, of course, be written directly in the architecture, but the example is intended to illustrate the principles involved. To make these clear, a small example has been chosen:

```
Library ieee;
Use ieee.std_logic_1164.ALL;

Entity ex1 is
  port( a,b,c,d:  in    std_logic;
        q:        out   std_logic);
end;
```

Architecture with internal signal *int* as a signal:

```
Architecture sig of ex1 is
signal int:std_logic;
begin,
  process(a,b,c,d,int)
  begin
    int<=a and b and c;
    q<=int and d;
  end process;
end;
```

Architecture with internal signal *int* as a variable:

```
Architecture var of ex1 is
begin
  process(a,b,c,d)
  variable int:std_logic;
  begin
    int:=a and b and c;
    q<=int and d;
  end process;
end;
```

The difference between these two examples can be seen in the sensitivity list for the process, among other things. In the signal example signal *int* is included, while in the variable example only the real input signals (*a,b,c,d*) are on the sensitivity list. Signal *int* has to be included in the signal example because it is only updated after a delta delay, i.e. when the line q<=int and d; is executed, signal *int* still has the old (incorrect) value. To get around this, *int* has to be on the sensitivity list, as the process will then be activated again (one delta after the signal assignment of *int*). The disadvantage of having *int* as a signal is quite clear in this example. Not only is it difficult to remember to assign all the signals which contain just an intermediate result in the sensitivity list, but also the signal example will take longer to simulate because the process must be gone through twice, compared with just once in the variable example. The advantage of the signal example is that signal *int* can be used as a waveform in the simulator. This is not possible if *int* is declared as a variable, as no time is linked to a variable. This makes it harder to debug the variable example than the signal example.

From the logic synthesis point of view, the result will be identical irrespective of whether the signal or variable description method is chosen in the above examples. If, on the other hand, the signal example had been synthesized without being included in the sensitivity list, the result from the majority of synthesis tools would have been "correct" as synthesis tools usually ignore sensitivity lists. We would then have had a mismatch between the RTL simulation and gate level behaviour, something which often has serious consequences (see next section).

It is recommended that you use variables only when you want to store a value temporarily.

14.2. Logic synthesis and sensitivity lists

Some synthesis tools do not pay any attention to what is in the sensitivity list. This can cause mismatches between simulation at RTL level and gate level. For example:

```
Library ieee;
Use ieee.std_logic_1164.ALL;

Entity ex2 is
 port( clk1,clk2,resetn,din,en:  in    std_logic;
       q:                        out   std_logic);
end;
Architecture bad of ex2 is
begin
 process(clk1,resetn)
 begin
  if resetn='0' then
    q<='0';
```

```
    elsif clk2'event and clk2='1' then
      if en='1' then
        q<=din;
      end if;
    end if;
  end process;
end;
```

After synthesis the result of the above example will be a D-type flip-flop with clk enable which is clocked by clk2, i.e. the synthesis program has ignored the sensitivity list with clk1. This description method is incorrect. The synthesis program usually also warns that clk2 is not included in the sensitivity list. This warning is easy to miss if the example forms part of a larger component. The error should, however, be detected before synthesis, i.e. in the simulation at RT level. It is impossible to stress too often how important the simulation at RT level is when it comes to finding functional design errors (see Chapter 11, "Design methodology").

14.3. Buffers and internal dummy signals

When you have to reread the value of an output signal, there are three possible solutions:

- Declare the signal to be of mode buffer in the entity
- Use an internal dummy signal in the architecture
- Use the new signal attribute 'driving_value (VHDL-93)

Examples of the three options are given below.

Example (buffer):

```
Library ieee;
Use ieee.std_logic_1164.ALL;
Entity ex3a is
 port( clk,resetn,din1: in      std_logic;
       q1:              buffer  std_logic;
       q2:              out     std_logic);
end;
Architecture buf of ex3a is
begin
 process(clk,resetn)
 begin
```

```
    if resetn='0' then
      q1<='0';
      q2<='0';
    elsif clk'event and clk='1' then
      q1<=din1;
      q2<=q1;         -- Reading signal q1.
    end if;
  end process;
end;
```

Example (intern "dummy-signal"):

```
Library ieee;
Use ieee.std_logic_1164.ALL;
Entity ex3b is
  port( clk,resetn,din1:  in    std_logic;
        q1:               out   std_logic;
        q2:               out   std_logic);
end;
Architecture buf of ex3b is
signal q1_b: std_logic;  -- The "dummy signal"
begin
  q1<=q1_b;
  process(clk,resetn)
  begin
    if resetn='0' then
      q1_b<='0';
      q2<='0';
    elsif clk'event and clk='1' then
      q1_b<=din1;
      q2<=q1_b;
    end if;
  end process;
end;
```

Example ('driving_value, VHDL-93):

```
Library ieee;
Use ieee.std_logic_1164.ALL;
Entity ex3c is
  port( clk,resetn,din1:  in    std_logic;
        q1:               out   std_logic;
```

```
          q2:              out   std_logic);
end;

Architecture buf of ex3c is
begin
 process(clk,resetn)
 begin
  if resetn='0' then
    q1<='0';
    q2<='0';
  elsif clk'event and clk='1' then
    q1<=din1;
    q2<=q1'driving_value;
  end if;
 end process;
end;
```

Alternative 3 is only valid in the new VHDL-93 standard. Synthesis tools do not yet support this new attribute, so it can only be used for simulation models. From the logic synthesis point of view the result will be identical irrespective of whether alternative 1 or 2 is chosen. Problems can arise during simulation, however, if the component is instanced in structural VHDL in the case of the buffer method because the designer will often find himself in a position where he is trying to use the **port map** command between a signal of mode **out** and one of mode **buffer** (see Figure 14.1), which is in breach of the VHDL standard.

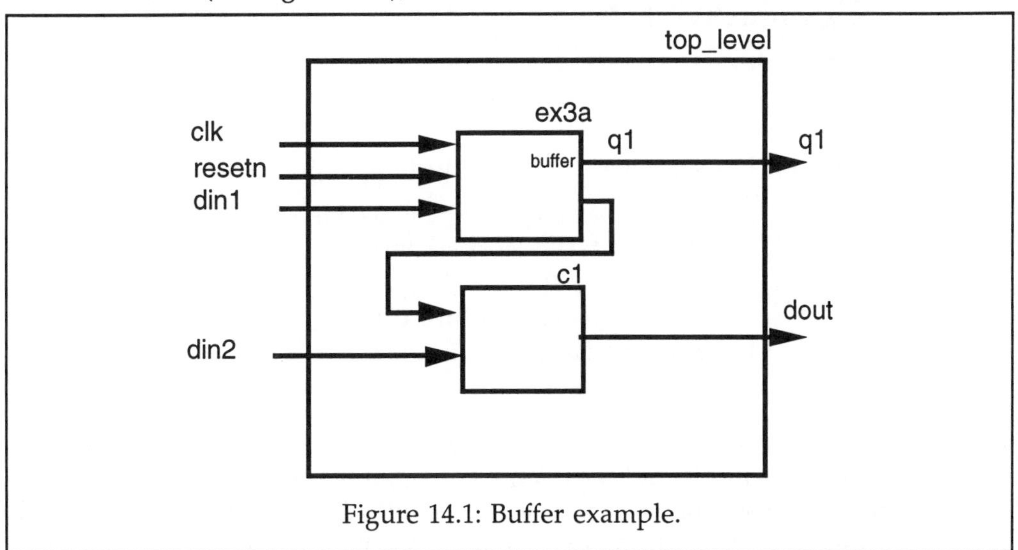

Figure 14.1: Buffer example.

Suppose that component ex3a, which has output signal *q1* defined as mode buffer, is to be instanced in component top_level. Suppose also that signal *q1* goes directly to the output on component top_level. If output signal *q1* in component top_level is defined as mode out, an error will occur during compilation, as you would be trying to map a signal of mode buffer to one of mode out, which is not permitted. Output signal *q1* in component top_level would also have to be defined as mode buffer. This does not work particularly well from a documentation point of view, as output signal *q1* is a pure output from top_level's perspective.

Example:

```
Library ieee;
Use ieee.std_logic_1164.ALL;

Entity top_level is
  port( clk,resetn,din1,din2:  in  std_logic;
        q1, dout:              out   std_logic);
end;

Architecture bad of top_level is
  component ex3a
   port( clk,resetn,din1:  in      std_logic;
         q1:               buffer  std_logic;
         q2:               out     std_logic);
   end component;
  component c1
  port(...
       ...
  end component;

  Signal i1: std_logic;

  For U1: ex3a Use entity  work.ex3a(buf);
  For U2: c1 Use entity  work.c1(rtl);
  begin
    U1: ex3a port map(clk,resetn,din1,q1,i1); --  Error
    U2: c1 port map(i1,din2,dout);
  end;
```

This problem can be solved by using the internal dummy signal in accordance with alternative 2. Using mode inout is not correct; it is certainly possible to both write and reread the signal, but others are also entitled to write to it. Inout should only be used for bidirectional signals or wired logic (wired-or, etc.). The schematic for the component using inout will certainly be correct, but problems can arise with the port map command. It is also bad design style to use inout when the signal is not bidirectional, as others may then believe that it is permissible to drive the signal, which may result in two drivers for a signal.

Sticking to the dummy signal alternative is recommended if structural VHDL is used. If, on the other hand, structural VHDL is not being used and the hierarchy is being drawn on a schematic tool (see Chapter 11, "Design methodology"), the choice does not matter (unless the component is instanced with structural VHDL in future).

14.4. Declaring vectors with downto or to

There are two ways of declaring vectors in VHDL:

```
signal a:std_logic_vector(0 to 3);
signal b:std_logic_vector(3 downto 0); -- Recommended
```

It is recommended that vectors should always be declared with downto. This is because if vectors are declared with **downto**, the most significant bit (MSB) will always be the one with the highest index number and the vector which is furthest to the left. If, however, **to** is used, the vector with index number (0) will correspond to the MSB.

14.5. Incompletely defined combinational processes

When a combinational process (not a memory element) is being designed, the output signals must always be assigned a value, i.e. it should not be possible to pass the process without having assigned all the output signals a value at least once.

An example of an incomplete process might be:

```
Architecture bad is
begin
 process(a,b)
 begin
  if a>b then
    q<='0';
  elsif a<b then
    q<='1';
  end if;
 end process;
end;
```

As figure 14.2 shows, a latch is obtained for the output signal, which is the sign of an incomplete combinational process.

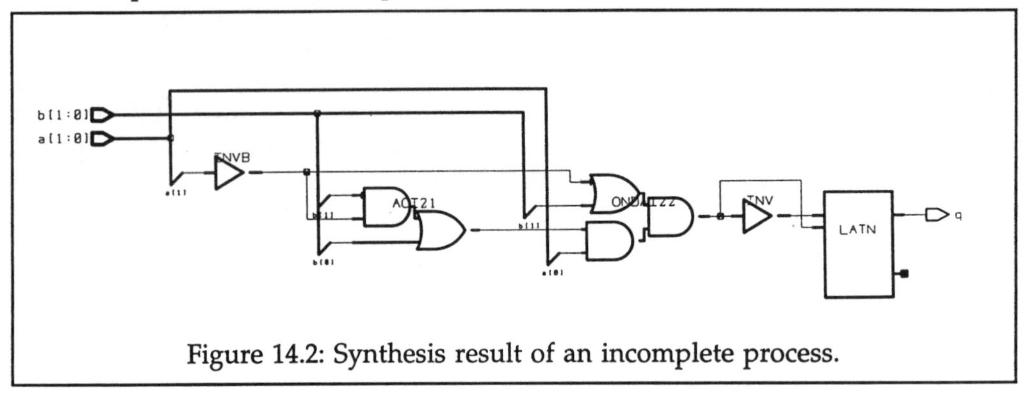

Figure 14.2: Synthesis result of an incomplete process.

A correct description method for the process would be:

```
Architecture good is
begin
  process(a,b)
  begin
    if a>b then
      q<='0'
    else
      q<='1'
    end if;
  end process;
end;
```

The above process assigns output signal *q* a value each time that the process is activated. This means that the synthesis tool does not have to infer a latch for the output. The synthesis result is shown in Figure 14.3.

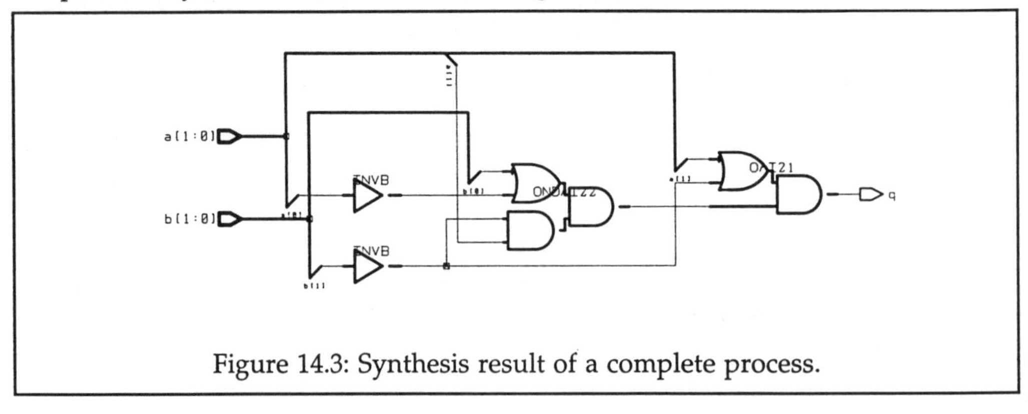

Figure 14.3: Synthesis result of a complete process.

15 Design examples and design tips

All the examples which follow are small. A large design has hundreds of lines of VHDL in the architecture. However, such a design normally contains a combination of the following examples (see the laboratory experiment entitled "A complete design example" on page 402).

The recommendation with regard to how large an entity (component) should be is dependent on which synthesis tool is being used and how much memory is available. Normally, however, the size should be between 500 and 6000 NAND equivalents (gates). If the components are smaller than 500 gates, describing the connections between them will take a long time in relation to the size of the components, irrespective of whether structural VHDL or a graphics tool is used. If the components are larger than 6000 gates, they will be awkward to handle and take a long time to synthesize.

The following examples were synthesized with a Synopsys synthesis tool. No special syntax for Synopsys was used, with the IEEE standard being satisfied in full by all the examples, which means that it will normally be possible to synthesize the design examples on the majority of "advanced" synthesis tools without modification. The one change which may have to be made is as a result of some of the following examples using package ieee.std_logic_unsigned. If the user's synthesis tool does not support this package, it can be swapped for the package which the chosen synthesis tool supports (see Chapter 5, "Library, package and subprograms").

Examples 15.1-15.11 were synthesized using Motorola's HDC library. Motorola is one of the leading ASIC suppliers and also has one of the most user-friendly ASIC design packages on the market.

15.1. Adders

15.1.1. One-bit adder with carry in

As explained in Chapter 6, "Structural VHDL", it is possible to describe a one-bit adder by instancing two ands, two xors and two ors. This description method was only given as an example to illustrate the principle of structural VHDL and is not suitable for a one-bit adder in design because it is awkward to use structural VHDL, so it also takes longer. If Synopsys is being used, the Design Ware option is also lost, i.e. constraint-controlled implementation of the adder (see Chapter 16, "Development tools").

In order to add std_logic_vectors, "+" has to be overloaded with input and output arguments std_logic and std_logic_vector. This functions is in packages ieee.std_logic_unsigned and ieee.std_logic_signed, among others.

Briefly, the difference is that package std_logic_unsigned adds without the sign, while package std_logic_signed adds with the sign (see "IEEE packages" in Appendix B). The reason for '0' & in some of the examples is that the function "+" has to return a vector of the right length: the function "+" returns a vector which is the same length as the longest of the input arguments for the function. For example:

```
Library ieee;
Use ieee.std_logic_1164.ALL;
Use ieee.std_logic_unsigned.ALL;

Entity add1 is
     port( a,b,cin:      in     std_logic;
              cout,sum:  out    std_logic);
end;

Architecture rtl of add1 is
signal s: std_logic_vector(1 downto 0);
begin
   s<=('0' & a)+b+cin;
   sum<=s(0);
   cout<=s(1);
end;
```

The synthesis result is shown in Figure 15.1.

Alternatively, package ieee.std_logic_arith could have been used. The addition would then have looked like this:

```
s<=unsigned(('0' & a)) + unsigned(b) + cin;
```

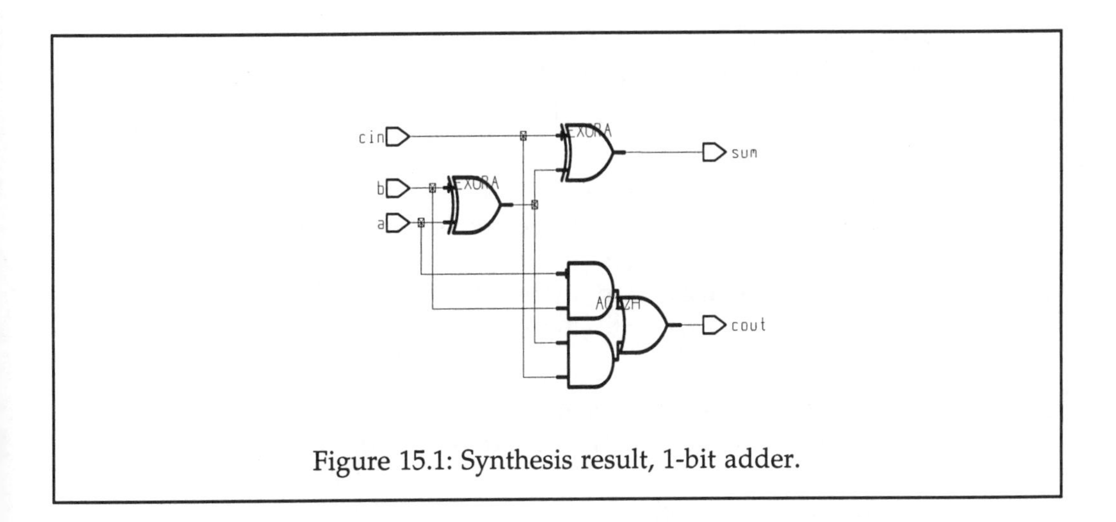

Figure 15.1: Synthesis result, 1-bit adder.

15.1.2. Eight-bit adder with carry in

```
Library ieee;
Use ieee.std_logic_1164.ALL;
Use ieee.std_logic_unsigned.ALL;

Entity add8 is
  port( a,b:  in  std_logic_vector(7 downto 0);
        cin:  in  std_logic;
        sum:  out std_logic_vector(7 downto 0);
        cout: out std_logic);
end;

Architecture rtl of add8 is
signal s: std_logic_vector(8 downto 0);
begin
  s<=('0' & a) + b + cin;
   sum<=s(7 downto 0);
  cout<=s(8);
end;
```

15.1.3. Generic adder with carry in

The following component adds two vectors with the generic length N. The value of N is determined when the component is instanced.

```
Library ieee;
Use ieee.std_logic_1164.ALL;
Use ieee.std_logic_unsigned.ALL;

Entity g_add is
   generic (N:positive:=4);
   port( a,b:  in   std_logic_vector(N-1 downto 0);
         cin:  in   std_logic;
         sum:  out  std_logic_vector(N-1 downto 0);
         cout: out  std_logic);
end;

Architecture rtl of g_add is
signal s:std_logic_vector(N-1 downto 0);
begin
  s<=('0' & a) + b + cin;
  sum<=s(N-1 downto 0);
  cout<=s(N);
end;
```

The synthesis result is shown in Figure 15.2.

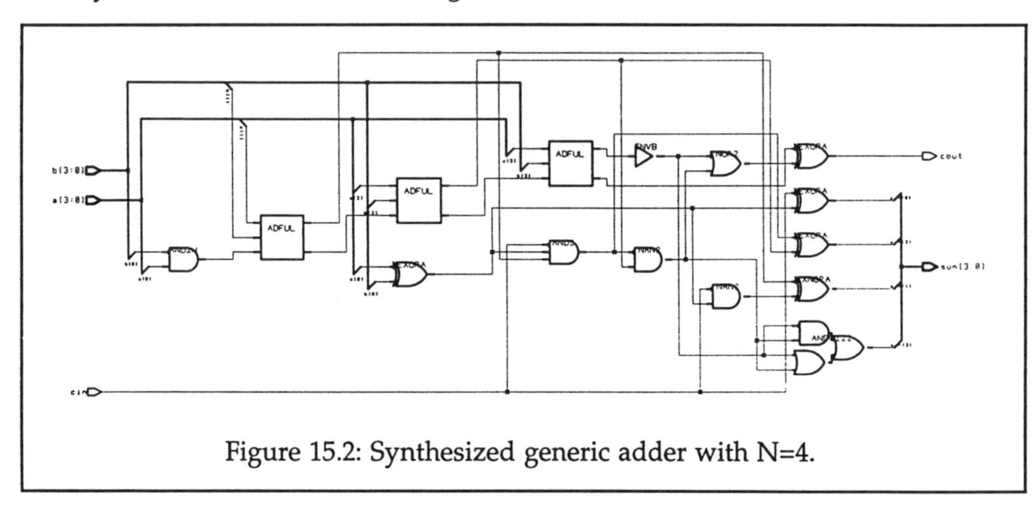

Figure 15.2: Synthesized generic adder with N=4.

15.1.4. Four-bit vector adder/subtractor

```
Library ieee;
Use ieee.std_logic_1164.ALL;
Use ieee.std_logic_unsigned.ALL;
```

```
Entity add_sub is
  port( add_subn: in   std_logic;
        a,b:      in   std_logic_vector(3 downto 0);
        q:        out  std_logic_vector(4 downto 0));
end;

Architecture rtl of add_sub is
begin
  process(a,b,add_subn)
  begin
    if add_subn='1' then
      q<=('0' & a) + b;
    else
      q<=('0' & a) - b;
    end if;
  end process;
end;
```

The synthesis result is shown in Figure 15.3.

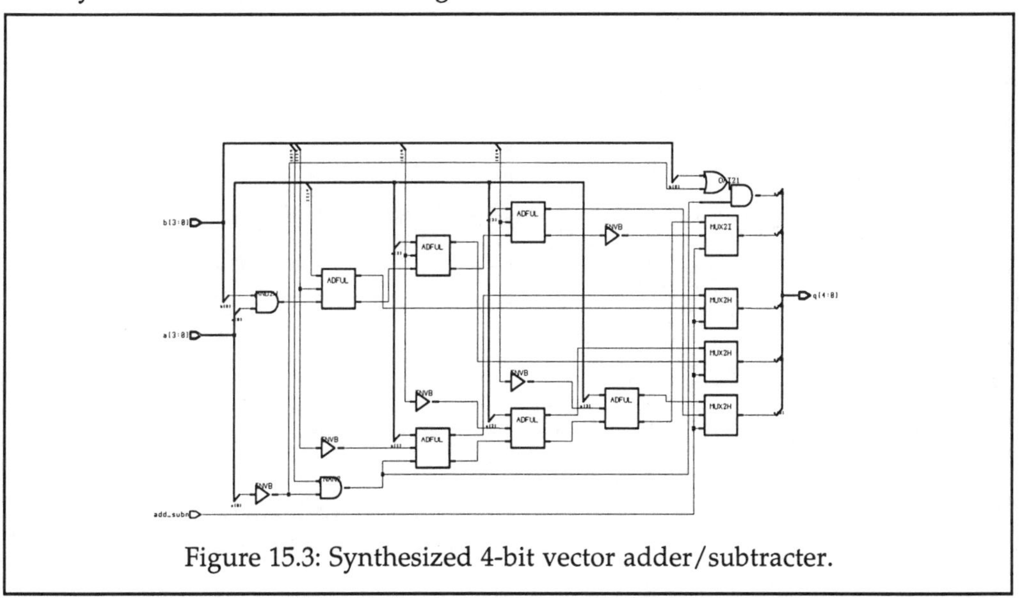

Figure 15.3: Synthesized 4-bit vector adder/subtracter.

15.2. Vector multiplication

```
Library ieee;
Use ieee.std_logic_1164.ALL;
Use ieee.std_logic_unsigned.ALL;
```

```
Entity v_mult is
  port(  a,b:  in   std_logic_vector(3 downto 0);
         c:    out  std_logic_vector(7 downto 0));
end;

Architecture rtl of v_mult is
begin
  c<=a * b;
end;
```

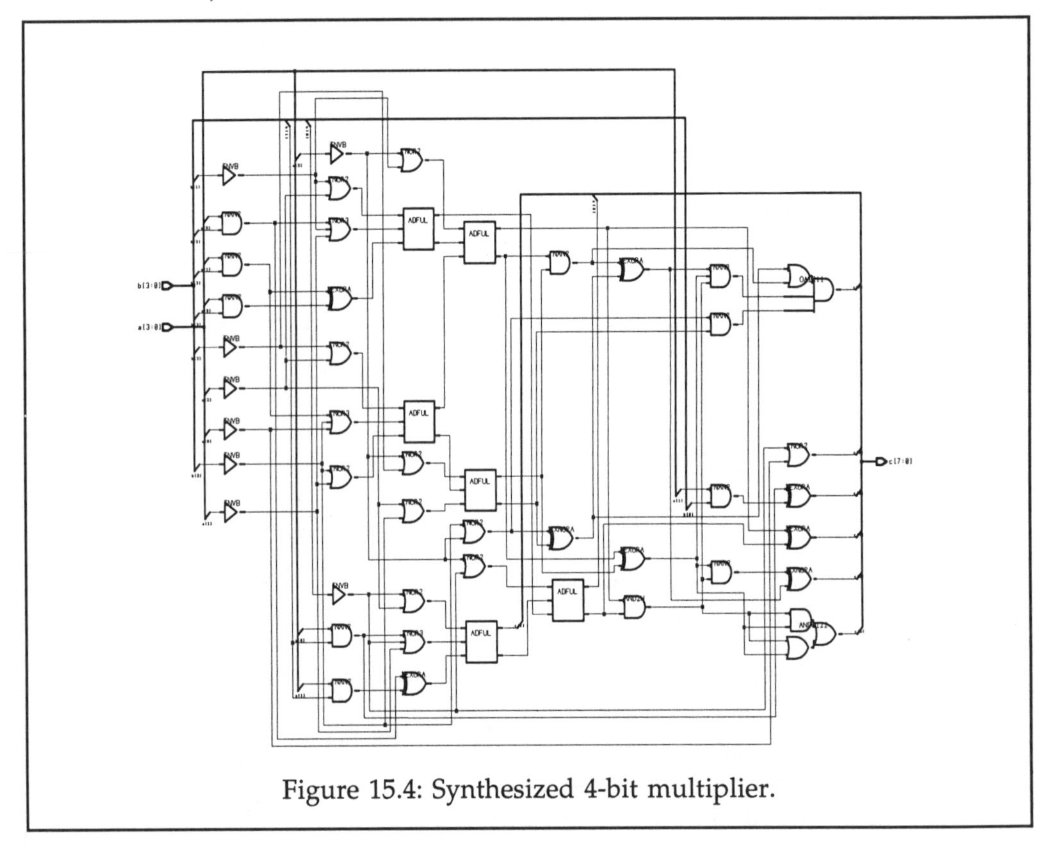

Figure 15.4: Synthesized 4-bit multiplier.

15.3. Resource sharing

15.3.1. Example when resource sharing of an adder is possible

The advanced synthesis tools contain what is usually called high-level optimization. This means, among other things, that the synthesis tool can share resources, e.g. an adder, among different vector adders. If this refinement is to be exploited, it is important to write the VHDL code correctly. The conditions for resource sharing are that the vectors must never utilize the adder at the same time and that this can be understood from the VHDL code. For example:

```
Library ieee;
Use ieee.std_logic_1164.ALL;
Use ieee.std_logic_unsigned.ALL;

Entity share is
 port( a,b,c,d: in   std_logic_vector(4 downto 0);
       sel:     in   std_logic;
       q:       out  std_logic_vector(4 downto 0));
end;

Architecture rtl of share is
begin
  process(a,b,c,d,sel)
  begin
    f sel='0' then
       q<=a+b;
    else
       q<=c+d;
    end if;
  end process;
end;
```

As the two additions are described in the same process and they can never be executed simultaneously (**if-then-else** statement) the synthesis tool can optimize in such a way that the adder is shared by the two additions.

15.3.2. Example when resource sharing of an adder is not possible

```
Library ieee;
Use ieee.std_logic_1164.ALL;
Use ieee.std_logic_unsigned.ALL;

Entity share is
  port(  a,b,c,d: in   std_logic_vector(4 downto 0);
         q1,q2:   out  std_logic_vector(4 downto 0));
end;

Architecture rtl of share is
begin
  q1<=a+b;
```

```
    q2<=c+d;
  end;
```

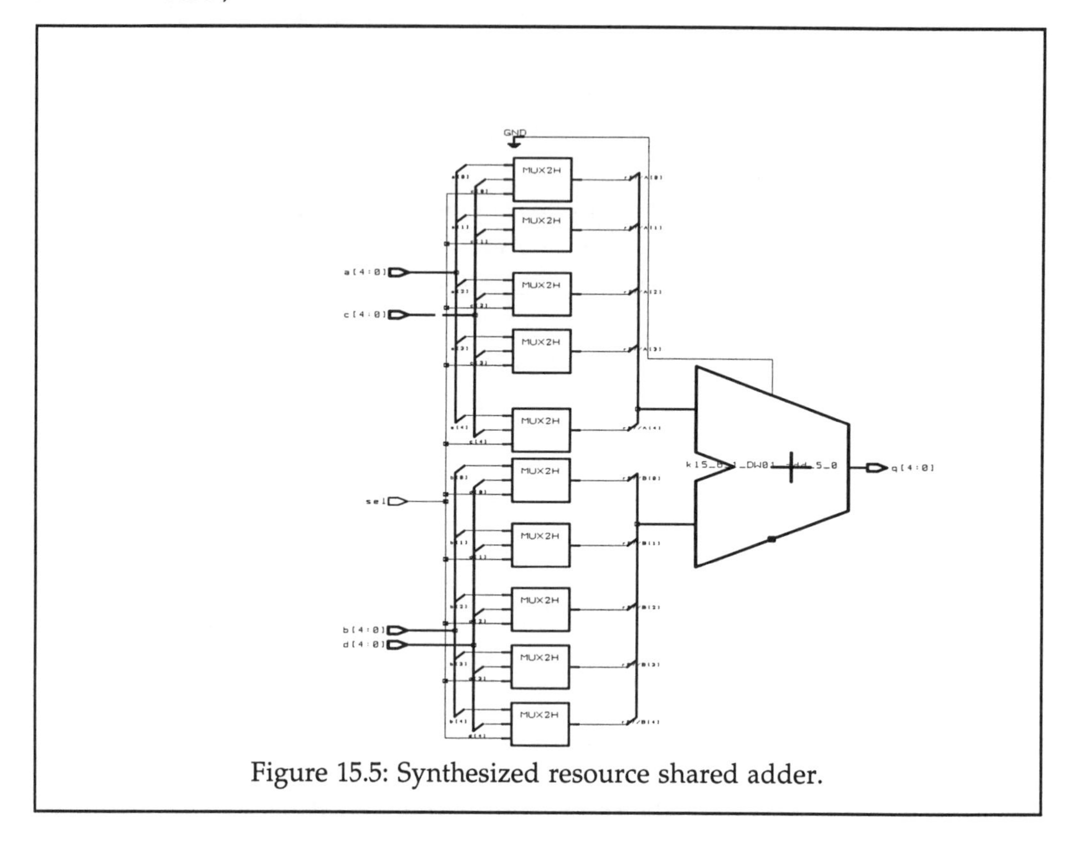

Figure 15.5: Synthesized resource shared adder.

As we can see from Figure 15.5, just one adder is obtained. The synthesis tool looked through the VHDL code and saw that the two additions in the code can never be executed simultaneously. Just because the adder is not used simultaneously, the above solution with one adder does not have to be the optimum one. The synthesis tool will choose the optimum solution on the basis of the area and timing constraints set for the design. In some cases, synthesis may result in two adders with multiplexors on the right of the adders. As explained in Chapter 10, "RTL synthesis", the synthesis result in the advanced synthesis tools is controlled entirely by the area and timing constraints which the designer sets.

The synthesis tools at RT level will never be able to manage resource sharing for the adders in example 15.3.2. This is also shown by the synthesis result in Figure 15.6. On the other hand, a behavioural synthesis tool can manage resource sharing of the adders (see Chapter 17, "Behavioural synthesis).

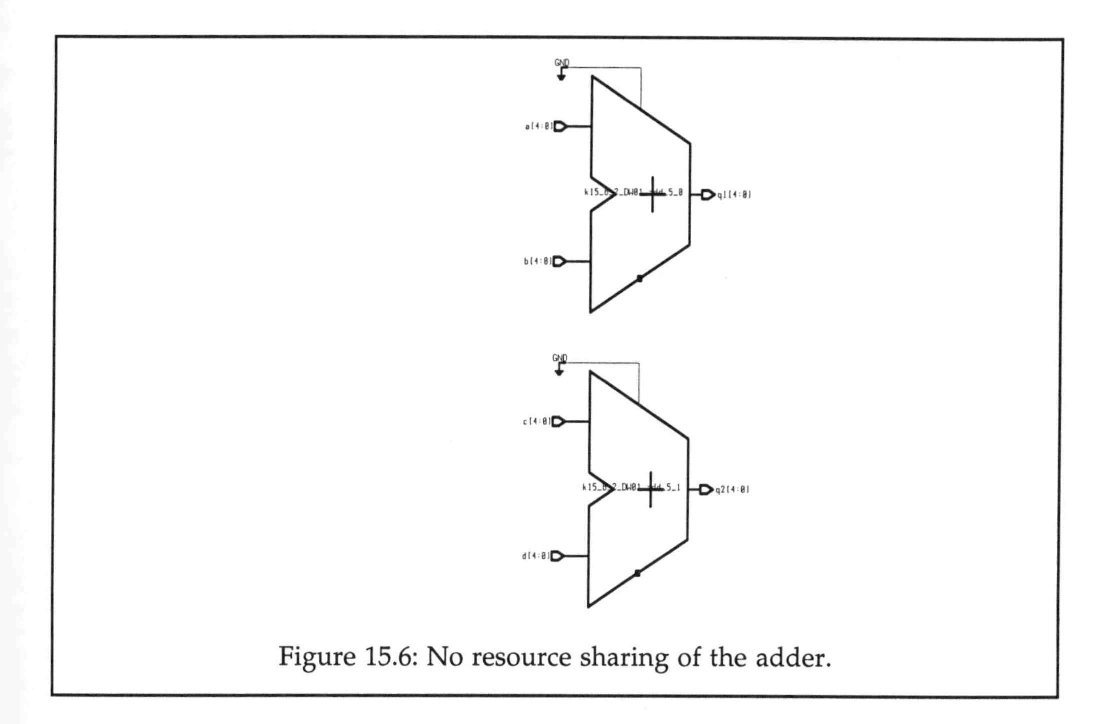

Figure 15.6: No resource sharing of the adder.

15.4. Comparators

Comparators are a useful component as it is often necessary to compare the sizes of vectors. The component, described below, compares two three-bit vectors and creates output signal *comp*. The function of output signal *comp* is determined with input signal *sel_f*.

sel_f	*Output*
00	a=b
01	a<b
10	a>b
11	output signal comp='0'

```
Library ieee;
Use ieee.std_logic_1164.ALL;
Use ieee.std_logic_unsigned.ALL;

Entity comp is
 port( a,b:   in   std_logic_vector(2 downto 0);
       sel_f: in   std_logic_vector(1 downto 0);
       q:     out  boolean);
end;
```

```
Architecture rtl of comp is
begin
  process(sel_f,a,b)
  begin
    case sel_f is
      when "00"     => q<= a=b;
      when "01"     => q<= a<b;
      when "10"     => q<= a>b;
      when others    => q<= false;
    end case;
  end process;
end;
```

The synthesis result is shown in Figure 15.7.

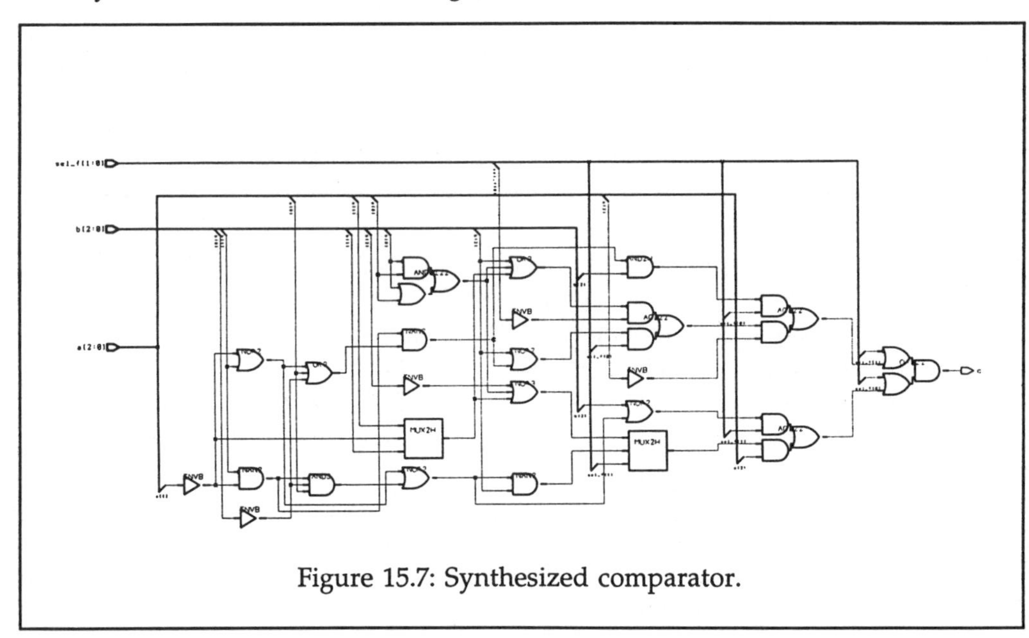

Figure 15.7: Synthesized comparator.

15.5. Multiplexors and decoders

15.5.1. Two-to-one multiplexor

There are many different ways of describing a 2-to-1 multiplexor, e.g. with **if-then-else** in a process, **case** in a process, a **with select** construction or with structural VHDL. The recommended method is to use a **when else** construction. This is because a 2-to-1 multiplexor is a very small structure and normally just one of many lines in a VHDL design. The multiplexor can be described in one line with a **when else** construction, making it easy to understand and read. For example:

```
Library ieee;
Use ieee.std_logic_1164.ALL;

Entity mux2 is
  port( a,b,sel:  in   std_logic;
        q:        out  std_logic);
end;

Architecture rtl of mux2 is
begin
  c<=a when sel='0' else b;
end;
```

15.5.2. Eight-to-one multiplexor

An 8-to-1 multiplexor cannot be described in a single line in the architecture. It is possible to use a **when else** construction, just as with the 2-to-1 multiplexor, but in order to make the VHDL design easier to understand, use of a **case** statement in a process is recommended. The result after synthesis is the same irrespective of which description method is used. For example:

```
Library ieee;
Use ieee.std_logic_1164.ALL;

Entity mux8 is
  port( a,b,c,d,e,f,g,h:  in    std_logic;
        sel:              in    std_logic_vector(2 downto 0);
        q:                out   std_logic);
end;

Architecture rtl of mux8 is
begin
  process(a,b,c,d,e,f,g,h,sel)
  begin
    case sel is
      when "000" => q<=a;
      when "001" => q<=b;
      when "010" => q<=c;
      when "011" => q<=d;
```

```
            when "100" => q<=e;
            when "101" => q<=f;
            when "110" => q<=g;
            when others => q<=h;
         end case;
      end process;
   end;
```

The synthesis result is shown in Figure 15.8.

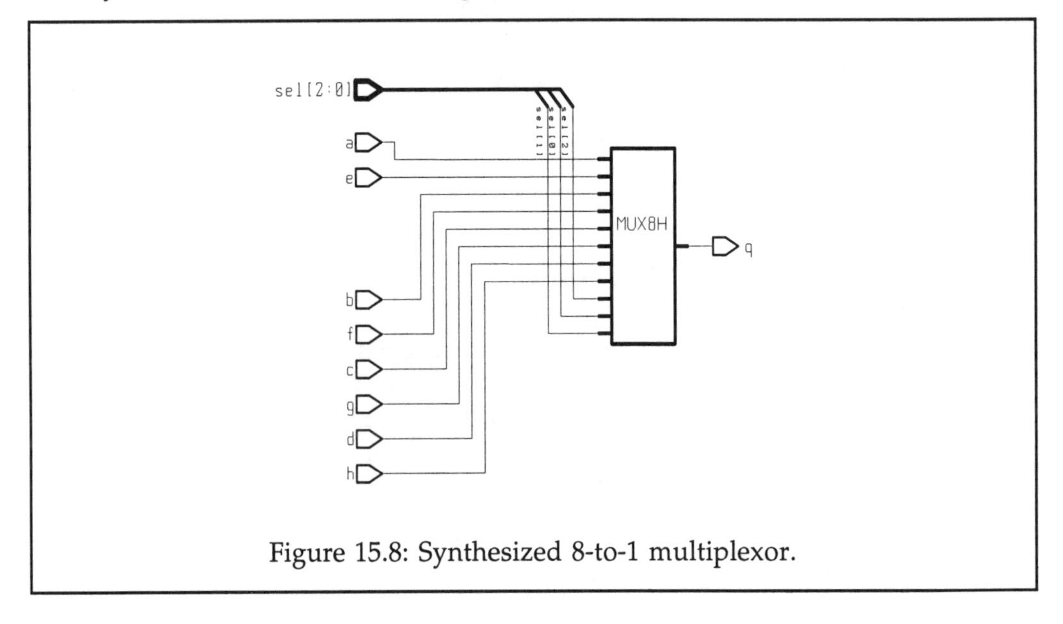

Figure 15.8: Synthesized 8-to-1 multiplexor.

15.5.3. Three-to-eight decoder

```
Library ieee;
Use ieee.std_logic_1164.ALL;
Use ieee.std_logic_unsigned.ALL;

Entity decoder is
 port( a,b,c,g1,g2_n: in    std_logic;
       y_n:           out   std_logic_vector(7 downto 0));
end;

Architecture rtl of decoder is
begin
  process(a,b,c,g1,g2_n)
  begin
```

```
        y_n<=(others=>'1');
        if g1='1' and g2_n='0' then
           y_n(conv_integer(c & b & a))<='0';
        end if;
     end process;
end;
```

The synthesis result is shown in Figure 15.9.

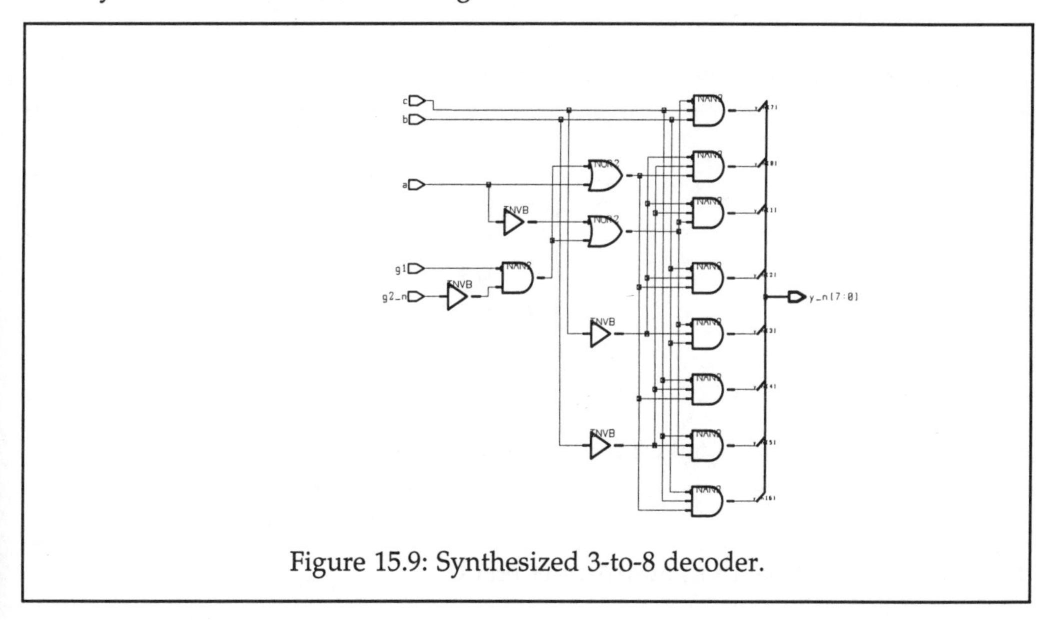

Figure 15.9: Synthesized 3-to-8 decoder.

15.6. Register

15.6.1. Flip-flop with asynchronous reset

```
Library ieee;
Use ieee.std_logic_1164.ALL;

Entity d_mem is
  port( clk,resetn,d_in:  in   std_logic;
        d_out:            out  std_logic);
end;

Architecture rtl of d_mem is
begin
```

```
    process(clk,resetn)
    begin
      if resetn='0' then
        d_out<='0';
      elsif clk'event and clk='1' then
        d_out<=d_in;
      end if;
    end process;
  end;
```

The synthesis result is shown in Figure 15.10.

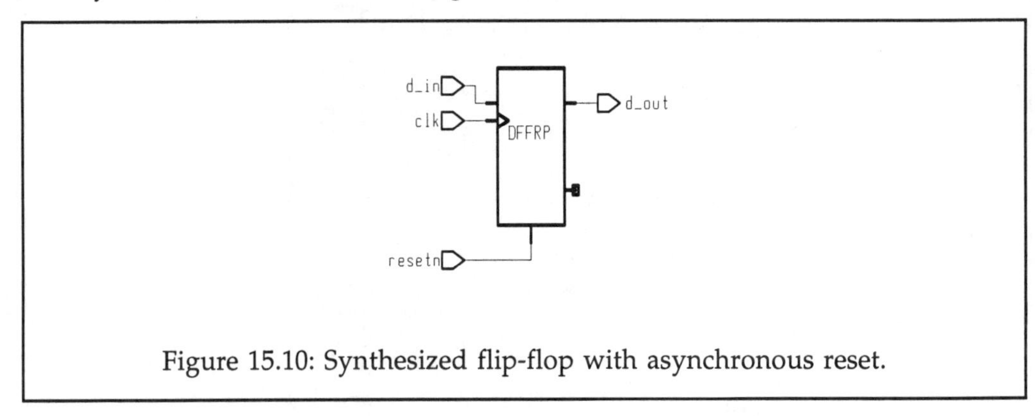

Figure 15.10: Synthesized flip-flop with asynchronous reset.

15.6.2. Flip-flop with synchronous reset

```
Library ieee;
Use ieee.std_logic_1164.ALL;

Entity d_mem is
  port( clk,resetn,d_in: in   std_logic;
        d_out:           out  std_logic);
end;

Architecture rtl of d_mem is
begin
  process(clk)
  begin
    if clk'event and clk='1' then
      if resetn='0' then
        d_out<='0';
      else
```

```
            d_out<=d_in;
         end if;
      end if;
   end process;
end;
```

The synthesis result is shown in Figure 15.11.

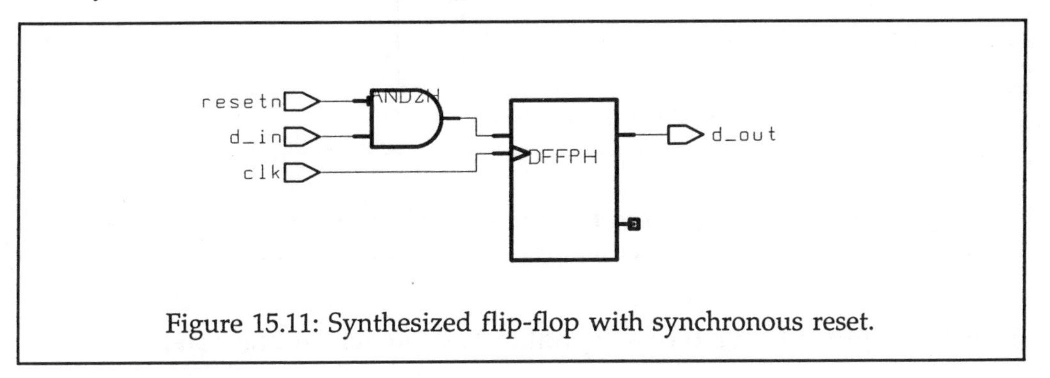

Figure 15.11: Synthesized flip-flop with synchronous reset.

15.6.3. Flip-flop with asynchronous reset and set

```
Library ieee;
Use ieee.std_logic_1164.ALL;
Entity d_mem is
   port( clk,resetn,presetn,d_in:  in     std_logic;
         d_out:                    out    std_logic);
end;
Architecture rtl of d_mem is
begin
   process(clk,resetn,presetn)
   begin
      if resetn='0' then
         d_out<='0';
      elsif presetn='0' then
         d_out<='1';
      elsif clk'event and clk='1' then
         d_out<=d_in;
      end if;
   end process;
end;
```

The synthesis result is shown in Figure 15.12.

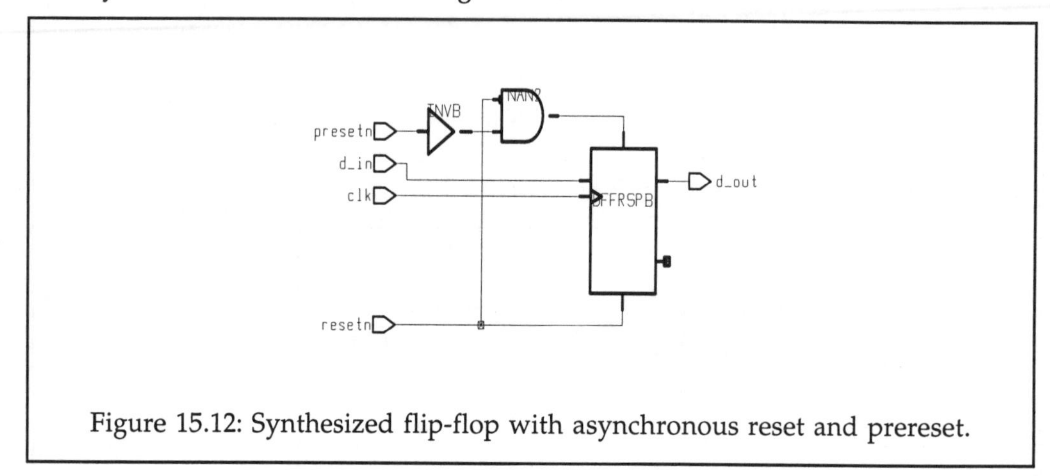

Figure 15.12: Synthesized flip-flop with asynchronous reset and prereset.

15.6.4. Eight-bit register with enable and asynchronous reset

```
Library ieee;
Use ieee.std_logic_1164.ALL;

Entity d_mem is
  port( clk,resetn,en:  in   std_logic;
        d_in:           in   std_logic_vector(7 downto 0);
        d_out:          out  std_logic_vector(7 downto 0));
end;

Architecture rtl of d_mem is
begin
  process(clk,resetn)
  begin
    if resetn='0' then
      d_out<=(others=>'0');

    elsif clk'event and clk='1' then
      if en='1' then
        d_out<=d_in;
      end if;
    end if;
  end process;
end;
```

The synthesis result is shown in Figure 15.13.

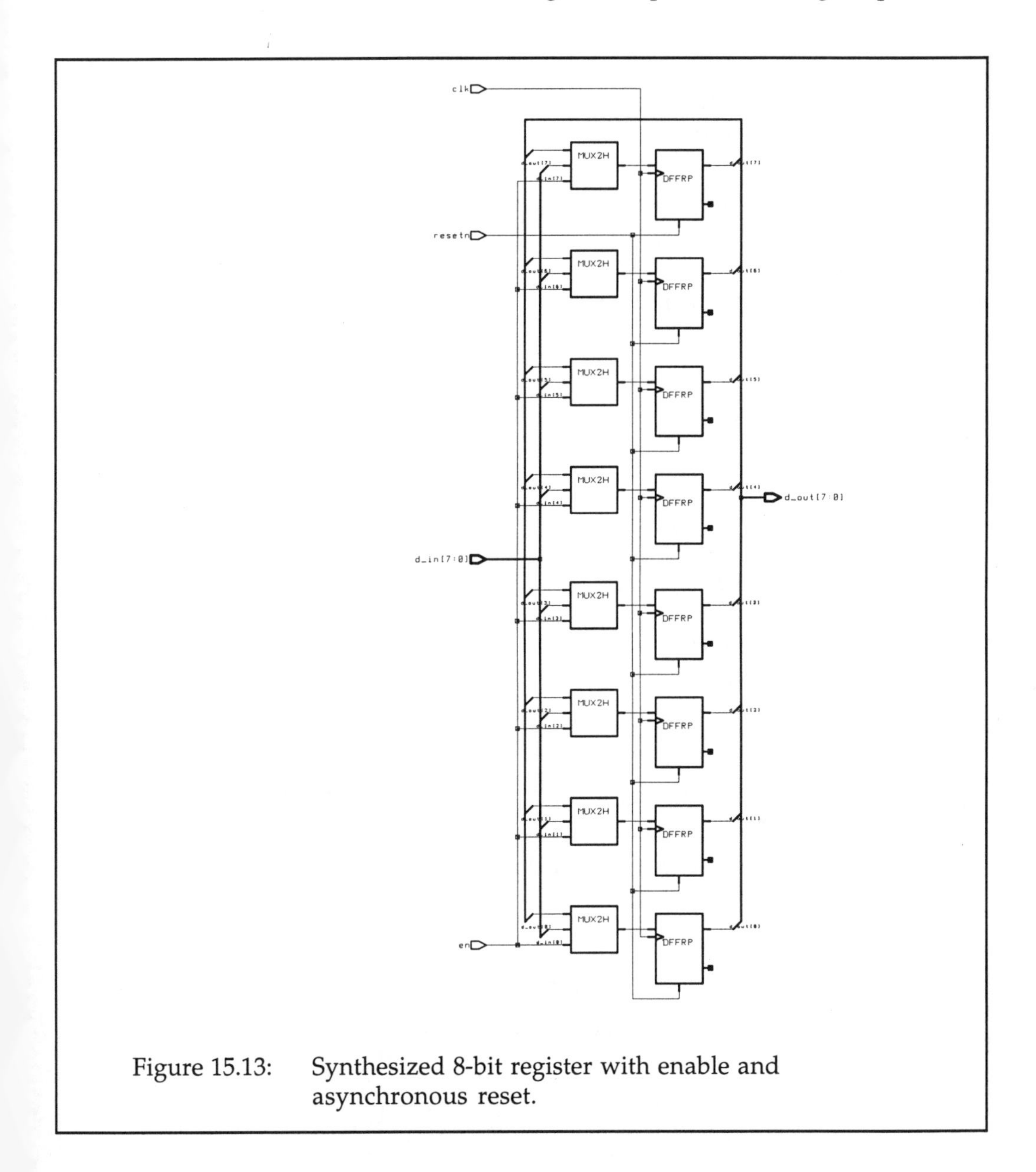

Figure 15.13: Synthesized 8-bit register with enable and asynchronous reset.

15.7. Edge-controlled pulse generator

The function of the following component is to generate a pulse for a clock when input din has a rising or falling edge. The choice of rising or falling edge is controlled by the pos_negn signal.

```
Entity puls_g is
  port( din,pos_negn,clk,resetn:  in   std_logic;
        puls:                     out  std_logic);
end;
Architecture rtl of puls_g is
signal din2:std:logic;
begin
  process(clk,resetn)
  begin
    if resetn='0' then
      din2<='0';
      puls<='0';
    elsif clk'event and clk='1' then
      din2<=din;
      if din=pos_negn and din2=not pos_negn then
        puls<='1';
      else
        puls<='0';
      end if;
    end if;
  end process;
end;
```

The synthesis result is shown in Figure 15.14.

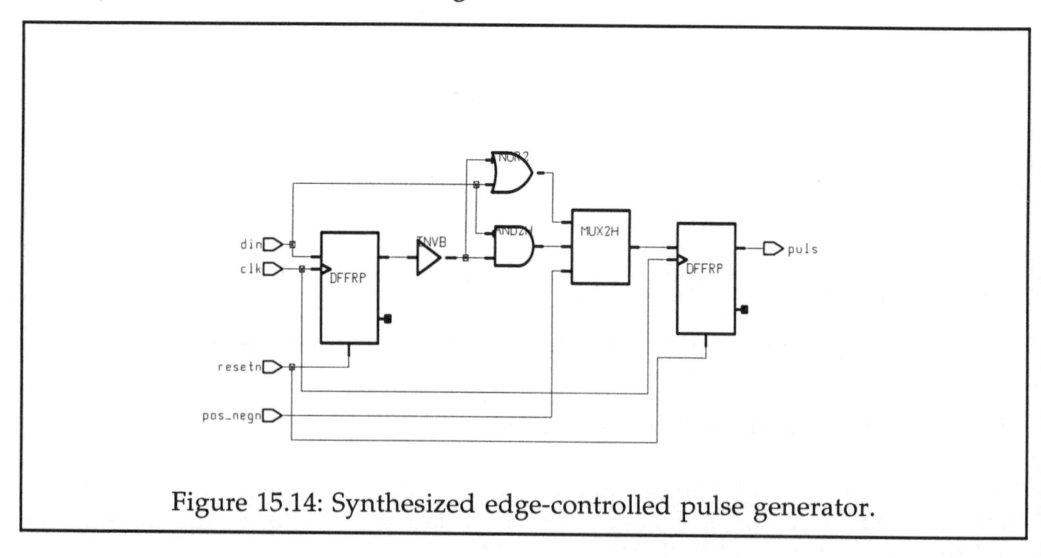

Figure 15.14: Synthesized edge-controlled pulse generator.

15.8. Counters

15.8.1. Three-bit counter with enable and carry out

If package ieee.std_logic_unsigned is being used, it is enough to use the "+" or "-" function when the counter value has to be increased or decreased. If the counter has been declared as a vector, the counter will change automatically when all the bits have the value '1' (in the case of "+"). If the counter is to stop at "111", for example, the value has to be tested by the counter before adding with +1. Below are two different architectures: one without a limit check and one with. The synthesis result is similar in the two cases, as the **if** statement which checks the limit does not add any new functionality. From the documentation point of view, the example with limit check is slightly easier to understand, but there are slightly more lines to write, increasing the risk of error. If the counter has been declared as type integer, there must always be a limit check, otherwise an error will occur in the simulator when count=7 and +1 is executed.

```
Library ieee;
Use ieee.std_logic_1164.ALL;
Use ieee.std_logic_unsigned.ALL;

Entity count3 is
  port( clk,resetn,count_en: in   std_logic;
        sum:                  out std_logic_vector(2 downto 0);
        cout:                out  std_logic);
end;

Architecture rtl of count3 is
signal count:std_logic_vector(2 downto 0);
begin
   process(clk,resetn)
   begin
      if resetn='0' then
         count<=(others=>'0');
      elsif clk'event and clk='1' then
         if count_en='1' then
            count<=count+1;
         end if;
      end if;
   end process;

   sum<=count;
   cout<='1' when count=7 and count_en='1' else '0';
end;
```

```
Architecture rtl of count3 is
signal count:std_logic_vector(2 downto 0);
begin
  process(clk,resetn)
  begin
    if resetn='0' then
      count<=(others=>'0');
    elsif clk'event and clk='1' then
      if count_en='1' then
        if count/=7 then
          count<=count+1;
        else
          count<=(others=>'0');
        end if;
      end if;
    end if;
  end process;
  sum<=count;
  cout<='1' when count=7 and count_en='1' else '0';
  end;
```

The synthesis result is shown in Figure 15.15.

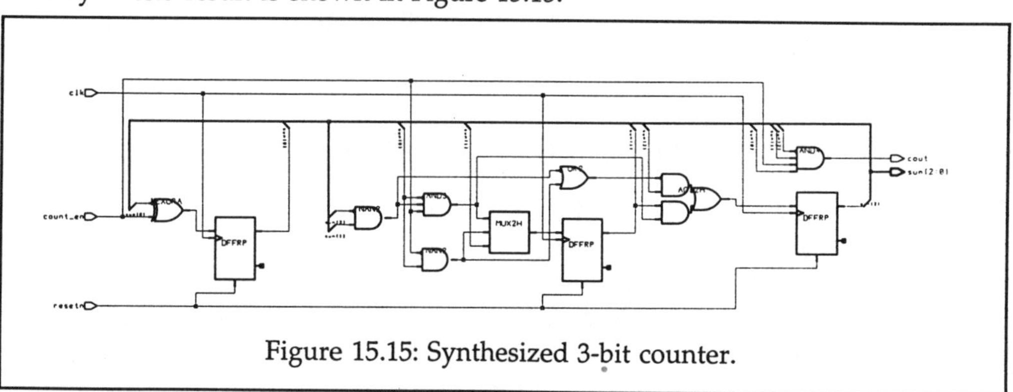

Figure 15.15: Synthesized 3-bit counter.

15.8.2. Three-bit up/down counter

The following counter counts up if input signal up is equal to 1, otherwise the counter counts down. The choice between using **case** or **if** statements normally has no effect on the synthesized result, so the decision should normally be made on the basis of readability.

```
Library ieee;
Use ieee.std_logic_1164.ALL;
Use ieee.std_logic_unsigned.ALL;

Entity up_down is
 port( clk,resetn:     in   std_logic;
       count_en,up:    in   std_logic;
       sum:            out  std_logic_vector(2 downto 0);
       cout:           out  std_logic);
end;

Architecture rtl of up_down is
signal count:syd_logic_vector(2 downto 0);
begin
   process(clk,resetn)
   begin
     if resetn='0' then
       count<=(others=>'0');
     elsif clk'event and clk='1' then
       if count_en='1' then
         case up is
           when '1'         => count<=count+1;
           when others      => count<=count-1;
         end case;
       end if;
     end if;
   end process;
   sum<=count;
 cout<='1' when count_en='1' and ((up='1' and count=7) or
                 (up='0' and count=0)) else '0';
end;
```

The synthesis result is shown in Figure 15.16.

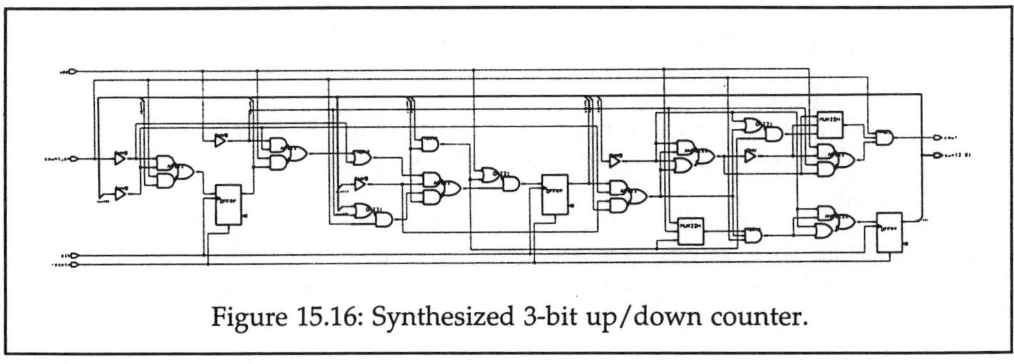

Figure 15.16: Synthesized 3-bit up/down counter.

15.8.3. Parallel loadable generic up/down counter

```
Library ieee;
Use ieee.std_logic_1164.ALL;
Use ieee.std_logic_unsigned.ALL;

Entity count_par is
 generic (N:positive:=3);
 port( clk,resetn: in   std_logic;
       count_en:   in   std_logic;
       load,up:    in   std_logic;
       data_in:    in   std_logic_vector(N-1 downto 0);
       sum:        out  std_logic_vector(N-1 downto 0);
       cout:       out  std_logic);
end;

Architecture rtl of count_par is
signal count:std_logic_vector(N-1 downto 0);
begin
  process(clk,resetn)
  begin
    if resetn='0' then
      count<=(others=>'0');
    elsif clk'event and clk='1' then
      if load='1' then
        count<=data_in;

      elsif count_en='1' then
        case up is
          when '1' =>       count<=count+1;
          when others =>    count<=count-1;
        end case;
      end if;
    end if;
  end process;

  sum<=count;
  cout<='1' when count_en='1' and ((up='1' and count=(2**N)-1)
        or (up='0' and count=0)) else '0';
end;
```

The synthesis result is shown in Figure 15.17.

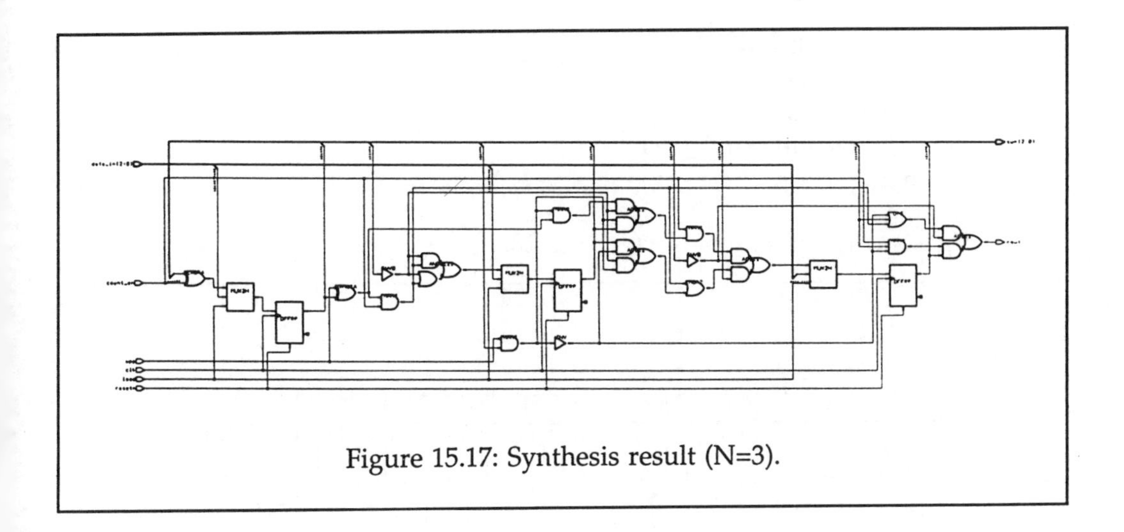

Figure 15.17: Synthesis result (N=3).

15.9. Shift register

15.9.1. Four-bit shift register with serial input data and parallel output data

```
Library ieee;
Use ieee.std_logic_1164.ALL;
Use ieee.std_logic_unsigned.ALL;

Entity shift_r is
 port( clk,resetn,d_in:  in       std_logic;
       shift_en:         in       std_logic;
       shift_out:        out  std_logic_vector(3 downto 0));
end;

Architecture rtl of shift_r is
signal shift_reg:std_logic_vector(3 downto 0);
begin
  process(clk,resetn)
  begin
    if resetn='0' then
       shift_reg<=(others=>'0');
    elsif clk'event and clk='1' then
       if shift_en='1' then
          shift_reg<=shl(shift_reg,"1");
          shift_reg(0)<=d_in;
```

```
                end if;
            end if;
        end process;
        shift_out<=shift_reg;
    end;
```

The synthesis result is shown in Figure 15.18.

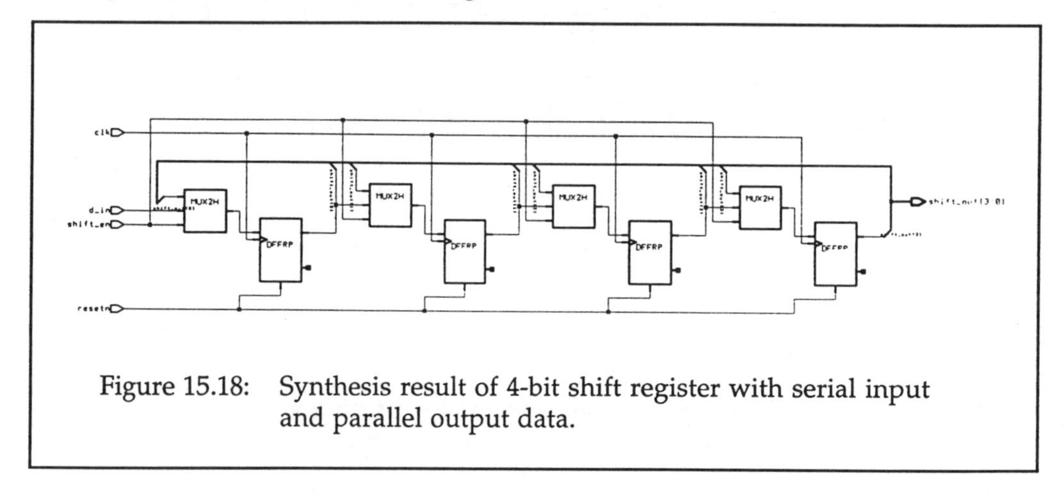

Figure 15.18: Synthesis result of 4-bit shift register with serial input and parallel output data.

15.9.2. Four-bit shift register with parallel load and serial output

```
Library ieee;
Use ieee.std_logic_1164.ALL;
Use ieee.std_logic_unsigned.ALL;

Entity shift_p is
 port( clk,resetn:    in   std_logic;
       shift_en,load: in   std_logic;
       d_in:          in   std_logic_vector(3 downto 0);
       shift_out:     out  std_logic);
end;

Architecture rtl of shift_p is
signal shift_reg:std_logic_vector(3 downto 0);
begin
  process(clk,resetn)
  begin
     if resetn='0' then
        shift_reg<=(others=>'0');
```

```
            elsif clk'event and clk='1' then
               if load='1' then
                  shift_reg<=d_in;
               elsif shift_en='1' then
                  shift_reg<=shl(shift_reg,"1");
                  shift_reg(0)<='0';
               end if;
            end if;
         end process;
         shift_out<=shift_reg(3);
      end;
```

The synthesis result is shown in Figure 15.19.

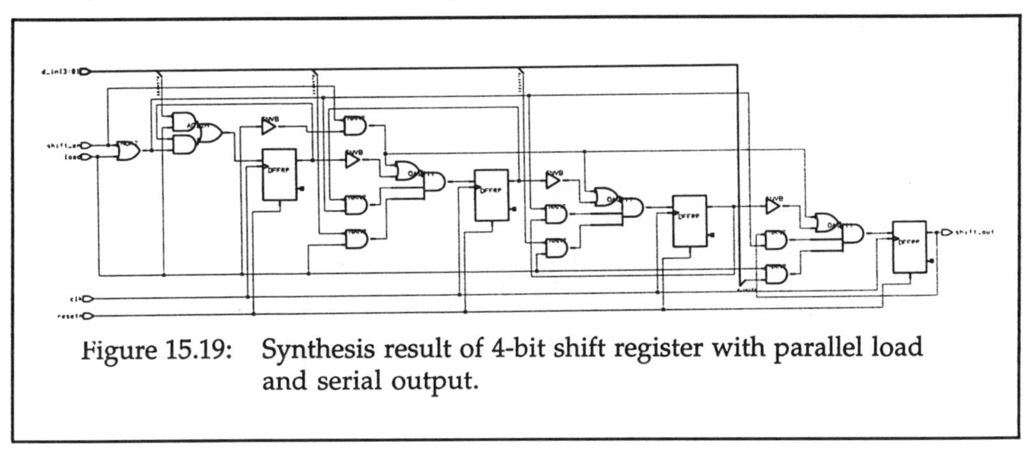

Figure 15.19: Synthesis result of 4-bit shift register with parallel load and serial output.

15.10. Filters

Below are some examples of elementary filters.

15.10.1. Four-input digital majority-voting filter

The filtered inputs are equal to the value sampled the most times in the previous three samples, i.e. if (d_in(N-2) + d_in(N-1) + d_in(N))>=2, then d_filtr='1', otherwise the filtered value will be '0'.

```
Library ieee;
Use ieee.std_logic_1164.ALL;
Use ieee.std_logic_unsigned.ALL;
```

```
Entity majority is
 port( clk,resetn:  in     std_logic;
       d_in:        in     std_logic_vector(3 downto 0);
       d_filtr:     out    std_logic_vector( 3 downto 0));
end;

Architecture rtl of majority is
type array_4 is array (3 downto 0) of
  std_logic_vector(2 downto 0);

signal shift_data: array_4;
signal d_filtr_i:std_logic_vector(3 downto 0);
begin
  process(clk,resetn)
  begin
    if resetn='0' then
      for i in 0 to 3 loop
        shift_data(i)<=(others=>'0');
      end loop;
    elsif clk'event and clk='1' then
      for i in 0 to 3 loop
        shift_data(i)<=shl(shift_data(i),"1");
        shift_data(i)(0)<=d_in(i);
      end loop;
    end if;
  end process;

  process(shift_data)
  type array_42 is array (3 downto 0) of
    std_logic_vector(1 downto 0);

  variable count: array_42;
  begin
    for i in 0 to 3 loop
      count(i):=shift_data(i)(2) + ('0' & shift_data(i)(1)) +
                shift_data(i)(0);
      if count(i)>=2 then
        d_filtr_i(i)<='1';
      else
        d_filtr_i(i)<='0';
      end if;
    end loop;
  end process;

  process(clk,resetn);
```

```
begin
  if resetn='0' then
    d_filtr<=(others=>'0');
  elsif clk'event and clk='1' then
    d_filtr<=d_filtr_i;
  end if;
end process;
end;
```

The synthesis result is shown in Figure 15.20.

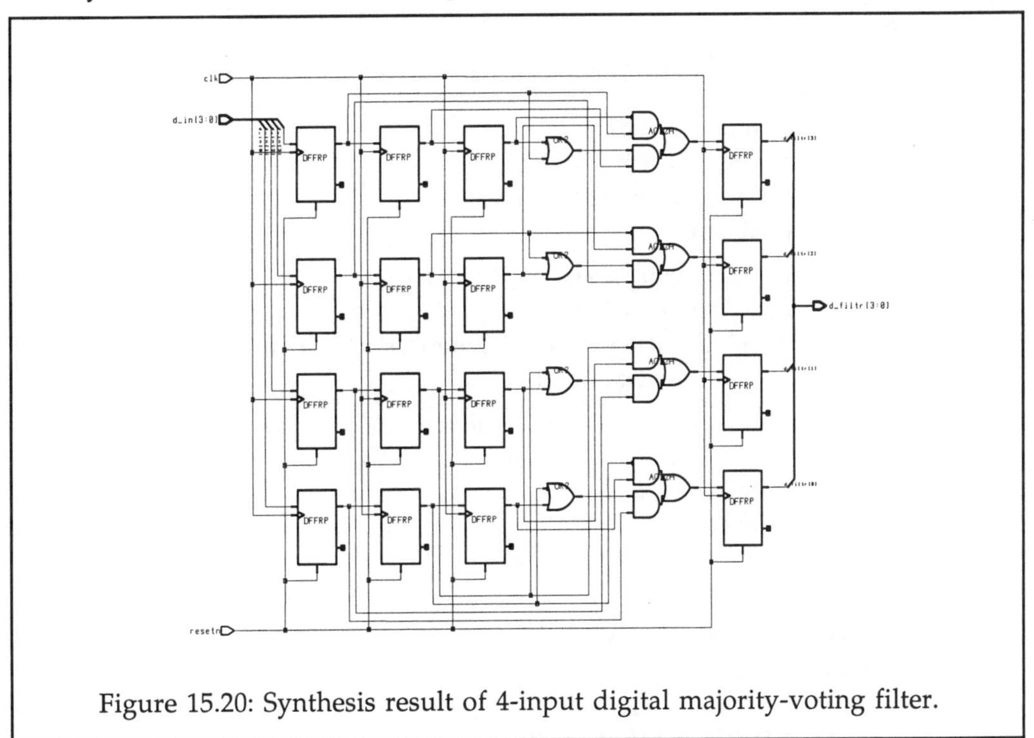

Figure 15.20: Synthesis result of 4-input digital majority-voting filter.

15.10.2. Four-input digital addition filter

In the following filter the inputs do not change over until the internal counter has counted up to 3 or down to 0. The counter adds 1 when d_in=1 and subtracts 1 when d_in=0.

```
Library ieee;
Use ieee.std_logic_1164.ALL;
Use ieee.std_logic_unsigned.ALL;
```

```
Entity add_filter is
 port( clk,resetn:  in    std_logic;
       d_in:        in    std_logic_vector(3 downto 0);
       d_filtr:     out   std_logic_vector(3 downto 0);
end;

Architecture rtl of add_filter is
type array_4 is array (3 downto 0) of
    std_logic_vector(1 downto 0);
signal count: array_4;
begin
  process(clk,resetn)
  begin
    if resetn='0' then
      for i in 0 to 3 loop
        count(i)<=(others=>'0');
      end loop;
    elsif clk'event and clk='1' then
      for i in 0 to 3 loop
        if d_in(i)='1' and count(i)/=3 then
          count<=count+1;
        elsif d_in(i)='0' and count(i)/=0 then
          count<=count-1;
        end if;
      end loop;
    end if;
  end process;

  process(clk,resetn)
  begin
    if resetn='0' then
      d_filtr<=(others=>'0');
    elsif clk'event and clk='1' then
      for i in 0 to 3 loop
        if count(i)=3 then
          d_filtr(i)<='1';
        elsif count(i)=0 then
          d_filtr(i)<='0';
        end if;
      end loop;
    end if;
```

```
    end process;
end;
```

The synthesis result is shown in Figure 15.21.

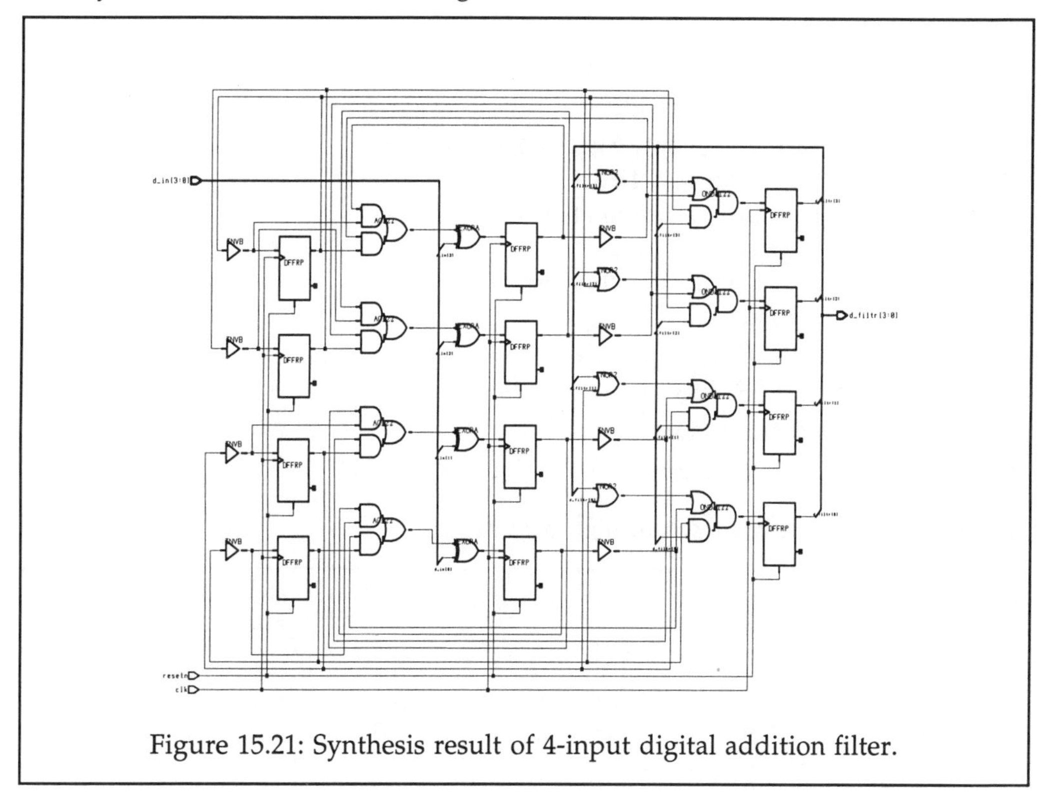

Figure 15.21: Synthesis result of 4-input digital addition filter.

15.11. Frequency dividers

The following component divides the clock frequency by 3. Output signal clk3 is directly from a register (i.e. spike-free), so it should be possible to use it as a clock in other parts of the design. If another frequency is required, the example is easy to modify.

```
Library ieee;
Use ieee.std_logic_1164.ALL;
Use ieee.std_logic_unsigned.ALL;

Entity div3 is
 port( clk,resetn: in    std_logic;
       clk3:       out   std_logic);
end;
```

```
Architecture rtl of div3 is
signal count:std_logic_vector(1 downto 0);
begin
  process(clk,resetn)
  begin
    if resetn='0' then
      count<=(others=>'0');
    elsif clk'event and clk='1' then
      if count<2 then
        count<=count+1;
      else
        count<=(others=>'0');
      end if;
    end if;
  end process;
  clk3<=count(1);
end;
```

The synthesis result is shown in Figure 15.22.

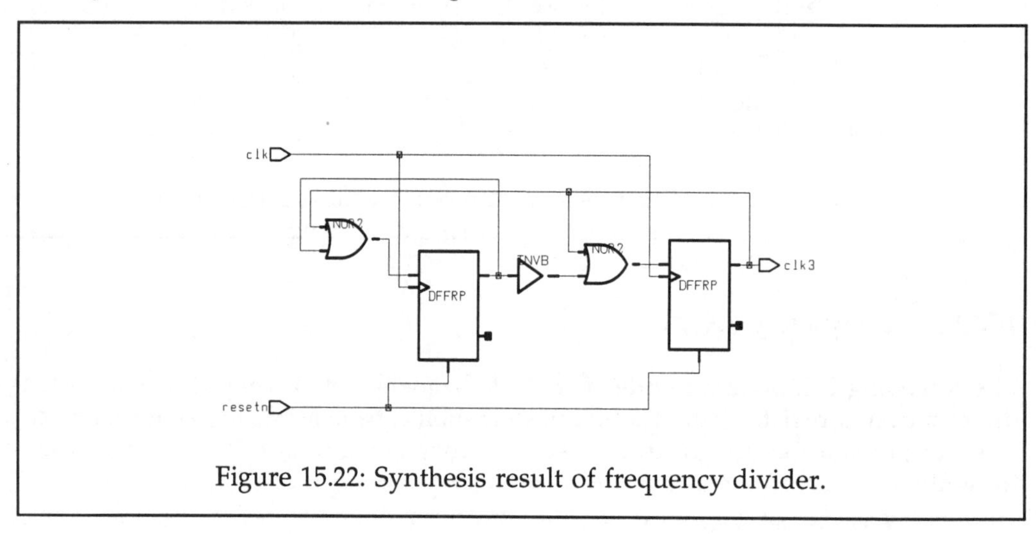

Figure 15.22: Synthesis result of frequency divider.

16
Development tools

There are now many suppliers of VHDL tools, primarily VHDL simulators. As VHDL is an IEEE standard, all VHDL simulators must be able to cope with the entire standard. What makes the VHDL simulators different from each other is their compilation and simulation speeds, plus integration with other tools. When it comes to synthesis, there is no standard which defines what subset of the VHDL standard should be synthesizable. The language constructions which are supported for synthesis vary widely from tool to tool. As far as synthesis is concerned, it is much more than just what subset of VHDL can be synthesized that is important (see Chapter 10, "RTL synthesis"). The optimization result, with regard to both area and speed, for example, differs greatly from tool to tool. As things stand, the market is relatively sound, i.e. the more you pay for a synthesis tool, the better the tool is. When it comes to synthesis for ASICs, Synopsys has been the market leader in the 1990s in terms of both number of licences and performance. Mentor Graphics, among others, is now trying to break this market dominance by making a synthesis tool which accepts the same subset (99 per cent) of VHDL code as Synopsys. This is very good from the user's point of view, as VHDL code can then be moved between these leading synthesis tools.

Several complex tools are required to be able to design ASICs/FPGAs quickly and efficiently, e.g. VHDL simulator, synthesis tool, timing analyser and ATPG tool. Apart from Synopsys and Mentor Graphics, there are several suppliers who offer the whole product range. Many tools can also work together with parts of other suppliers' tools. To illustrate present-day development tools, there follows a brief presentation of parts of the Synopsys product range.

16.1. Synopsys

Synopsys synthesis tools have been and still are the the market leaders. Synopsys offers a lot more than just synthesis tools: its tools cover virtually the entire area of ASIC design. Figure 16.1 shows its product range and how the products are related.

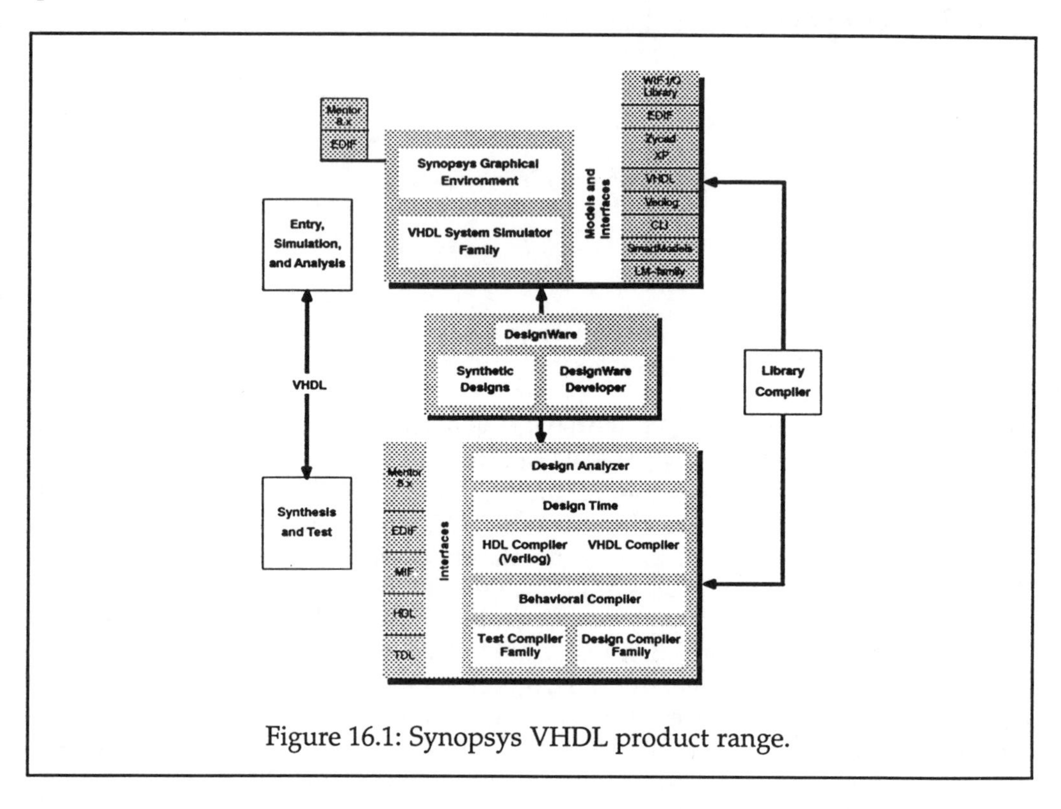

Figure 16.1: Synopsys VHDL product range.

16.1.1. VHDL Compiler and Design Analyzer

The Synopsys VHDL Compiler translates VHDL code to a generic schematic. The VHDL Compiler supports a large part of the VHDL standard (IEEE-1076) for synthesis. When translating from VHDL to the generic schematic, the VHDL Compiler identifies flip-flops, latches, multiplexors, adders, etc. On the basis of the generic schematic, the Design Compiler then optimizes the schematic to a chosen technology. For optimization to produce a good result, constraints (timing requirements, etc.) and design requirements have to be set. Those most commonly used in Synopsys are (V3.3a):

Clock frequency
Multi clock cycles
Clock skew and clock uncertainty
Maximum input delay
Maximum output delay
Maximum delay

Minimum delay
Point to point timing
Maximum power (ECL designs)
Maximum transition time
Maximum fan-out
Maximum area
Driving strength
Load
Wire load
Operating conditions
False paths

The above constraints can be set either via a script, directly on the command line or graphically in the Design Analyzer.

Setting up design requirements may seem like a big job. This is not usually the case, however. If, for example, you have a fully clocked design (which is highly recommended), you only have to give the command create_clock -period 25 clk to indicate that all the flip-flops in this design are clocked at 40 MHz.

The Design Analyzer is the Synopsys graphics tool. From the graphical environment it is possible, among other things, to use such tools as VHDL Compiler, Design Compiler, FPGA Compiler, Test Compiler, Test Compiler plus, Design Ware and Design Time.

Setting up a complete constraint picture for a design allows the synthesis tool to do a good job. The design is also easier to analyse after optimization. Synopsys has a static timing analyser: Design Time. Using this it is possible to generate report files which say whether the timing constraints have been met, for example. It is also possible to generate reports which show whether other constraints have been met. If there is a critical data path through the design which does not meet the timing constraints, Synopsys can be asked to mark it in the schematic. This often makes the problem easy to identify. It is also virtually always possible to click on generated report files from Synopsys and then view the causes graphically.

It is also possible to see the link between the VHDL code and the schematic by marking either a line of VHDL or a gate. If, for example, a line of VHDL is marked, Synopsys will mark which gates the line has produced and vice versa.

16.1.2. Design Ware

Design Ware is a Synopsys tool which, to put it simply, contains a collection of libraries of different implementations of components. Synopsys currently (V3.1a) has five Design Ware libraries:

ALU family
Advanced math family
Basic sequential family
Fault tolerant family
Control logic family

Examples of the components included are adders, subtracters, multipliers and multiplexors. It is possible to develop your own Design Ware components in addition to these libraries. Many manufacturers also have their own Design Ware components which are included in their design package.

The VHDL Compiler identifies the Design Ware components, e.g. an adder, during synthesis. This adder can be implemented in different ways. Design Ware will test implementation of all the possible adders in one of the Design Ware libraries. Depending on which design requirements have been set, Synopsys then chooses the adder which takes up the least area while meeting the set timing constraints.

Example:

```
Library ieee;
Use ieee.std_logic_1164.ALL;
Use ieee.std_logic_unsigned.ALL;

Entity add is
  port(  a,b: in   std_logic_vector(4 downto 0);
         q:   out  std_logic_vector(4 downto 0));
end;

Architecture rtl of add is
begin
  q<=a + b;
end;
```

Suppose, first, that you have a design for which the following design requirements and attributes have been set in Synopsys:

```
Set_max_delay 5 -from a -to q
Set_max_delay 5 -from b -to q
Set_max_area 200
Set_operating_conditions "WCCOM" -library HDCM
Set_wired_load "1k" -library HDCM -mode segmented
```

As the timing constraint of 5 ns is much tougher than the area constraint of 200 gates, Synopsys will choose to implement the adder in the fastest way, e.g. a fast carry look-ahead adder (Figure 16.2).

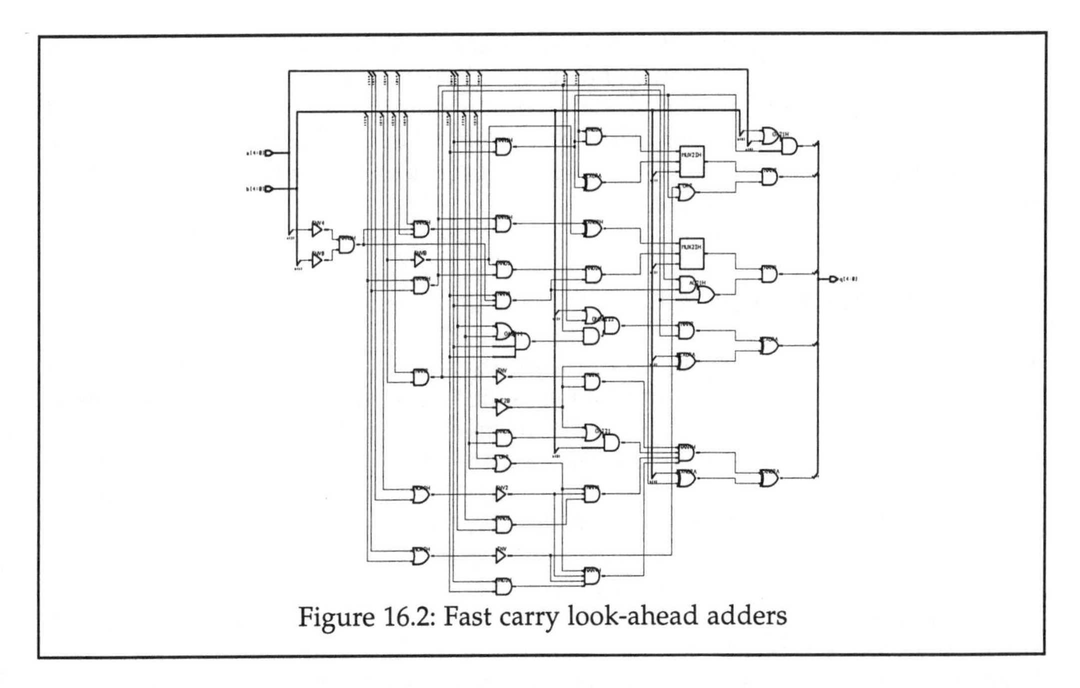

Figure 16.2: Fast carry look-ahead adders

If, instead, you set up the design requirements so that Synopsys optimizes primarily on the basis of area, Synopsys will choose a slower adder which takes up far less area, e.g. a ripple carry adder (Figure 16.3).

Design requirements/attributes set in Synopsys are:

Set_max_delay **15** -from {a,b} -to q
Set_max_area **50**
Set_operating_conditions "WCCOM" -library HDCM
Set_wired_load "1k" -library HDCM -mode segmented

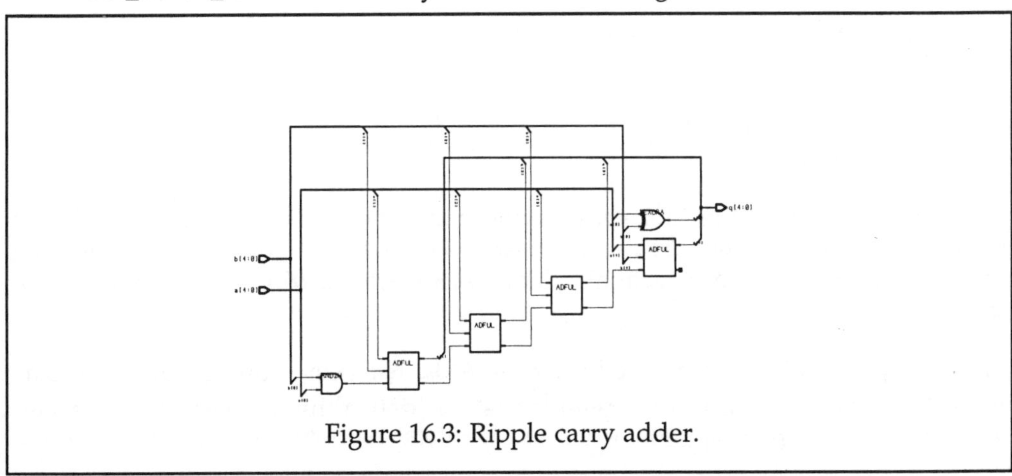

Figure 16.3: Ripple carry adder.

The fast carry look-ahead adder is about twice as large and twice as fast as the ripple carry adder. This example shows the different synthesis results which can be obtained depending on what design requirements are set for the synthesis tool. Note that both the adders are fully compatible from a functional point of view. This must be the case, as the same VHDL code has been used as the point of departure in both the examples.

16.1.3. Design Compiler

Design Compiler is the Synopsys optimization tool. Optimization is normally controlled by the design requirements which have been set. Based on these requirements, the Design Compiler tries to meet all the timing constraints in the smallest possible area. The design requirements are of differing priority: design rules have the highest priority, e.g. maximum transition and maximum fan-out, followed by the timing constraints and, finally, the area constraints.

During optimization, Synopsys may share resources among adders, for example (see Chapter 15, "Design examples and design tips"). Synopsys makes the decision to share a resource or not on the basis of the design requirements which have been set. The structure which best meets the requirements will be selected.

The implementation of Design Ware components is also chosen during optimization.

During optimization it is possible to specify several parameters:

Incremental mapping
Prioritize minimum path
Only design rule optimization
No design rule
Map effort (low, medium or high)
In-place optimization
Ungroup all
Boundary optimization
Verify design

Mapping means that the tool tries to find components from the chosen library which either take up less area or are faster. The tool does not examine just one component at a time, but normally looks at several nearby components at the same time.

In-place optimization can be used after the ASIC has been "routed" and problems with the driving strength, for example, have arisen. With in-place optimization, Synopsys can swap components with an identical footprint in order to solve the problem. Just swapping to components with the same "footprint" means that the ASIC does not have to be "rerouted", with the component merely having to be changed in the layout tool.

Apart from the above optimization parameters it is also possible to choose different algorithms during optimization, depending on whether the design is area- or timing-critical. An acceptable result is normally obtained with the default setting on the tool.

The Design Compiler also includes a state machine optimizer which can perform state coding, state minimization and area optimization for state machines (see Chapter 9, "State machines").

As Synopsys is the leading tool in this area, virtually all ASIC suppliers of any importance support Synopsys Design Compiler for synthesis.

Synopsys can also read in an ordinary schematic for optimization or convert the schematic to another technology. This means that technology can be changed smoothly and easily even if you do not have access to a VHDL description.

16.1.4. ATPG tools

Synopsys has two ATPG tools: Test Compiler and Test Compiler plus. Test Compiler is a full scan ATPG tool, while Test Compiler plus is a partial scan ATPG tool (see Chapter 12, "Test Methodology").

As the various ASIC suppliers support different test vector formats, Synopsys can convert the test vectors to different test vector formats.

In addition to generating test vectors, it is also possible to perform boundary scan synthesis. If you mark the in/out pads for your design, Test Compiler can then insert boundary scan cells which comply with the IEEE-1179.1 standard.

Both tools are well integrated with the other tools from Synopsys.

16.1.5. FPGA Compiler

The FPGA Compiler is the Synopsys tool for optimizing to an FPGA, and most FPGAs can be designed with the tool. However, Synopsys has a five-year co-operation agreement with Xilinx. This means that the FPGA Compiler can both optimize to Xilinx CLB (Configurable Logic Block) structure and at the same time optimize its timing.

The FPGA Compiler also supports the Xilinx X-Blox library. The X-Blox library is regarded by the FPGA Compiler as an extra Design Ware library. During synthesis to Xilinx 4000, the FPGA Compiler will investigate whether it should choose a component from the X-Blox library to serve either CLBs or timing.

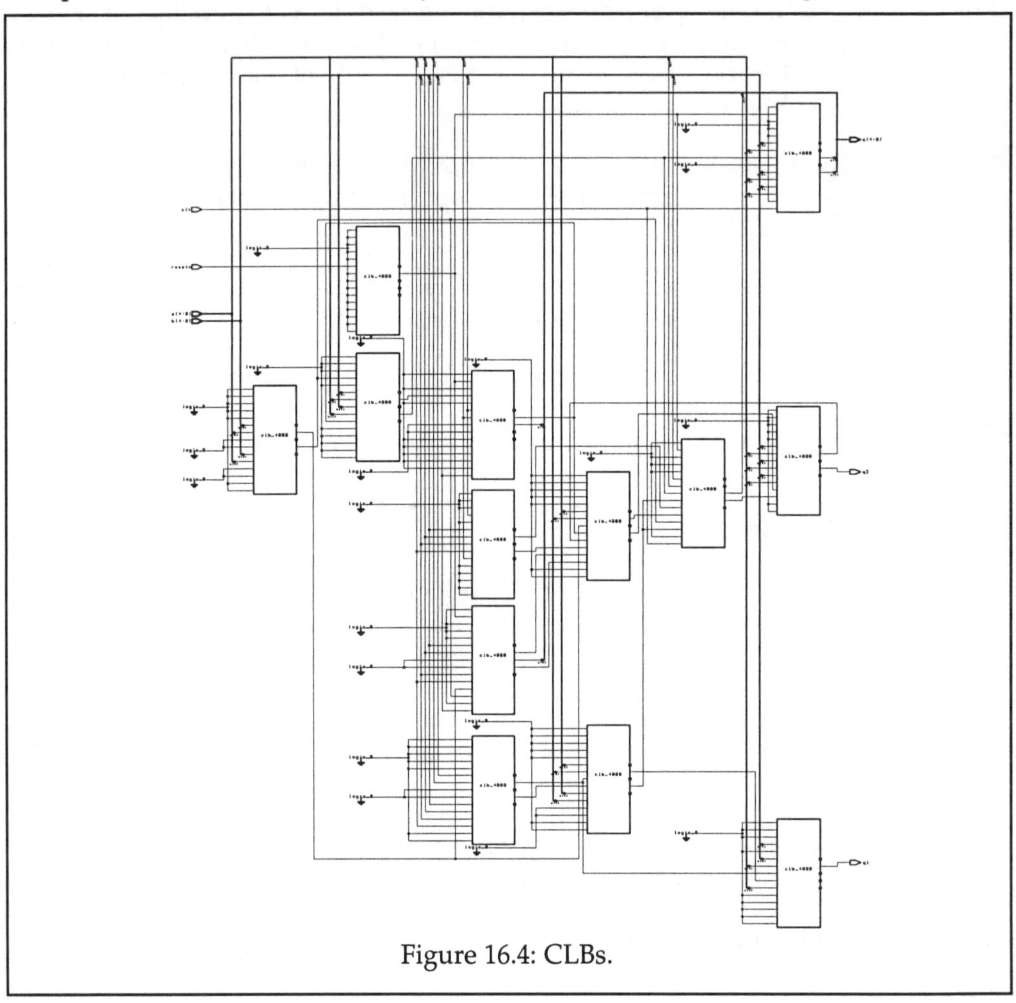

Figure 16.4: CLBs.

Figure 16.4 shows how the FPGA Compiler can map to Xilinx CLBs. It is already possible at this stage to analyse the design's timing. Synopsys will then estimate what delay will be obtained after routing. This estimate is normally about 10 per cent away from the real delay.

It is also possible to look at what the various CLBs contain in terms of logic. Both the equations and a graphical image can be viewed (Figure 16.5).

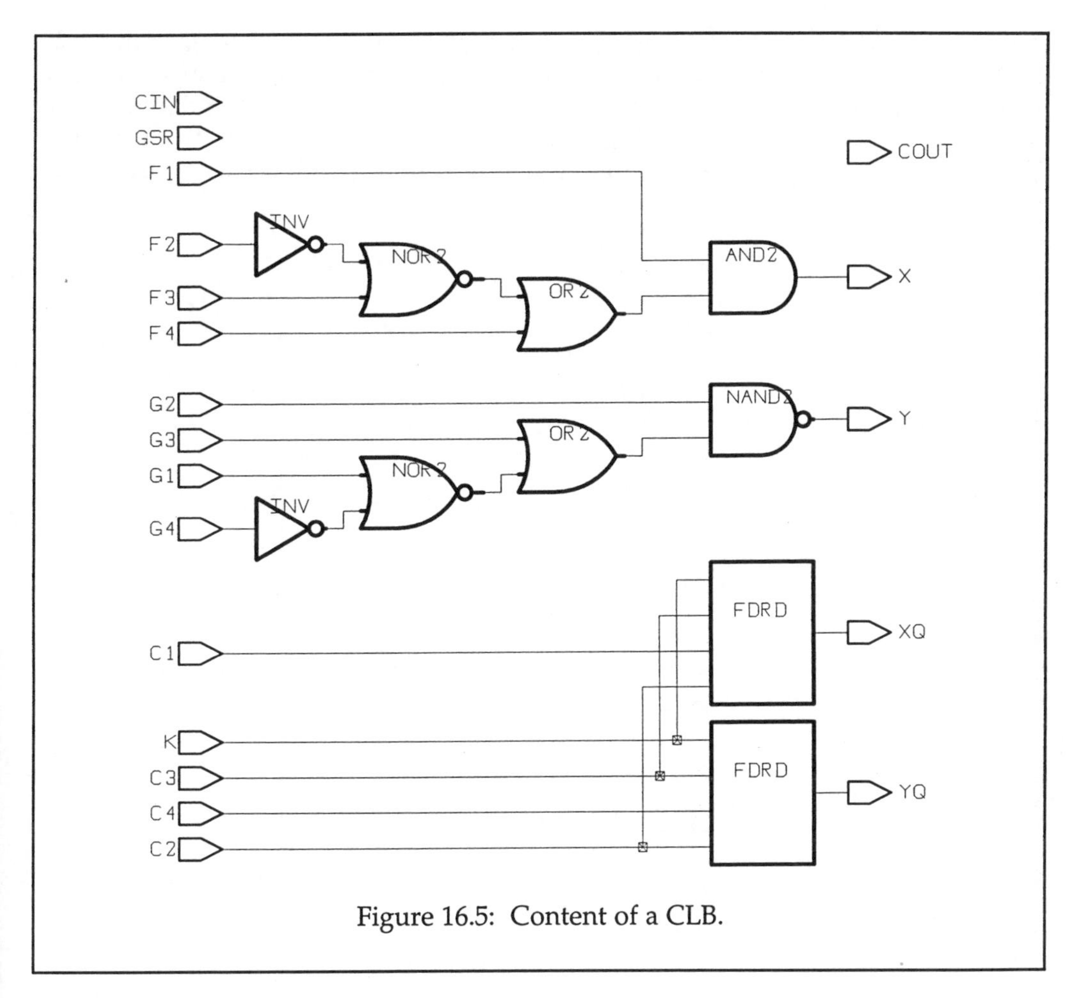

Figure 16.5: Content of a CLB.

The FPGA Compiler can also print out an XNF netlist containing the set timing constraints. Xilinx can then use these for routing, i.e. timing driven routing. The FPGA Compiler can also read in an XNF netlist for back annotation, optimization or conversion to a new technology.

16.1.6. VHDL simulator

The Synopsys VSS simulator is a VHDL simulator which supports the VHDL-87 standard in its entirety. VSS is one of the fastest VHDL simulators on the market. Several ASIC manufacturers have also started to support what is known as sign-off from VSS. This means that it is possible both to simulate the VHDL code and to verify the circuit timing at gate level in the same simulator.

To cope with the various simulation tasks, VSS has three different "motors" built into it: interpreting, compiling and gate level. The different motors are not visible to the user, but VSS uses the most suitable for each simulation. This makes it possible to achieve high performance in simulation from VHDL behavioural level right down to timing simulation at gate level using the same simulator.

17
Behavioural synthesis

17.1. Introduction

In most cases, finding a good design entails an iterative process. Working at as high a level of abstraction as possible makes it simpler to iterate, resulting in a more reliable analysis. Behavioural synthesis provides estimates of the area, speed and cost of an ASIC.

The other important aspect is that behavioural synthesis will compete with machine code for CPUs in future. Behavioural synthesis can in many cases create smaller, cheaper and more reliable solutions for small, integrated systems than a CPU solution can achieve.

Behavioural synthesis is what many designers want. It means that work such as designing control machines for sharing functions, RAM interfaces, etc., can be done automatically.

Definition:
Behavioural synthesis (high-level, algorithm and architectural synthesis) means going from an algorithm specification to an RT level which implements that behaviour.

17.1.1. Terminology

Data path
The data path consists of a number of interconnected components at RT level. Components such as multipliers, adders, registers and subtractors, for example.

Control unit
The control unit is a state machine which generates control signals for the data path.

Scheduling
Scheduling means determining in which clock cycle resources (operators), I/O read and write operations and memory access are to be executed. A timetable is set for optimizing the utilization rate of resources without violating the timing constraints set for the functions. Different functions can use the same resources in order to reduce the number of resources on the chip, i.e. to reduce the area and with it the price of the ASIC. The scheduler is implemented in a state machine (control unit).

Resource allocation
Resource allocation is used to allocate the operators which have been scheduled in different clock cycles by the scheduler, a hardware resource. It is the allocator which takes the final decision as to whether an operator is to be shared or not.

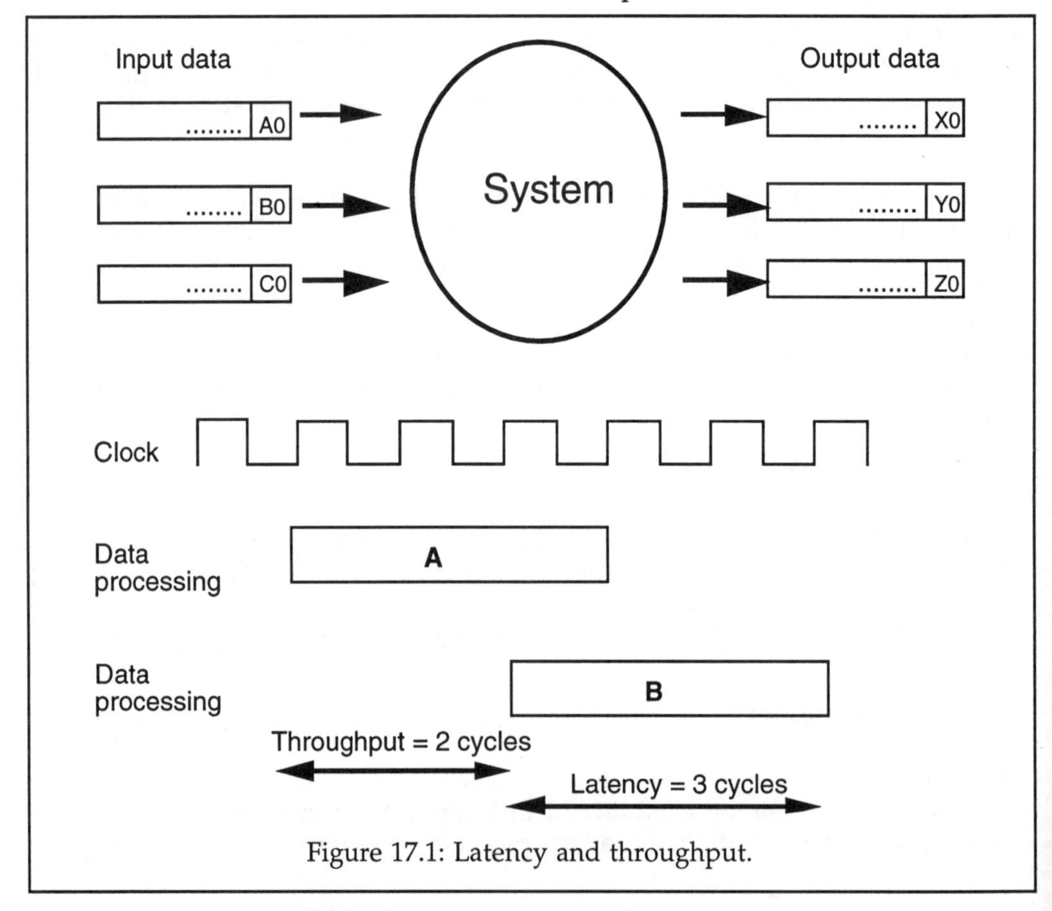

Figure 17.1: Latency and throughput.

Latency
Latency is the number of clock cycles required to execute the function (Figure 17.1).

Throughput
Throughput defines how often new data can be clocked into the system (Figure 17.1).

The following example uses two multiplications and three additions at RT level.

```
a <= (b * c) + (e * 34) + 56
```

Behavioural synthesis can be used to share resources. This means that, following behavioural synthesis, the hardware can be shrunk down to one multiplier and one adder. The gain is in terms of area, while the loss is that it takes longer to carry out the calculation, i.e. latency and throughput increase.

At RT level, then, timing constraints are written locally for operations, while for behavioural level they are written for processes (globally). Optimization at behavioural level is done across clock stages and on RTL components (type multiplier), while RT level optimizes within a clock cycle and with Boolean functions (gates). The designer has to specify resource utilization at RT level whereas it is done automatically at behavioural level.

17.2. Handshaking

As the behavioural synthesis tool can change the cycle-to-cycle behaviour of the I/O signals, problems arise during design and verification. If a testbench is to be used for all the levels (behavioural, RT and gate level, Figure 17.2), handshaking must be used. When the behavioural VHDL code is simulated, the data will be read and written exactly as defined in the VHDL code. After behavioural synthesis, this cycle-to-cycle behaviour has normally been changed. If the testbench expects the RTL/gate model to read and write in exactly the same clock cycles as defined in the behavioural synthesis code, the test will not be passed. The solution is to use handshake signals. These signals should say when the model reads/writes to the I/O signals. Depending on whether the block which is to send data has a fixed response time or not, a one-way handshake or a two-way handshake should be used.

Handshaking makes it possible to use the same testbench for all the levels (Figure 17.2).

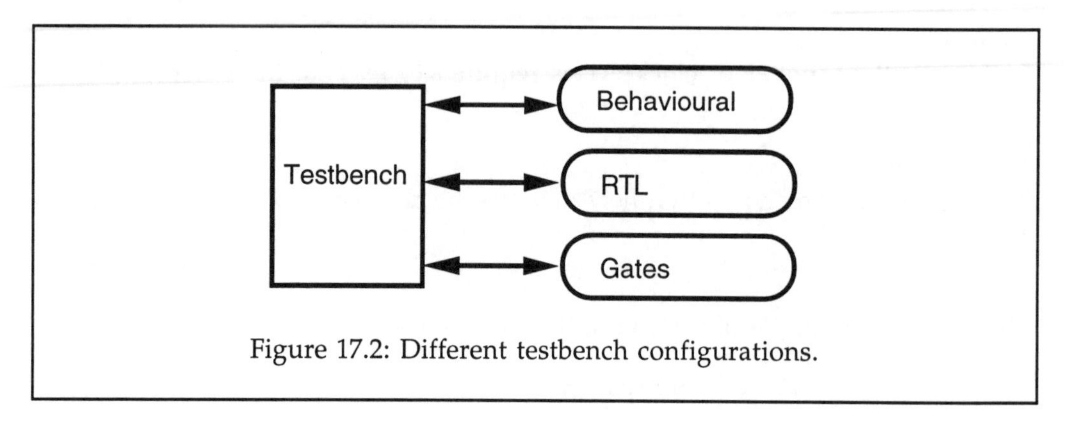

Figure 17.2: Different testbench configurations.

Exactly the same problem applies when data are to be transferred between hierarchical blocks in a design if any of the blocks have been synthesized using a behavioural synthesis tool. Handshaking must then be used to transfer data between such blocks.

17.2.1. One-way handshake protocol

A one-way handshake can be used when the response time is fixed. Only one signal is required to implement this handshake: *ready_for_data* (Figure 17.3).

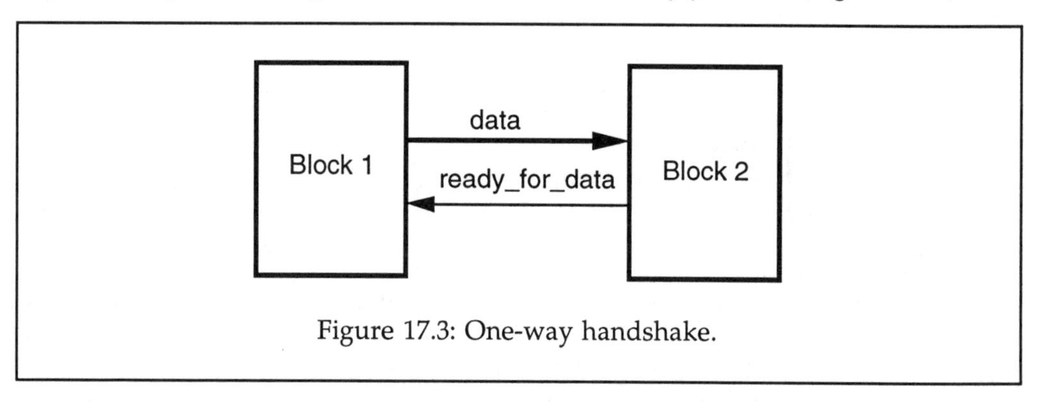

Figure 17.3: One-way handshake.

The requirement for block 1 in Figure 17.3 is that it should have a fixed response time after ready_for_data has been activated. This normally applies to testbenches, but not to hierarchical blocks where behavioural synthesis has been used. For blocks which do not have a fixed response time, a two-way handshake must be used.

The following example represents block 1 and is of the one-way handshake type. The process waits until ready_for_data is activated before it writes to I/O (data).

Example:

```
process
begin
  for i in 0 to 7 loop
    wait until clk='1';
    while ready_for_data='0' loop;
      wait until clk='1';
    end loop;
    wait until clk='1';
    data<=data_out(i);
    wait until clk='1';
  end loop;
end process;
```

17.2.2. Two-way handshake protocol

A two-way handshake can be used when the response time is not fixed. Two signals are required to implement this handshake: ready_for_data and data_valid (Figure 17.4).

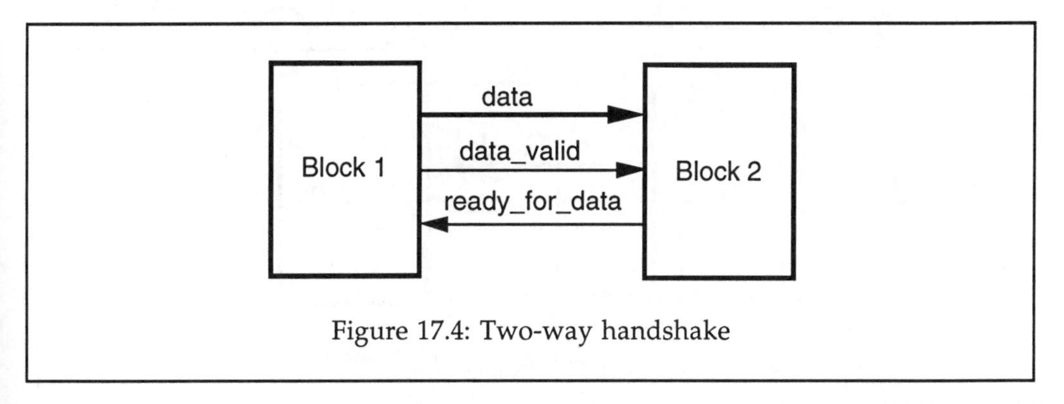

Figure 17.4: Two-way handshake

The following example represents block 2 and is of the two-way handshake type. The process activates ready_for_data when it is ready to process new data. Then the process waits for data_valid before it reads I/O (data).

Example:

```
process
begin
  main: loop
```

```
        wait until  clk='1';
        ready_for_data<='1';
        while data_valid='0' loop;
           wait until  clk='1';
        end loop;
        wait until  clk='1';
        ready_for_data<='0';
        data_in<=data;
        wait until  clk='1';
        q:=data_in * b;
        ...
    end process;
```

17.3. Example of behavioural/RTL synthesis - FIR filter

This section explains how behavioural synthesis can be used to design an FIR filter (see Figure 17.5).

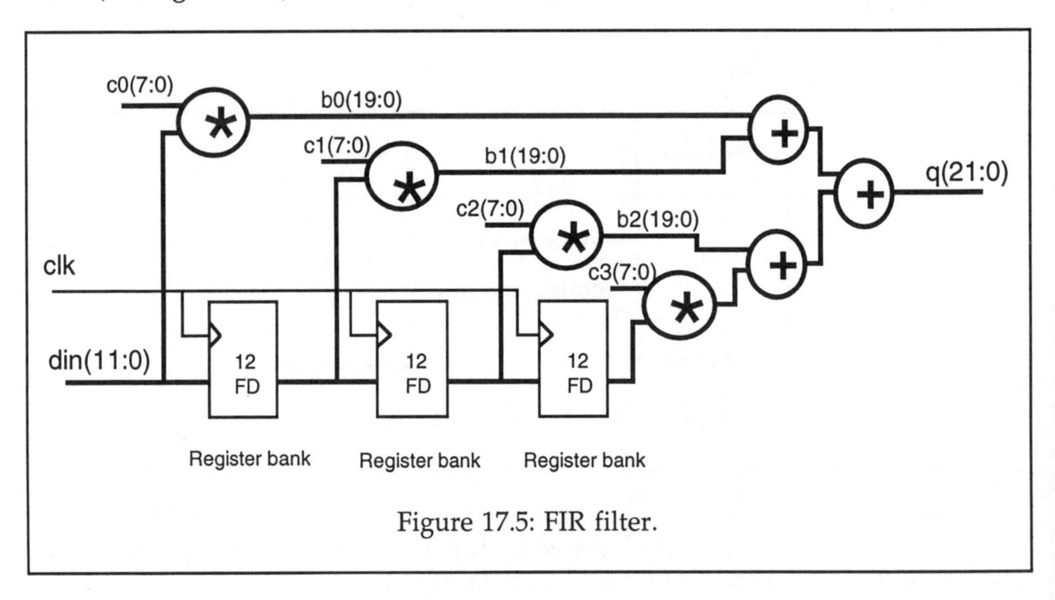

Figure 17.5: FIR filter.

The filter is to filter four digital channels, each with 12 bits. The FIR coefficients c0-c3 should be configurable, i.e. inputs to the filter component.

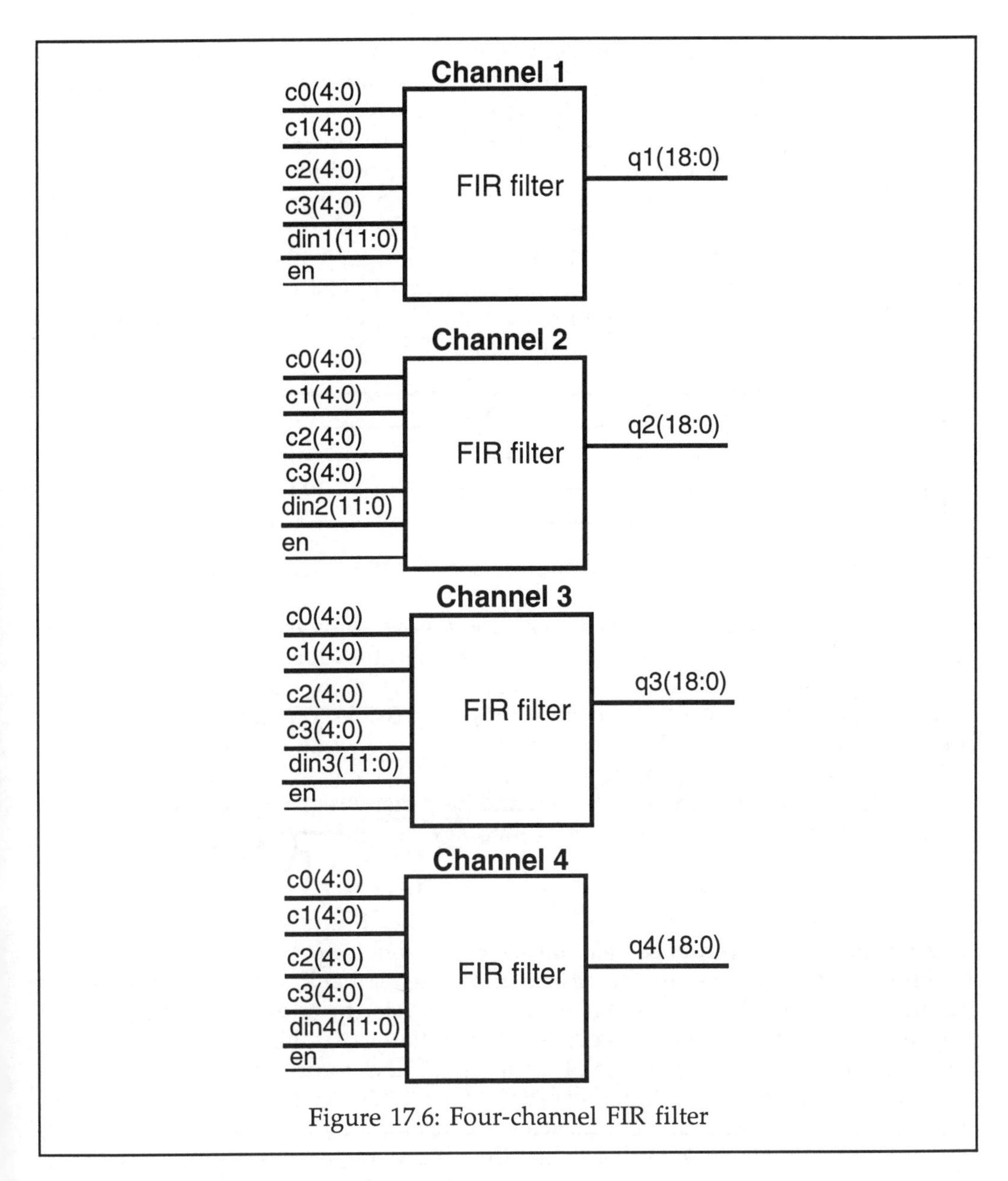

Figure 17.6: Four-channel FIR filter

If the four-channel FIR filter in Figure 17.6 is designed at RT level, any of the following three options can be chosen:

1. Instance four FIR filters
2. Share one FIR filter for all four channels
3. Share one FIR filter and share the multipliers and adders

Alternative 1 is the easiest: the designer only needs to design a filter of the type in Figure 17.5 and then instance it four times, i.e. one per channel. This design method produces 4 * 4, i.e. 16, multipliers and twelve adders. The same filter can be used for all four channels (alternative 2) if it is time-shared among the four channels. This solution produces four multipliers and three adders. In addition, a control unit has to be designed to control the resource sharing in the time between the channels. Alternative 3 is the most complex: as well as sharing the FIR filter between the four channels, the designer also has to share the multipliers and adders. This alternative can result in just one multiplier and one adder being required. A control unit for controlling the resource sharing also has to be designed. The results are summarized in Table 17.1.

	Multiplier	**Adder**	**Control unit**
Alt 1	16	12	0
Alt 2	4	3	1
Alt 3	1	1	1

Table 17.1: Number of resources used by the three alternatives.

At RT level the designer has to calculate how long each data path takes. Depending on whether a data path is longer than the clock period, the designer may have to insert a pipeline in the data path, i.e. store the result in a register bank temporarily. In one technology it may be possible to accommodate both the multiplication and the addition in one clock period, while in another technology several clock periods may be required. The VHDL code at RT level has to define what is to be done in each clock period and how many register banks are required. This means that the design is partly technology-dependent at RT level.

Suppose that we have two technologies, A and B. In technology A, multiplication and the two adders take 34 ns, while in technology B multiplication and the two adders take 17 ns. Suppose that the clock period is to be 18 ns. In this case technology A has to store the result in a register bank temporarily, whereas technology B can do everything in one clock cycle. In addition, if there is any sharing, the designer has to keep track manually of when the resources are available at RT level. In the case of resource sharing, information on latency and throughput are important design constraints if the RTL architecture is to be good. Latency refers to how many clock cycles it takes for a data value to be calculated, i.e. the number of clock cycles from filter input to filter output. Throughput refers to how often a new value is produced by the filter. If resources are shared, the filter cannot produce a new value every clock cycle. These two constraints affect the architecture at RT level, while a behavioural synthesis tool can retain the same VHDL code. The behavioural synthesis tool will automatically create different architectures to satisfy the set latency and throughput constraints.

If the designer has access to a behavioural synthesis tool, the VHDL code can be written as follows:

```
Library ieee;
Use ieee.std_logic_1164.ALL;

Package types is
  Constant L_in:   positive:=12;      -- Number of bits in input data.
  Constant L_mult:positive:=8;        -- Number of bit in mult.
  Constant L_ch:   positive:=4;       -- Number of channels.

  Constant L_reg: positive:=L_IN+L_mult; -- Number of bits in
                                         -- register after mult.

  Constant L_data:positive:=L_reg+2;     -- Number of bits in
                                         -- output.

  subtype data is std_logic_vector(L_data-1 downto 0);
                                         -- data out

  type data_array is array (integer range 0 to L_ch-1) of data;

  subtype reg is std_logic_vector(L_reg-1 downto 0);
                                         -- reg after mult

  type data_reg is array (integer range 0 to L_ch-1) of reg;

  subtype reg_in is std_logic_vector(L_in-1 downto 0);
                                         -- input reg

  type data_reg_in is array (integer range 0 to L_ch-1) of reg_in;
end;


Library ieee;
Use ieee.std_logic_1164.ALL;
Use ieee.std_logic_unsigned.ALL;
Use work.types.ALL;

Entity fir is
  port( d_in:           in   reg_in;
        c0,c1,c2,c3:    in   std_logic_vector(L_mult-1 downto 0);
        clk,resetn,en:  in   std_logic;
        done:           out  std_logic;
        d_out:          out  data_array);
end;
```

```
Architecture behv of fir is
begin
  p1:process
 variable d_in_i:          reg_in;
 variable  d1,d2,d3:       data_reg_in;
 variable  b0,b1,b2,b3:    reg;
 variable  i1,i2:          std_logic_vector(L_reg  downto 0);
 variable  d_out_i:        data;
 variable ch:              std_logic_vector(1  downto 0);
 variable ch_nr:           integer  range 0 to  L_ch;
 begin
   res_loop:loop
     done<='0';
     d_out_i:=(others=>'0');
     ch:=(others=>'0');
     d_out<=(others=> (others=>'0'));

     main:loop
       wait until  clk='1';
       exit res_loop  when  resetn='0';

       ch_nr:=conv_integer(ch);
       done<='0';

       wait until  clk='1';
       exit res_loop  when  resetn='0';

       while  en='0'  loop
         wait until  clk='1';
         exit res_loop  when  resetn='0';
       end loop;

       d_in_i:=d_in;
       b3:=  d3(ch_nr)  *  c3;
       b2:=  d2(ch_nr)  *  c2;
       b1:=  d1(ch_nr)  *  c1;
       b0:=  d_in_i  *  c0;

       i1:=('0'  &  b0)  +  b1;
       i2:=('0'  &  b2)  +  b3;

       wait until  clk='1';
       exit res_loop  when  resetn='0';

       d_out_i:=('0'  &  i1)  +  i2;

       done<='1';
```

```
          d3(ch_nr):=d2(ch_nr);
          d2(ch_nr):=d1(ch_nr);
          d1(ch_nr):=d_in_i;

          wait until clk='1';
          exit res_loop when resetn='0';

          d_out(ch_nr)<=d_out_i;
          ch:=ch+1;

          wait until clk='1';
          exit res_loop when resetn='0';
        end loop;
      end loop;
    end process;
  end;
```

The above VHDL code does not define which resources are to be shared. That decision is left to the behavioural synthesis tool, allowing the designer to concentrate on getting the algorithms, i.e. the behaviour, right. The resources to be shared and the latency and throughput that the designer wants the system to have are defined as constraints for the behavioural synthesis tool. Describing the same thing at RT level is time-consuming, as the resource sharing has to be described in the VHDL code. The designer also has to define latency and throughput in the VHDL code by selecting an architecture which he/she believes will pass the constraints. If he/she selects the wrong architecture, the VHDL code has to be rewritten.

Different architectures, and therefore different results, are obtained depending on what latency and throughput constraints are set for the behavioural synthesis tool.

The above VHDL code can be synthesized with an ordinary RTL synthesis tool. In this case the result will be precisely the cycle-to-cycle behaviour defined in the VHDL code, i.e. no latency or throughput constraints can be specified for the tool.

If the designer chooses to design at RT level, the VHDL code can be written as shown below. In order to manage the clock frequency of 40 ns with a good margin in the chosen technology (Texas Instruments TGC2000, 0.65 μm), the design has a pipeline, i.e. signals *b0-b3* are stored in register banks. A pipeline has also been inserted between the adders for the same reason.

Example (RT level):

```
  Library ieee;
  Use ieee.std_logic_1164.ALL;
```

```
Package types is
   Constant L_in:  positive:=12;  -- Number of bits in input data.
   Constant L_mult:positive:=8;   -- Number of bit in mult.
   Constant L_ch:  positive:=4;   -- Number of channels.

   Constant L_reg: positive:=L_IN+L_mult;   -- Number of bits in
                                            -- reg. after mult.
   Constant L_data:positive:=L_reg+2;       -- Number of bits in
                                            -- output.
   subtype data is std_logic_vector(L_data-1 downto 0);
                                            -- data out

   subtype reg is std_logic_vector(L_reg-1 downto 0);
                                            -- reg after mult

   subtype reg_in is std_logic_vector(L_in-1 downto 0);
                                            -- input reg
end;


Library ieee;
Use ieee.std_logic_1164.ALL;
Use ieee.std_logic_unsigned.ALL;
Use work.types.ALL;

Entity fir_rtl is
  port( d_in:           in  reg_in;
        c0,c1,c2,c3:    in  std_logic_vector(L_mult-1 downto 0);
        clk,resetn,en:  in  std_logic;
        q:              out data);
end;

Architecture rtl of fir_rtl is
signal d1,d2,d3:      reg_in;
signal b0,b1,b2,b3:   reg;
signal d_out_i:       data;
signal i1,i2:         std_logic_vector(L_reg downto 0);
begin
  p1:process(clk,resetn)
  begin
    if resetn='0' then
      d1<=(others=>'0');
      d2<=(others=>'0');
```

```
      d3<=(others=>'0');
      b0<=(others=>'0');
      b1<=(others=>'0');
      b2<=(others=>'0');
      b3<=(others=>'0');
      q<=(others=>'0');
    elsif clk'event and clk='1' then
      if en='1' then
        b0<=d_in * c0;      -- Register bank
        b1<=d1 * c1;        -- Register bank
        b2<=d2 * c2;        -- Register bank
        b3<=d3 * c3;        -- Register bank

        d3<=d2;             -- Shift data
        d2<=d1;
        d1<=d_in;
      end if;

      i1<=('0' & b0) + b1;   -- Register bank
      i2<=('0' & b2) + b3;   -- Register bank
      q<=('0' & i1) + i2;

    end if;
  end process;

end;


Library ieee;
Use ieee.std_logic_1164.ALL;
Use ieee.std_logic_unsigned.ALL;
Use work.types.ALL;

Entity fir_4_rtl is
  port( d_in1,d_in2:     in   reg_in;
        d_in3,d_in4:     in   reg_in;
        c0,c1,c2,c3:     in   std_logic_vector(L_mult-1 downto 0);
        clk,resetn,en:   in   std_logic;
        q1,q2,q3,q4:     out  data);
end;

Architecture rtl of fir_4_rtl is
component fir_rtl
  port( d_in:         in  reg_in;
        c0,c1,c2,c3:  in  std_logic_vector(L_mult-1 downto 0);
```

```
        clk,resetn,en: in  std_logic;
        q:             out  data);
   end component;
   For ALL:fir_rtl Use entity  work.fir_rtl(rtl);
   begin
     CH1:fir_rtl port map(d_in1,c0,c1,c2,c3,clk,resetn,en,q1);
     CH2:fir_rtl port map(d_in2,c0,c1,c2,c3,clk,resetn,en,q2);
     CH3:fir_rtl port map(d_in3,c0,c1,c2,c3,clk,resetn,en,q3);
     CH4:fir_rtl port map(d_in4,c0,c1,c2,c3,clk,resetn,en,q4);
   end;
```

If we synthesize the above examples of the FIR filter, the following results are obtained. One synthesis result is shown for RT level and two for behavioural level.

RTL: The VHDL code which instances four FIR filters, i.e. without resource sharing, was used in the RTL synthesis.

Behavioural 1: The behavioural VHDL code was used. The constraints were set to minimize the number of resources, i.e. the area of the design.

Behavioural 2: The behavioural VHDL code was used. The constraints were set to reduce latency in the design.

Synthesis tool	Throughput (#clock cycles)	Latency (#clock cycles)	Area (#gates)	Resources (#mult/#add)
RTL	1	3	22,000	16/12
Behavioural 1	8	4x2=8	6000	2/2
Behavioural 2	40	4x10=40	5000	1/2

Table 17.2: Number of resources.

Below are parts of the report from the first behavioural synthesis (area controlled). These are followed by an analysis. Three different parts of the report are shown: Summary, Operation schedule and Register usage of process p1.

```
*************************************
*  Summary report for process p1:   *
*************************************
```

```
---------------------------------------------------------------------------
    Timing Summary
---------------------------------------------------------------------------
 Clock period 40.00
  Loop timing information:
    p1                                10 cycles (cycles 0 - 10)
       res_loop                       10 cycles (cycles 0 - 10)
         main                         10 cycles (cycles 0 - 10)
           _L0                        1 cycle  (cycles 2 - 3)
              (exit) EXIT_L64         (cycle 3)
---------------------------------------------------------------------------
    Area Summary
---------------------------------------------------------------------------
  Estimated combinational area          2251
  Estimated sequential area             3440
 TOTAL                                  5691

 11 control states
 12 basic transitions
 2 control inputs
 15 control outputs

---------------------------------------------------------------------------
    Resource types
---------------------------------------------------------------------------
       Register Types
 ============================================
       1-bit  register                    1
       2-bit  register                    1
       8-bit  register                    2
       12-bit  register                   2
       20-bit  register                   1
       21-bit  register                   2
       48-bit  register                   3

       Operator Types
 ============================================
       (2_2->2)-bit DW01_add              1      (Channel counter)
       (12_8->20)-bit DW02_mult           1
       (21_21->21)-bit DW01_add           1
       (22_22->22)-bit DW01_addsub        1

       I/O Ports
 ============================================
       1-bit input port                   1
       1-bit registered output port       1
       8-bit input port                   4
       12-bit input port                  1
       88-bit output port                 1

---------------------------------------------------------------------------
```

As the above summary report shows, from the scheduling each main loop takes 10 clock cycles. As the design consists of four channels, this means that loop main has to be executed four times to process the data for all the channels. Hence the latency will be 4 x 10 = 40 cycles. The latency is relatively high because the constraint on the behavioural synthesis tool was only to minimize the area, regardless of the cost in terms of latency.

The behavioural synthesis tool also estimates the area after scheduling and allocating the design. This area estimate (5691 gates) is before optimization at RT level by an RTL synthesis tool. If the RTL synthesis tool optimizes the design with medium effort and the hierarchies created are retained, the area estimate is normally quite accurate. The behavioural synthesis tool (Synopsys) creates a hierarchy for the control machine, operators and register banks (mainly to produce a readable schematic). If the RTL synthesis is done with high effort and with a switch which says that the hierarchy should be flattened out, a slightly smaller area is usual obtained. The area after optimization of this type was 5000 gates (see Table 17.2). This is more than four times smaller than the synthesis result for the RTL VHDL code. Latency and throughput, on the other hand, have increased considerably. This is completely in line with the constraints we set for the behavioural synthesis, i.e. area constraints only. When the behavioural synthesis tool was given a more balanced set of constraints, much lower throughput and latency were obtained (behavioural 2). The area did not increase very much either, only 1000 gates, while latency and throughput were reduced from 40 to 8 clock cycles.

The above report also shows that the control machine which controls resource sharing consists of 11 states. It is also possible to read the number of operators from the report. As we can see, there is one multiplier, as expected. It is surprising to see, on the other hand, that there are two adders. This is because the behavioural synthesis tool has calculated that it is not worth sharing these two adders from the point of view of area. Sharing would require a larger control machine and multiplexors (21) in the data path to supply correct data values to the adder. A multiplexor takes three gates. An extra register bank is also required to store the next value to be calculated. If a register bit with asynchronous reset is assumed to take eight gates, this will result in an area cost of 21x8 + 21x3 + larger control machine = 231 gates (minimum). This meant that the two adders were not shared in this case. As twelve adders are required if no resource sharing is done, the two adders obtained are shared with many other resources, contributing to the low area consumption of 5000 gates.

```
*************************************
*  Operation schedule of process p1:  *
*************************************
```

Resource types

c0	8-bit input port
c1	8-bit input port
c2	8-bit input port
c3	8-bit input port
d_in	12-bit input port
done	1-bit registered output port
en	1-bit input port
loop	loop boundaries
p0	88-bit output port d_out
r46	(12_8->20)-bit DW02_mult
r52	(22_22->22)-bit DW01_addsub
r57	(21_21->21)-bit DW01_add
r411	(2_2->2)-bit DW01_add

```
cycle | loop | d_in | c1 | c2 | c3 | c0 | p0 | en | r46 | r52 | r57 | r411 | done
-----------------------------------------------------------------------------------
  0   |..L6..|......|....|....|....|....|.W1.|....|.....|.....|.....|......|..W0..
      |..L3..|......|....|....|....|....|....|....|.....|.....|.....|......|......
      |..L0..|......|....|....|....|....|....|....|.....|.....|.....|......|......
  1   |......|......|....|....|....|....|....|....|.....|.....|.....|......|..W4..
  2   |..L9..|......|....|....|....|....|....|.R6.|.....|.....|.....|......|......
  3   |.L12..|..R0..|.R1.|.R2.|.R3.|....|....|....|.o5..|.....|.....|......|......
      |.L11..|......|....|....|....|....|....|....|.....|.....|.....|......|......
      |.L10..|......|....|....|....|....|....|....|.....|.....|.....|......|......
  4   |......|......|....|....|....|.R4.|....|....|.o7..|.....|.....|......|......
  5   |......|......|....|....|....|....|....|....|.o3..|.....|.....|......|......
  6   |......|......|....|....|....|....|....|....|.o4..|.....|.o9..|......|..W3..
  7   |......|......|....|....|....|....|....|....|.....|.....|.o8..|......|......
  8   |......|......|....|....|....|....|....|....|.....|.o6..|.....|......|......
  9   |......|......|....|....|....|....|.W2.|....|.....|.o6..|.....|......|......
      |......|......|....|....|....|....|.R5.|....|.....|.....|.....|......|......
 10   |..L8..|......|....|....|....|....|....|....|.....|.....|.....|.o10..|......
      |..L7..|......|....|....|....|....|....|....|.....|.....|.....|......|......
      |..L5..|......|....|....|....|....|....|....|.....|.....|.....|......|......
      |..L4..|......|....|....|....|....|....|....|.....|.....|.....|......|......
      |..L2..|......|....|....|....|....|....|....|.....|.....|.....|......|......
      |..L1..|......|....|....|....|....|....|....|.....|.....|.....|......|......
```

Operation name abbreviations

L0	loop boundaries	p1_design_loop_begin
L1	loop boundaries	p1_design_loop_end
L2	loop boundaries	p1_design_loop_cont
L3	loop boundaries	res_loop/res_loop_design_loop_begin
L4	loop boundaries	res_loop/res_loop_design_loop_end
L5	loop boundaries	res_loop/res_loop_design_loop_cont
L6	loop boundaries	res_loop/main/main_design_loop_begin
L7	loop boundaries	res_loop/main/main_design_loop_end
L8	loop boundaries	res_loop/main/main_design_loop_cont
L9	loop boundaries	res_loop/main/_L0/_L0_design_loop_begin
L10	loop boundaries	res_loop/main/_L0/_L0_design_loop_end
L11	loop boundaries	res_loop/main/_L0/_L0_design_loop_cont
L12	loop boundaries	res_loop/main/_L0/EXIT_L64

```
R0      12-bit read res_loop/main/d_in_72
R1      8-bit read res_loop/main/c1_75
R2      8-bit read res_loop/main/c2_74
R3      8-bit read res_loop/main/c3_73
R4      8-bit read res_loop/main/c0_76
R5      88-bit read res_loop/main/U4
R6      1-bit read res_loop/main/_L0/en_64

W0      1-bit write res_loop/done_49
W1      88-bit write res_loop/d_out_52
W2      88-bit write res_loop/main/U5
W3      1-bit write res_loop/main/done_86
W4      1-bit write res_loop/main/done_59

o3      (12_8->20)-bit MULT_UNS_OP res_loop/main/mul_73/mult/mult
o4      (12_8->20)-bit MULT_UNS_OP res_loop/main/mul_75/mult/mult
o5      (12_8->20)-bit MULT_UNS_OP res_loop/main/mul_74/mult/mult
o6      (22_22->22)-bit ADD_UNS_OP res_loop/main/add_84/plus/plus
o7      (12_8->20)-bit MULT_UNS_OP res_loop/main/mul_76/mult/mult
o8      (21_21->21)-bit ADD_UNS_OP res_loop/main/add_78/plus/plus
o9      (21_21->21)-bit ADD_UNS_OP res_loop/main/add_79/plus/plus
o10     (2_2->2)-bit ADD_TC_OP res_loop/main/add_95/plus/plus

******************************************************
```

The above report shows how the operators have been scheduled in the 10 clock cycles. We can see when I/O ports are read or written to: output signal *d_out* is written to in clock cycle 9 (p0, w2), for example. *d_out* is also written to in clock cycle 0, with this being the output reset.

The report also shows in which clock cycles the operators are used. The multiplier (r46) is used in clock cycles 3, 4, 5 and 6. This means that the utilization rate for the multiplier is 4/10 = 40 per cent. The 22-bit adder (r52) is used by operation o6 in clock cycles 8 and 9. This means that the adder operates across two clock cycles and that the behavioural synthesis tool will set up what are known as multicycle constraints for this data path automatically (see Chapter 10, "RTL synthesis"). Setting up such constraints manually at RT level can be time-consuming.

The operation name abbreviations part of the report explains the resource names. The figure included in some names, e.g. 1-bit write res_loop/main/done_86, refers to the line in the original VHDL code.

```
***********************************
*  Register usage of process p1:  *
***********************************

        Storage resource types
=================================
```

r21	48-bit register
r136	1-bit register
r145	2-bit register
r292	21-bit register
r319	12-bit register
r331	12-bit register
r429	8-bit register
r430	21-bit register
r434	8-bit register
r444	48-bit register
r445	20-bit register
r446	48-bit register

```
cycle| r444 | r21  | r446 | r292 | r430 | r445 | r331 | r319 | r429 | r434 | r145 |r136
---------------------------------------------------------------------------------------
 (48)| (48) | (48) | (21) | (21) | (20) | (12) | (12) | (8)  | (8)  | (2)  | (1)
=======================================================================================
  0  |. v0..|..v1..|..v2..|......|......|......|......|.v12..|......|......|.v17..|...
  1  |..v0..|..v1..|..v2..|......|......|......|......|.v12..|......|......|.v17..|...
  2  |..v0..|..v1..|..v2..|....  .|......|......|.v14..|.v12..|......|......|.v17..|v19
  3  |..v3..|..v1..|..v2..|..v9..|......|......|.v11..|.v12..|.v15..|.v16..|.v17..|...
  4  |..v3..|..v1..|..v2..|..v9..|......|.v10..|.v11..|.v12..|.v15..|.v16..|.v17..|...
  5  |..v3..|..v1..|..v2..|..v9..|..v7..|.v10..|.v11..|.v13..|.v15..|......|.v17..|...
  6  |..v3..|..v1..|..v2..|..v8..|..v6..|.v10..|.v11..|......|......|......|.v17..|...
  7  |..v3..|..v4..|..v2..|..v5..|..v6..|......|......|......|......|......|.v17..|...
  8  |..v3..|..v4..|..v2..|..v5..|..v6..|......|......|......|......|......|.v17..|...
  9  |..v3..|..v4..|..v2..|......|......|......|......|......|......|......|.v17..|...
 10  |......|......|......|......|......|......|......|......|......|......|......|...
```

Data value name abbreviations

v0	48-bit data value	res_loop/main/d2
v1	48-bit data value	res_loop/main/d1
v2	48-bit data value	res_loop/main/U1/net
v3	48-bit data value	res_loop/main/U2/net
v4	48-bit data value	res_loop/main/U3/net
v5	21-bit data value	res_loop/main/add_78/plus/plus/Z
v6	21-bit data value	res_loop/main/add_79/plus/plus/Z
v7	20-bit data value	res_loop/main/mul_73/mult/mult/Z
v8	20-bit data value	res_loop/main/mul_75/mult/mult/Z
v9	20-bit data value	res_loop/main/mul_74/mult/mult/Z
v10	20-bit data value	res_loop/main/mul_76/mult/mult/Z
v11	12-bit data value	res_loop/main/d_in_72/net
v12	12-bit data value	res_loop/main/mul_73/mult/mult/A
v13	12-bit data value	res_loop/main/mul_75/mult/mult/A
v14	12-bit data value	res_loop/main/mul_74/mult/mult/A
v15	8-bit data value	res_loop/main/c1_75/net
v16	8-bit data value	res_loop/main/c3_73/net
v17	2-bit data value	res_loop/main/ch
v19	1-bit data value	res_loop/main/_L0/U2/Z

The above report shows the register usage after behavioural synthesis has scheduled and allocated process p1. As we can see from the report, different data values are stored in the same register resource. This means that the behavioural synthesis tool also shares register banks when possible and worthwhile from the point of view of area. From the report it is possible to see that the pipeline registers after the multipliers (20-bit) and adders (21-bit) are being shared. Without resource sharing four 21-bit and two 20-bit registers per channel would have been required to store the values. The report shows that there are only two 21-bit and one 20-bit registers, i.e. the registers have been shared.

In behavioural synthesis 2, latency of two clock cycles per channel is achieved, i.e. eight clock cycles in total. If the example had been larger, latency could have been reduced by means of what is known as loop pipelining. Loop pipelining means that the next iteration of the loop starts before the previous one is complete (Figure 17.7).

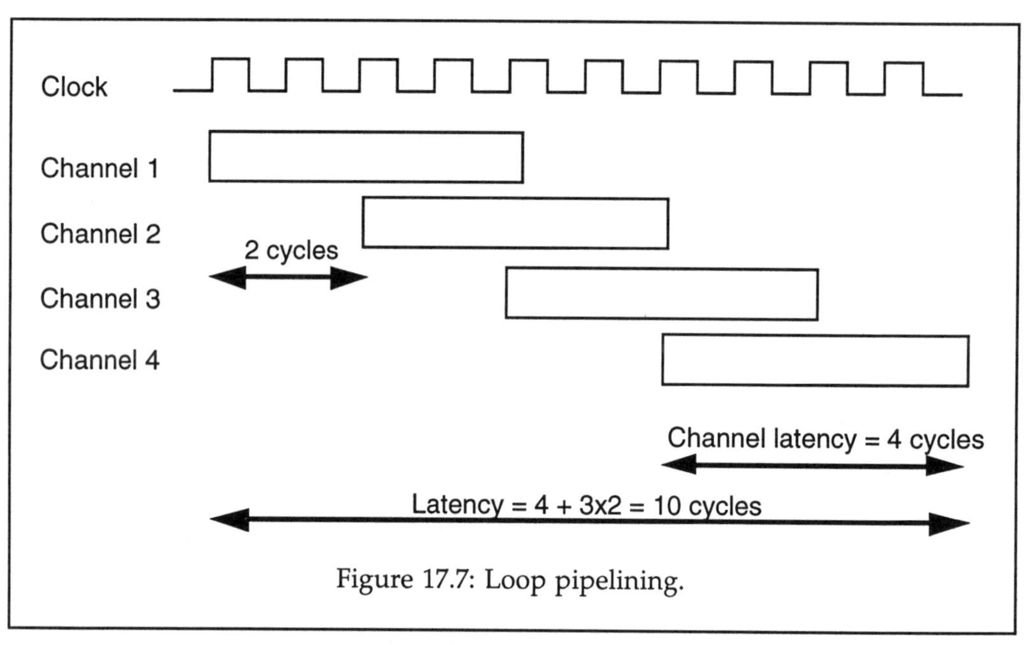

Figure 17.7: Loop pipelining.

Suppose that the minimum possible latency of a loop iteration had been four clock cycles. With loop pipelining, the next iteration of the loop (channel) could have been started after two clock cycles. Without loop pipelining the latency for the four channels would have been 4 x 4 = 16 clock cycles. With loop pipelining the latency for the example is 4 + 4 x 2 = 10 clock cycles (Figure 17.7). In larger designs, loop pipelining can greatly reduce latency in the design at the expense of slightly larger area. A behavioural synthesis tool can do loop pipelining automatically.

In the above synthesis result ordinary flip-flops have been used to implement the register banks. The behavioural synthesis tool can insert a RAM block automatically instead, if so desired. The behavioural synthesis tool will then also automatically build a control unit which generates read, write and enable strobes for the RAM. For the behavioural synthesis tool (Synopsys Behavioural Compiler) to include RAM instead of flip-flops, an attribute has simply to be included in the VHDL code. If, for example, variables *d1*, *d2* and *d3* are to be stored in RAM instead of register banks, the following attribute must be included in the VHDL code:

```
constant my_RAM:resource:=0;
attribute variable of my_RAM:constant is "d1 d2 d3";
attribute map to module of my_RAM:constant is <name_of_RAM>;
```

The present version of Synopsys (V3.3b) requires that the index range of the variables should not overlap, i.e. this also has to be modified in the VHDL code.

If RAM is wanted when designing at RT level, it has to be instanced, in addition to which the entire control unit which generates the strobes has to be designed, all of which is time-consuming. Instancing RAM at RT level also makes the code technology-dependent, as RAM from an ASIC library is instanced. At behavioural level the VHDL code is technology-independent. There may be an attribute for technology-specific RAM in the code, but it can simply be ignored or easily changed if another technology is required. Allowing a behavioural synthesis tool to insert RAM automatically in this way is very quick and efficient. In many cases it is also possible to reduce the area of the design considerably.

18
Laboratories

As procedures vary from synthesis tool to synthesis tool, the method for experiment 1 is shown for both ViewLogic and Synopsys (together with Mentor). In the case of ViewLogic we use both the ViewLogic VHDL simulator and synthesis tool, while in the case of Synopsys we use the Mentor Graphics VHDL simulator (Quick-HDL) and then do the synthesis with Synopsys.

VHDL is defined in the IEEE-1076 standard, but only the simulators can cope with the full standard. Synthesis tools can only synthesize a subset of it. This is partly because some VHDL commands are impossible to do synthesis on, which means that there will never be a tool which can synthesize the full standard. There are also differences between the various synthesis tools with regard to how much of the standard each can synthesize. As already mentioned, Synopsys is the leading logic synthesis company today, and the Synopsys synthesis supports a larger subset of the standard than most other tools (including ViewLogic). Consequently, the solutions to the experiments are given for both ViewLogic and Synopsys, whereby the entire spectrum of tools, from simple PC tools to advanced workstation tools, is covered.

As ViewLogic does not support as large a subset of the VHDL standard as Synopsys, some parts of VHDL code have to be rewritten to what ViewLogic supports (but still within the framework of the IEEE standard).

18.1. "Hands on" for ViewLogic

Lab 1: Design, simulate and synthesize a simple component

ViewLogic works partly under Windows on a PC. The instructions assume a certain familiarity with Windows. If you use SUN, contact the authors by E-mail.

1. Start the development system

ViewLogic is started by double-clicking on the Cockpit icon. Cockpit is in the WVPLUS program group.

A window called Workview PLUS Cockpit opens. The next step is to start a project.

Creating a project
A project is an aid to keeping track of all the files for a job. All the laboratory experiments can be called a project. A project creates a directory structure. The root directory is given the same name as the project and creates a number of subdirectories.

c:\<project_name> Top directory
c:\<project_name>\behv: VHDL files (<filename>.VHD)
c:\<project_name>\wir: Synthesis results (gate descriptions)
c:\<project_name>\sch: Schematics
c:\<project_name>\sym: Symbols
c:\<project_name>\cmd: Commando files

The Create command is found in the project menu. Create a project, e.g. C:\lab. If you have already created a project, you can activate it by pressing Current Project and "down arrow".

2. Enter VHDL code

Now a project directory has been created. Create a subdirectory for the project called "behv".

Design a component in VHDL with the following behaviour:

a	b	c
0	0	1
0	1	0
1	0	0
1	1	0

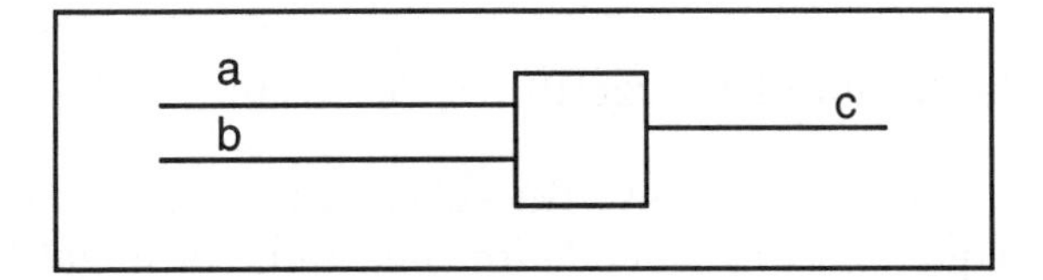

Signals *a*, *b* and *c* should be of type **vlbit**. The entity name should be **lab1**.

Use the word-processor you are familiar with and save the document just as text. Now the file is ready, so save it in the subdirectory (NB: text only), <project_directory>**behv**, as **lab1.vhd**.

3. Verify the component

Creating a simulatable model
The simulator cannot read the VHDL code directly. It has to be compiled first. This is done using **analyser**. Analyser is activated in the box on the right. Double-click on the "**analyser**" icon. A window will appear containing:

Enter the command line: (type) **VHDL behv\lab1.vhd -v -e=0**

-v means that a simulatable file will be created (lab1.vsm).
-e=0 means that all errors will be described in the file lab1.lis.

If you get syntax errors, open the file **lab1.lis** in the project directory (not in the subdirectory behv).

When the file is error-free, go to the next point.

Start the simulator and simulate the model
Double-click on the **viewsim** icon. The following question will appear:

Enter the command line: (type) **WVSIM lab1**

Description of some simple commands for the simulator

wave lab1.wfm a b c

Wave is a command for opening a graphics window for signals of interest. The above command opens a graphics window called lab1 and signals *a*, *b* and *c* are displayed.

a b 0
"a" (assign) sets the input signal to 0 for signal *b*.

s 10ns
"s" (simulate) starts the simulator for 10 ns. The result can be seen in the wave window.

An example of a sequence of commands for the simulator might be:

a a 0

```
a b 0
s 100ns
a b 1
s 100ns
```

(click on the wave window)

Command file for ViewLogic (cmd file) (supplementary study)
Below are descriptions of a few more commands which are useful to know in later experiments and of how to create a command file for the simulator. The advantage of a command file is that it only has to be written once. Then it can be executed every time the simulator is started.

The command file <file name>.cmd is placed in the CMD directory in your project. The name of the file should be the same as the VHDL code. For example, suppose that your component has the following interface:

```
din
bus_in(8 downto 0)
bus_out(6 downto 0)
clk
resetn
```

To avoid rewriting all the commands as vectors, for example, and assigning values to signals, etc., in ViewSim, a text file is created.

First of all you define which signals are to be represented as a bus with **VECTOR**. Then you specify **STEPSIZE**, which specifies the length of a simulation cycle. The next command is **CLOCK**, which specifies which signals are to be repeated periodically. The signals which are to be displayed in ViewWave should be defined before simulation. This is done with the following line:

wave<file name>.wfm clk resetn din bus_in bus_out

A complete file might look like this:

```
vector bus_in bus_in[8:0]
vector out_bus bus_out[6:0]
stepsize 50ns
clock clk 0 1
wave A.wfm clk reset_n din int_bus bus_in out_bus
a resetn 0
s 100ns
a resetn 1
s 100ns
```

```
a din 1
a bus_in  100110111
```

The next time ViewSim is started, select the following in the menu:

Misc - Execute - CmdFile
Type: **cmd\<file name>.cmd** followed by OK

If an internal signal inside the architecture is to be displayed in ViewWave, e.g. int_bus, type the following:

```
vector int_bus <entity-name >\int_bus.[3:0]
wave<file name>.wfm int_bus
```

Swap <entity name> for the name specified in the VHDL code.

4. Synthesize the component

Synthesizing the VHDL code to gates
A large main memory is needed to synthesize the design, so Windows sometimes has to be shut down, but usually the MS-DOS prompt can be used.

Change to the project directory.

Give the command "**VHDLDES BEHV\LAB1 -tec=xc4000**"

-tec=xc4000: This means that the target technology is the XILINX 4000 (xc4000) series. The synthesis result is in the file **lab1.rpt.**

For more advanced use of the synthesis software, enter VHDLDES and then help.

Creating a graphical schematic and analysing the answer
The next step is to generate a schematic in order to take a look at the result. This is done with the command **VIEWGEN LAB1 -makesym.**

-makesym: This creates a symbol for lab1.

Type exit (if Windows is running).

Start the schematic program by clicking on **VIEWDRAW** in the COCKPIT window. Click on **OK** in the window which appears. A ViewDraw window should be visible on screen. There are directories at the bottom, while the top part shows what is in the selected directory. Select the project directory and **schematics**. Now lab1.1 should be on top. Double-click on **lab1.1.**

The result should be a schematic of a NOR gate. There is a menu at the top of the schematic page (View is the most interesting), and parts of the menu can be activated by pressing the left button. The button functions are described as L (left), M and R in the register below the menu line.

Among other things, help can be found under the red box at the top right (Help - Help on Commands).

Try to press the left button and activate zoom. Indicate an area with the middle button (M:zoom). Try a few times and it will work well.

The NOR model can be studied in greater detail by marking the NOR symbol with select (left button). Press down the right button and select **Level - Push** schematic. A simulation model of the NOR gate should now be visible on screen. If you want to return to the preceding hierarchical level, select **Level - pop** in the menu.

18.2. "Hands on" for Synopsys synthesis tool and Mentor Graphics VHDL simulator

Lab 1: Design, simulate and synthesize a simple component

1. Start the development system

Create project directories:

<project>/rtl	: VHDL files (<file name>.vhd)
<project>/syn	: Synthesize result
<project>/sch	: Schematic
<project>/script	: Synthesis scripts

Example:

```
mkdir vhdl
cd vhdl
mkdir rtl
mkdir sch
mkdir syn
mkdir script
```

Go into the directory rtl with the command **cd<project>/rtl**.

Create a work library for the VHDL compiler (Quick-HDL) with the command: **qhlib work**

2. Enter VHDL code

Start any text editor and enter the VHDL code.

At the top of the file type:

```
Library ieee;
Use ieee.std_logic_1164.ALL;
```

Save the file under the name **lab1.vhd** in the **rtl** directory.

Design a component in VHDL with the following behaviour:

a	b	c
0	0	1
0	1	0
1	0	0
1	1	0

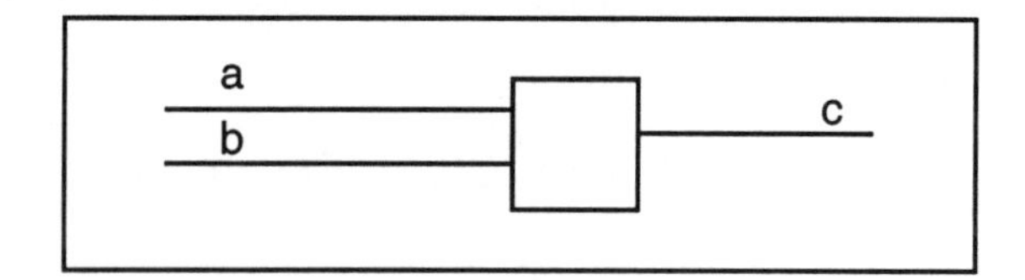

Signals *a*, *b* and *c* should be of type **std_logic**. The entity name should be **lab1**.

Compile the VHDL code with the command **qvhcom lab1.vhd -source.**

3. Simulate the component

Start the simulator by typing **qhsim lab1 &**
Enter signals in the wave window with the command:

```
wave a
wave b
wave c
```

Give input stimuli to the model
The input signals can be forced with the command force:

```
force a 0
force b 1
run 100
```

Once functionality has been verified, exit the simulator **without saving**.

4. Synthesize the component

Go to the <project> directory with the command **cd ..**
Enter the command **dc_shell -f script/lab1.scr**.The content of the script is shown later in this chapter. If no error has occurred, a file called **syn/lab1.db** should have been created.

Start the Synopsys graphical environment with the command
design_analyzer&

From the menu select **file-read** (format db, file lab1.db).
Press OK.
Double-click on the icon lab1.
Double-click on the symbol lab1.

Study the schematic.

18.3. Script for Synopsys users

This script can be used for laboratory experiment 1. By modifying the name in the script, it can also be used for the other laboratory experiments. Note that in some laboratory experiments a user-written package also has to be entered before the VHDL code itself (e.g. mypack in experiment 4).

Motorola's HDC library is used as the ASIC library. The library can be swapped for any ASIC/FPGA library by changing the set-up file for Synopsys.

The script: lab1.scr:

```
read -format VHDL rtl/lab1.vhd
set_max_area 0
compile -map_effort high -verify
write -format db -hierarchy -output syn/lab1.db
quit
```

18.4. Laboratory assignments

Note that types std_logic and std_logic_vector should be swapped for types vlbit and vlbit_vector for ViewLogic PC users.

Lab 1: Design, simulate and synthesize a simple component

1. Enter VHDL code

Design a component in VHDL with the following behaviour:

a	b	c
0	0	1
0	1	0
1	0	0
1	1	0

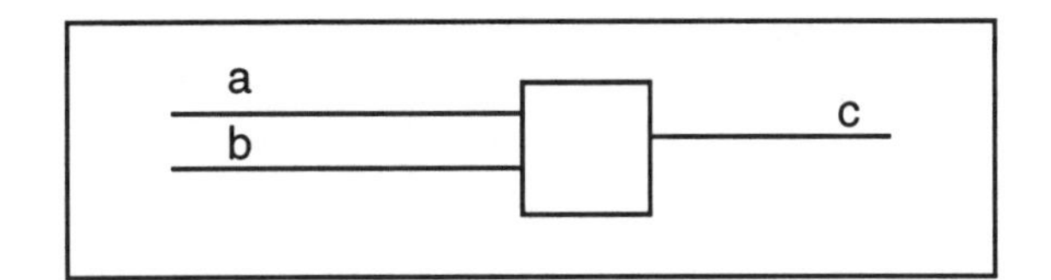

Signals *a, b* and *c* should be of type **std_logic**. The entity name should be **lab1.**

2. Simulate the component

3. Synthesize the component

Study the schematic.

Lab 2: Parallel VHDL

1. Enter VHDL code

Design a 4-input multiplexor with a generic delay on Tdelay ns with the following behaviour:

Sel (1:0)	q
00	a
01	b
10	c
11	d
X-	X
-X	X
Z-	X
-Z	X

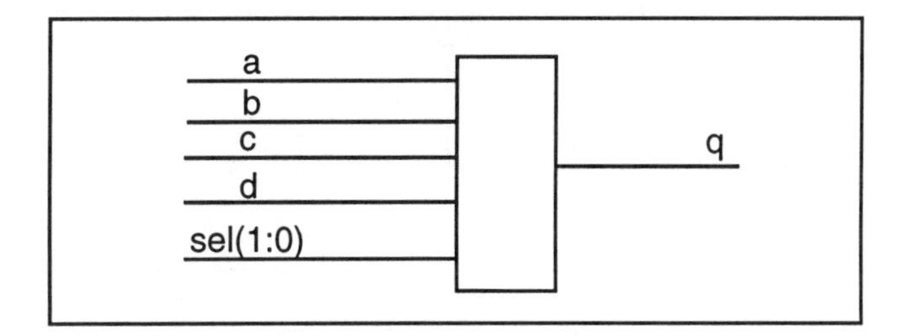

Signals *a*, *b* and *c* should be of type **std_logic**. The entity name should be **lab1.**

Signals *a*, *b*, *c*, *d* and *q* should be of type std_logic. Signal *sel* should be of type std_logic_vector(1 downto 0).

2. Simulate the component

3. Synthesize the component

Study the schematic.

Lab 2: Supplementary study

Do the following assignment if you have the time.

Modify the code in lab 2, so that the input signals *a*, *b*, *c* and *d* are of the type std_logic_vector with the length 4 bits.

- **Simulate**
- **Synthesize**
- **Study the schematic**

Lab 3: Sequential VHDL

1. Enter VHDL code

Design a component with the following behaviour in VHDL:

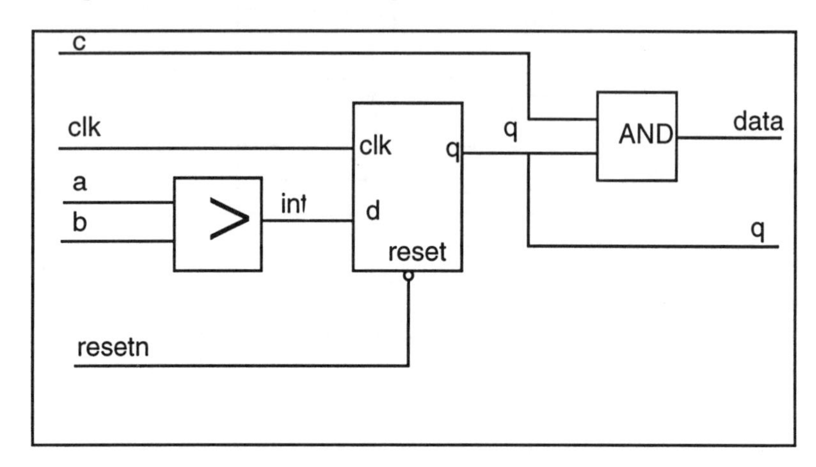

Signals *c, clk, int, q, resetn* and *data* should be of type **std_logic;**
Signals *a* and *b* should be of type **std_logic_vector(1 downto 0);**
Signal *int* should be equal to '1' if a>b else '0'.
The flip-flop should have synchronous reset.
The whole behaviour should be described in one architecture.

2. Simulate the component

3. Synthesize the component

Study the schematic.

Lab 3: Supplementary study

Do the following assignment if you have the time.

Modify the code in lab 3 so that the flip-flop gets asynchronous reset.

- **Simulate**
- **Synthesize**
- **Study the schematic**

Lab 4: Subprogram

1. Enter VHDL code

Create a package **mypack** which contain a function **max.** The function should return the largest vector. Then create an architecture (entity=lab4) which uses the function.

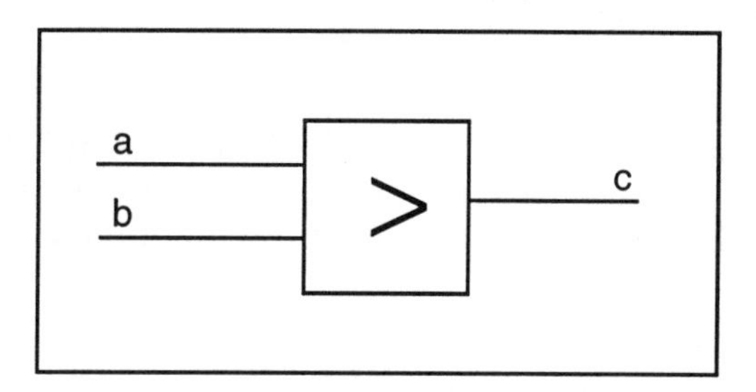

Signals *a*, *b* and *c* should be of type **std_logic_vector(2 downto 0).**

2. Simulate the component

3. Synthesize the component

Study the schematic.

Lab 4: Supplementary study

Do the following assignment if you have the time.

Design a component with the following behaviour in VHDL:

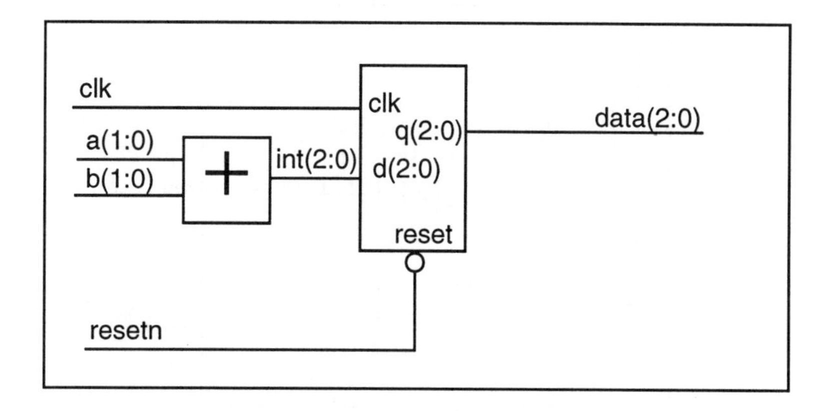

Signals *a* and *b* should be of type **std_logic_vector**.

ViewLogic users
For the addition use the function addum(v1,v2:vlbit_1d) return vlbit_1d. This function is ready-written and is in library std and package standard. For the function to be used, the library and package "have" to be declared.

```
Library  std;
Use   std.standard.ALL;
```

Synopsys users
For addition use the function "+"(a,b:std_logic_vector) return std_logic_vector. This function is ready-written and is in library ieee and package std_logic_unsigned. For the function to be used, the library and package have to be declared.

```
Library  ieee;
Use   ieee.std_logic_unsigned.ALL;
```

The flip-flop should have synchronous reset.

- **Simulate**
- **Synthesize**
- **Study the schematic**

Lab 5: Structural VHDL

1. Enter VHDL code

Design three components **K1, K2** and **K3** (K1=and, K2=or and K3=xor). Connect the component together to a 1-bit adder. Use structural VHDL.

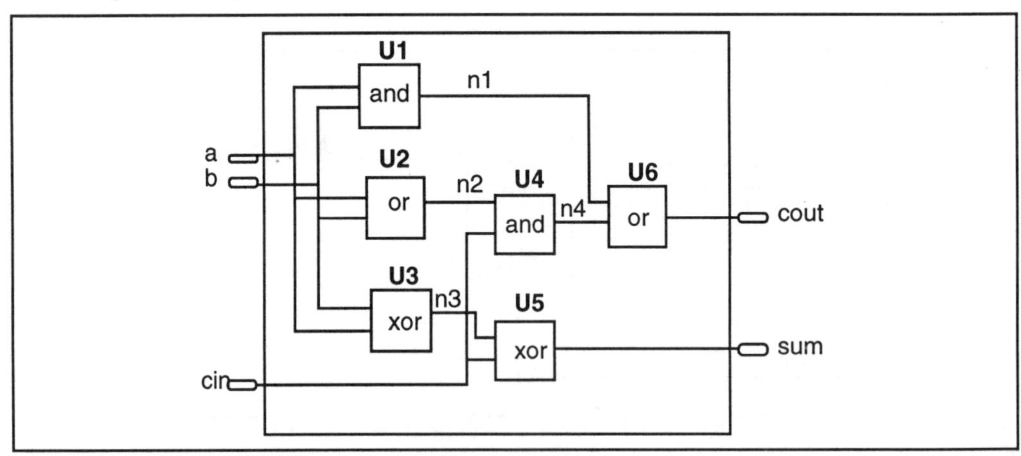

2. Simulate the component

Note that you have to compile four files before you can simulate the adder.

3. Synthesize the component

Study the schematic.

Lab 6: State machines

1. Enter VHDL code

Design a Mealy machine with the following behaviour:

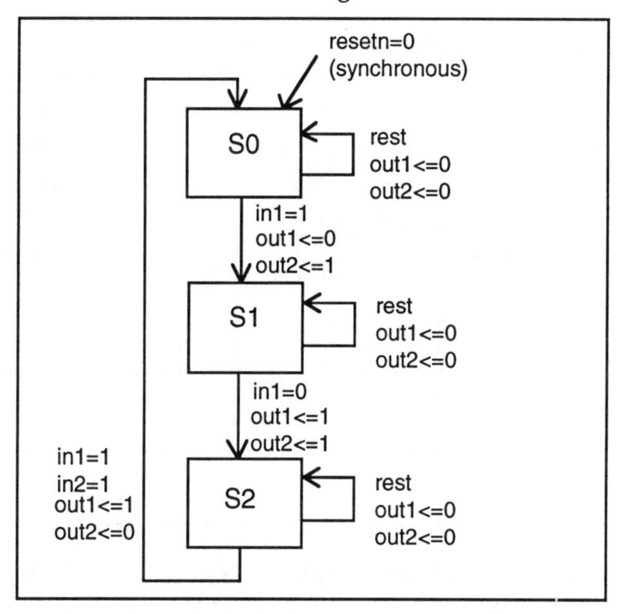

2. Simulate the component

3. Synthesize the component

Study the schematic.

Lab 6: Supplementary study

Do the following assignment if you have the time.

Redesign the Mealy machine to a clocked output Mealy machine.

- **Simulate**
- **Synthesize**
- **Study the schematic**

Lab 7: A complete design example

1. Enter VHDL code

Design a system with the following behaviour:

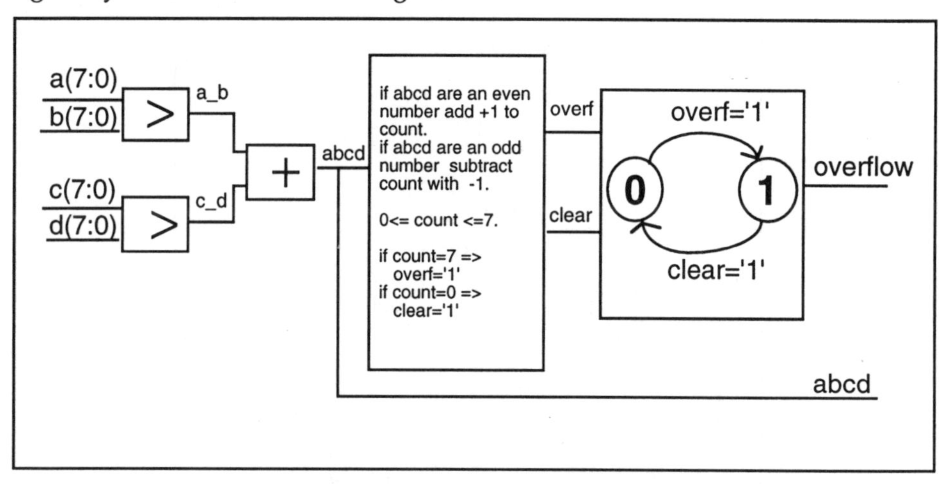

Signals *a, b, c, d, a_b, c_d* should be of type **std_logic_vector(7 downto 0).**
Signal *abcd* should be of type **std_logic_vector(8 downto 0).**
Signals *clk, resetn, overflow* should be of type **std_logic**.

Count is a counter which can have values between 0 and 7.
Declare *count* as a signal of type **std_logic_vector(2 downto 0).**

Use the function **max** in package **mypack** for "larger than" operators. For addition, use the function **addum** (ViewLogic) or "+" (Synopsys). For subtraction, use the function **subum** (ViewLogic) or "-" (Synopsys).

The state machine should be of type **output=state**, be clocked by *clk* and have asynchronous reset. Declare the state machine's state as type std_logic. Use two constants: **normal_state** and **of_state** as state names.

Make a process for the counter, which should be clocked by *clk* and have asynchronous reset. Make a separate process (combinational) for the comparators (=7 and =0).

ViewLogic users
Declare a variable *c_int* in the comparator of type integer. Assign it the value of *count* (c_int:=v1d2int(counter);). The conversion function **vld2int()** can also be used in the counter process. Declare *count* as a signal of type vlbitc_vector(3 downto 0).

2. Verify the components

Verify the component by designing a VHDL **testbench** of type 2, i.e. a testbench which generates input signals and checks the value of the output signals. The name of the testbench should be **tb_7**. The testbench should print out the result "**Test OK**" if the test is passed (use the assert command).

3. Synthesize the component

Study the schematic.

Lab 8: Controlling a stepping motor

This exercise is intended to provide practical training in designing Moore machines with VHDL.

Specification of the task

Task: To design and test a control system in a chip for a stepping motor.

Description of stepping motor

The stepping motor is a type of motor which can be controlled simply with ones and zeros. By applying different currents to the rotor windings, the rotor can be made to assume different positions. In this case we apply 5 V (1) or 0 V (0) to actuate the semiconductor switch.

Figure 18.1 shows the layout of the experiment board for the stepping motor.

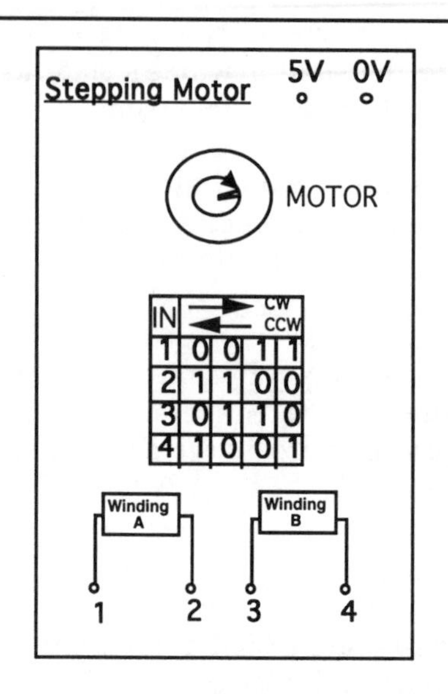

Figure 18.1: Stepping motor experiment board.

System design

Create a component (Figure 18.2) with:

inputs: *reset_n*, *clk*, *dir* and *puls*
outputs: *d_out*(3 downto 0)

clk: system clock approximately 20 Hz, connected the oscillator on the experiment board
reset_n: low, approximately 1 μs before the circuit is started, controlled manually
dir: determines the direction of the motor
puls: turns the shaft at the clk frequency

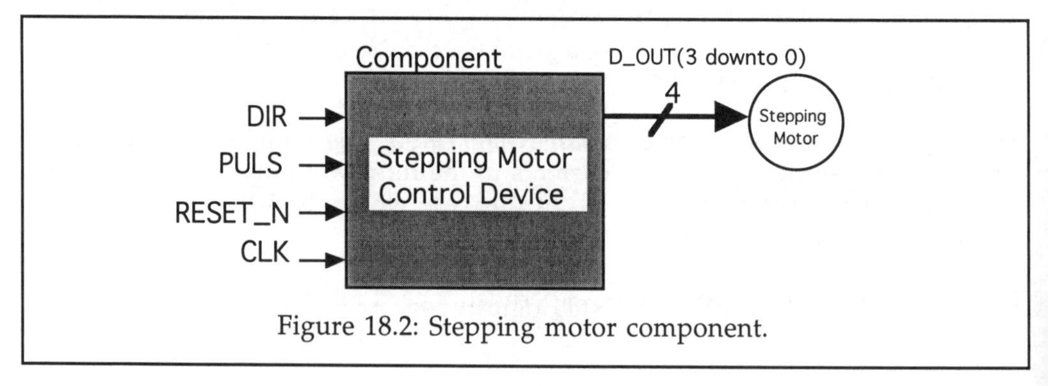

Figure 18.2: Stepping motor component.

Remember the following:

- The design is clocked with the clock.
- All internal signals and output signals should be set to a defined position by reset_n = '0'. See state machine in Figure 18.3.

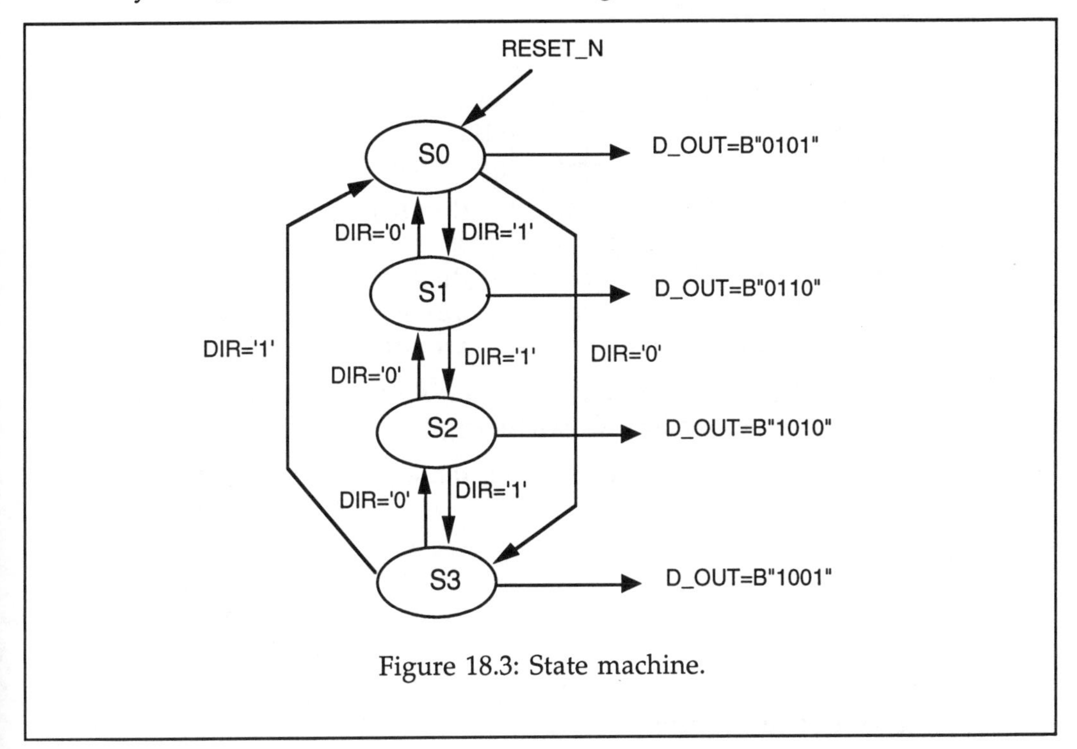

Figure 18.3: State machine.

1. Enter VHDL code

2. Verify the component

Verify the component by using the following VHDL testbench. The testbench is of type 2, i.e. it generates input signals and checks the value of the output signals. The testbench prints out the result "Test OK" if the test is passed.

Remember that ViewLogic users should replace std_logic with vlbit.

```
Library ieee;
Use ieee.std_logic_1164.ALL;

Entity tb_lab8 is
 port(Test_ok: out std_logic);
end;
```

```
Architecture rtl of tb_lab8 is
 component lab8
  port( clk,resetn,puls,dir: in   std_logic;
        d_out:               out  std_logic_vector(3 downto 0));
 end component;
 signal clk:              std_logic:='0';
 signal Test_ok_b:        std_logic:='H';
 signal resetn,puls,dir:  std_logic;
 signal d_out:            std_logic_vector(3 downto 0);
 constant S0:std_logic_vector(3 downto 0):="0101";
 constant length:integer:=26;
 constant S1:std_logic_vector(3 downto 0):="0110";
 constant S2:std_logic_vector(3 downto 0):="1010";
 constant S3:std_logic_vector(3 downto 0):="1001";
 type rom_table is array (integer range 0 to length) of
      std_logic_vector(3 downto 0);
 constant rom:rom_table:=rom_table'(  S0, S0, S0, S0, S3,
                                      S2, S1, S0, S1, S2,
                                      S3, S0, S0, S0, S1,
                                      S1, S1, S2, S2, S2,
                                      S3, S3, S3, S0, S3,
                                      S2, S0);
type d_type is (S_0, S_1,S_2, S_3);
signal dout:d_type;
begin
  U1:lab8 port map(clk,resetn,puls,dir,d_out);
  Test_ok<=Test_ok_b;
  clk<=not clk after 100 ns;
  resetn<= '0',
           '1' after 400 ns,
           '0' after 5050 ns;
  puls<= '0',
         '1' after 600 ns,
         '0' after 2200 ns,
         '1' after 2600 ns,
         '0' after 2800 ns,
         '1' after 3200 ns,
         '0' after 3400 ns,
```

```
                '1' after 3800 ns,
                '0' after 4000 ns,
                '1' after 4400 ns;

  dir<= '0',
              '1' after 1400 ns,
              '0' after 2400 ns,
              '1' after 2600 ns,
              '0' after 3000 ns,
              '1' after 3200 ns,
              '0' after 3600 ns,
              '1' after 3800 ns,
              '0' after 4200 ns,
              '1' after 4400 ns,
              '0' after 4600 ns;

  process
  variable i:integer range 0 to  length:=0;
  begin
    wait for 50 ns;

    while i<length loop
      if rom(i)/=d_out then
        Test_ok_b<='0';
      end if;
      i:=i+1;
      wait until clk='0';
    end loop;

    assert Test_ok_b/='H'
    report ("Design Lab8 passed test")
    severity note;

    wait;
  end process;

  dout<=   S_0 when d_out=S0 else
           S_1 when d_out=S1 else
           S_2 when d_out=S2 else
           S_3;

  Test_ok_b<=  'H' when (d_out="0101" or d_out="0110" or
                    d_out="1010" or d_out="1001") else
               '0';

end;
```

3. Synthesize the component

4. Instance buffers and pads

For implementation of the following part of the experiment it is assumed that you have access to ViewLogic, Xilinx software and experiment boards.

Create a new schematic (name CHIP)
Start the schematic program by clicking on **VIEWDRAW** in the COCKPIT window. Click on **OK** in the window.

In the ViewDraw window enter CHIP on the line. Enter name, then press OK. A new schematic called CHIP.1 should now be created.

Instancing pads and buffers (see Figure 18.4)
The next step is to fetch components for the schematic with **Add-Comp**. Press the project library symbol in the window, following which the components in your library should be visible at the top of the window. Select your symbol (lab 8) and click with the left button at the appropriate place in the schematic.

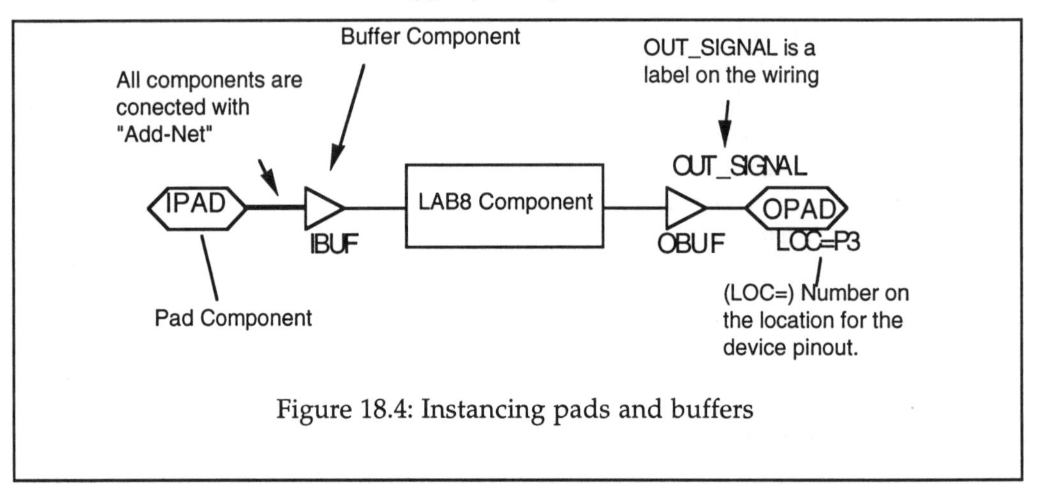

Figure 18.4: Instancing pads and buffers

Fetch the IBUF, OBUF, IPAD and OPAD components. They are in library xc4000. Press xc4000 when you have activated Add-Comp.

Buffers (IBUF and OBUF) are used to describe how an I/O block should be configured. Pads (IPAD and OPAD) are used to determine the pin number on the circuit.

Connect the components with interconnections
Connect all the components with **Add-Net**. Draw a network (interconnection) between pins on symbols. Start and finish by pressing the middle button.

Name the interconnections
The next step is to name the interconnections which are interesting to monitor during simulation. We recommend that you name those interconnections which go to the pads. This is done using **Add-Label**. Click on the interconnection with the left button. Then select ADD-LABEL. Enter the name after "**Label String:**". Position the text in a suitable place, e.g. above the interconnection, using the middle button.

Determine the pin numbering
The next step is to determine the pin number for the pads. This is done by marking a pad and then selecting Add-Attr. For example, write LOC=P3, which means pin 3. Position the text with the middle button. Use pins 3, 4, 13, 14, 15, 16, 17, 18 and 19.

Save the schematic
The final step is to save the design. This is done using File - Write.

5. Wiring and downloading to the real chip

Convert the design to a Xilinx loading module
Start the **MS-DOS prompt** (if you are short of memory, close Windows) and change to your project directory.

Start the Xilinx shell program **XDM**. XDM is structured with a menu at the top.
Activate **Profile - Family** in the menu, select **xc4000.**
Activate **Profile - Part** in the menu, select **4005pc84**. This means approximately 5000 gates and an 84-pin package. The Speed menu will appear automatically, select **-5.**

The chosen circuit has now been described. Start the translation with **Translate Xmake Done,** select **Lab8** and **Bitstream.**

There is a report file (Lab8.rpt) in the project directory. Study the contents of the file.

Creating a simulatable file with fan-out and interconnection capacitance (supplementary study)

Once the design has been mapped on the chip, a simulatable model of the chip can be created. This is done in XDM with menu selection: **Verify Xsimmake,** select **-F (flow..)**. Select **Viewlogic_timing.** The design can then be simulated in ViewLogic with real delays.

Downloading to a physical prototype
Connect a downloading cable from the PC's serial port to the contact on the experiment board and a power supply to the experiment board. Connect the power. Also connect external components such as stepping motor, contacts and LEDs. Give the following command in XDM: **Verify xchecker Done, lab8**

The download program now starts. Press <enter> to download. Finally the circuit is validated against the specification.

Lab 8: Supplementary study

Do the following assignment if you have time.

Describe the state machine in lab 8 as a clocked Mealy machine.

- **Simulate**
- **Synthesize**
- **Instance buffers and pads**
- **Wiring and downloading to the real chip**

19

Answers

19.1. Solutions to selected exercises

Chapter 1

1. **Advantages:**

- It makes it possible to verify electronic functions or models without knowing the implementation technology.
- It makes it possible to work at a higher level of abstraction than traditional gate level.
- It makes it possible to write programs which can also be used as a specification.
- It makes it possible to reuse components.
- It is possible to buy standard components in VHDL (not synthesizable) from a number of suppliers. These components can be used to verify the design.
- VHDL is a standardized language. The standard makes it possible to change development tools. VHDL is a text file which can easily be moved between different computer systems.
- VHDL supports both top-down and bottom-up methodology.

Disadvantages:

- VHDL is not yet standardized for analog electronics.
- VHDL for synthesis is not standardized.

2. The US military.

3. To obtain a uniform specification description for electronic systems.

4. See section 1.3.

5. (a)

(b)

6.

Exercise		Logic synthesis	RTL synthesis	Behavioural synthesis
(a)	a:= b and c;	x	x	x

(b)	if b='1' then
	c:='1';
	else
	c:='0';
	end if;
		x (1)	x	x

(c)	wait until clk='1'; d <= '1';		x (2)	x
(d)	a<= (b*d) + (c*e);		x (3)	x
(e)	wait for 1 ms;			x (4)

(1) This is a Boolean expression.

(2) In order to wait for a signal before a certain value is to be output on *d*, a flip-flop is required. Logic synthesis cannot synthesize this.

(3) Translating "a<=(b*d) + (c*e)" with RTL synthesis produces two multipliers. Behavioural synthesis produces only one multiplier, i.e. a much smaller design. One exception: very short time constraints which can "force" the behavioural synthesis to generate two multipliers. The symbol "<=" means that the value is saved in a register. Therefore it is not possible to synthesize the logic in the exercise.

(4) A 1 ms clock is created from a reference clock. If the reference clock is 1 μs, for example, a counter is instanced which creates the 1 ms clock. Behavioural synthesis should manage this automatically. RTL synthesis will not manage to create an extra counter.

Chapter 2

2. Behavioural or functional level. This is because it is faster to build and to simulate the models. The disadvantage is that the accuracy is worse.

3. The highest level of abstraction which can be synthesized from automatically should be chosen. This is currently RTL as well as some behavioural designs in VHDL.

4.

```
Entity John is
   port( a_in:   in    std_logic;    -- input
         b_in:   in    std_logic;    -- input
         c_in:   in    std_logic;    -- input
         d_out:  out   std_logic);   -- output
```

```
end;
```

Alternative:

```
entity John is
   port(  a_in, b_in, c_in:     in     std_logic;
          d_out:                out    std_logic);
end;

architecture rtl of John is
--* no architecture_declarative_part --
begin
      d_out <= (a_in and b_in) and c_in;
end;
```

5. See in Chapter 2.

6. See in Chapter 2.

7. It means that the rest of the line is a comment.

8. See example at the end of section 2.3.1.

Chapter 3

1. <=

2. Inertial and transport delay.

3.

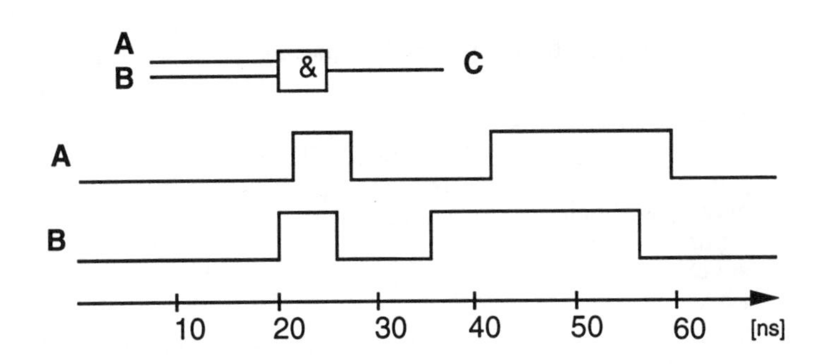

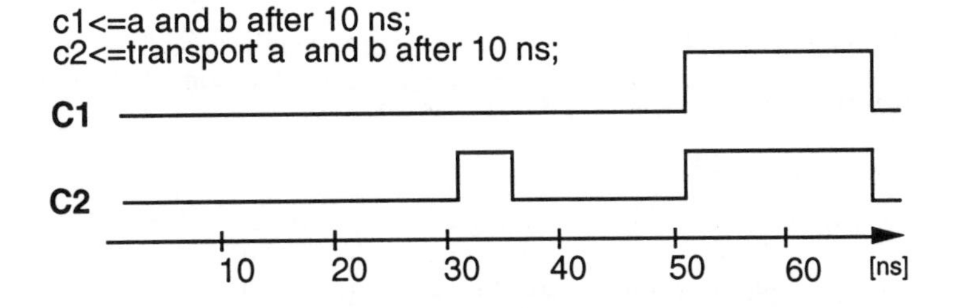

4.

```
Architecture rtl of ex is
begin
  b<=a;
  c<=b;
end;
```

	0	10	10+1delta	10+2delta	20	20+1delta	20+2delta
A	0	1	1	1	0	0	0
B	0	0	1	1	1	0	0
C	0	0	0	1	1	1	0

```
Architecture rtl of ex is
begin
  c<=b;
  b<=a;
end;
```

	0	10	10+1delta	10+2delta	20	20+1delta	20+2delta
A	0	1	1	1	0	0	0
B	0	0	1	1	1	0	0
C	0	0	0	1	1	1	0

5.

(a) **When else** must be ended with an **else** condition.
(b) q<= a when sel='0' else b;

(c) It is allowed to end the command with a **when** expression.

(d) Yes.

6. Note, Warning, Error, Failure. Default is Error

7.

```
Architecture rtl of ex is
signal a: std_logic_vector(2 downto 0);
begin
  assert a/="111"
  report "to many ones"
  severity error;
  a<=c and b;
end;
```

8.

(a) Constant, signals and variables.

(b) See section 3.11

(c) std_logic and std_ulogic, see section 3.11

9. See section 3.12.1.

10.

(a) Yes.

(b) Example: Architecture rtl of ex is

```
signal a: integer;
begin
  a<=16#FF#;
end;
```

11. See section 3.12.2.

12. Length-independent assignment.

13. *a*="01000", *b*="00010"

14.
(a) 'T'
(b) '0'
(c) 'U'
(d) Yes, see section 3.17.

Chapter 4

1.
(a) Combinational and clocked processes
(b) Combinational processes produce combinational logic and clocked processes give rise to memory elements.

2.
(a) A latch on the output.
(b) Include all signals which are "read" in the sensitivity list.

3.
(a) Different synthesis tools support different description methods. All are within the framework of the VHDL standard, however.
(b) See section 4.12.

4. See section 4.3.2.

5. 5

6.
(a) Clocked and asynchronous reset.
(b) Asynchronous reset.

7. Two asynchronous resets and one clocked reset.

8. 2

9.
(b)
```
Architecture rtl of c1 is
begin
  process(a,b,sel)
  begin
    case sel is
      when "00"    => q<=a xor b;
      when "01"    => q<=a or b;
      when "10"    => q<=a nor b;
      when "11"    => q<=a and b;
```

```
            when  others  =>q<="XX";
          end case;
        end process;
      end;
```

(d) If an advanced synthesis tool is used, the synthesis result will be identical.

10. (a) 4, 5 and 6.

(b) (1)

(2)

(3)

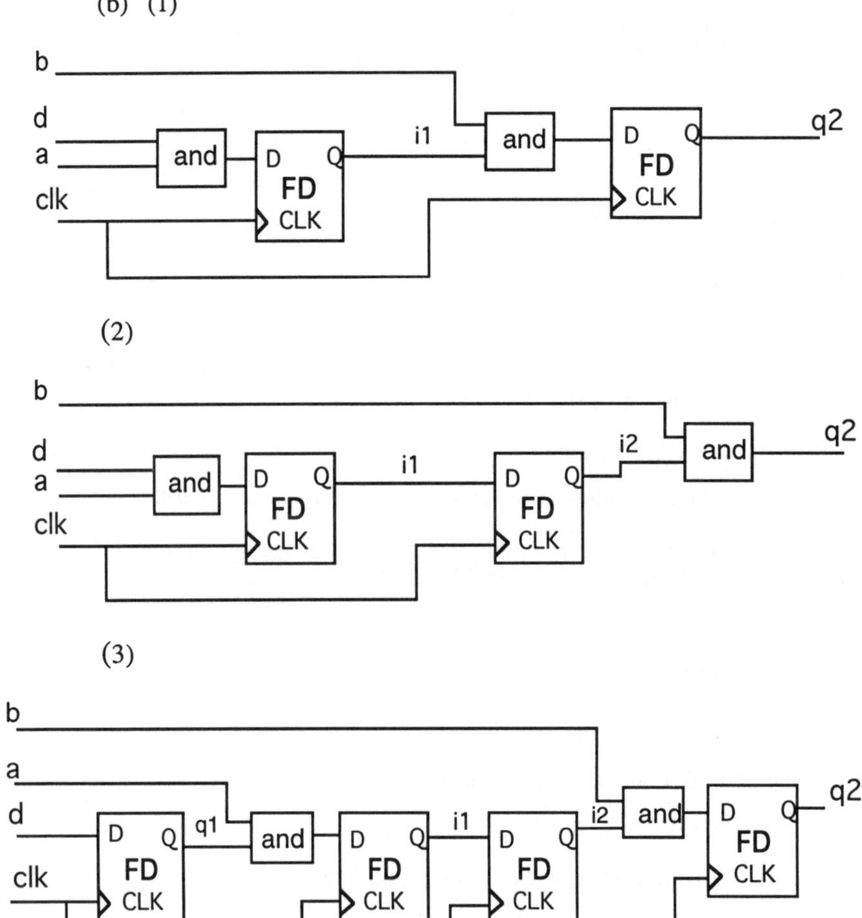

(4), (5) and (6)

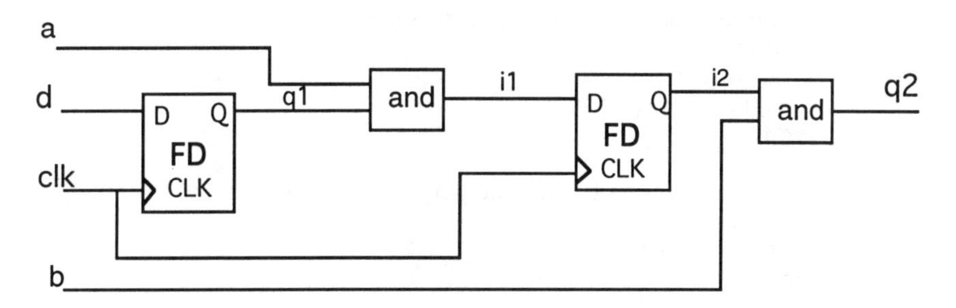

Chapter 5

1. (a) In library std and package standard.

(b) Library work, std;
Use std.standard.ALL;

2. Functions, procedures, types and constants.

3. (a) Package, architecture and process.

(b) Package

(c) Library <libary_name>;
Use <libary_name>.<package_name>.ALL;

4. See section 5.3.3.

5. See section 5.3.

6. To obtain length-independent functions.

7. See section 5.4.

8. (a)

```
Library ieee;
Use ieee.std_logic_1164.ALL;
Use ieee.std_logic_unsigned.ALL;

Package mypack is
function average1(a,b: in integer) return integer;
function average2(a,b: in std_logic_vector) return
  std_logic_vector;
function sum1(a,b: in integer) return integer;
function sum2(a,b: in std_logic_vector) return std_logic_vector;
end;

Package body of mypack is
function average1(a,b: in integer) return integer is
begin
  return (a+b)/2;
end;
```

```
function average2(a,b: in std_logic_vector) return
  std_logic_vector is
variable int: std_logic_vector(a'range);
begin
  int:=a+b;
  return shr(int,"1");
end;

function sum1(a,b: in integer) return integer is
begin
  return (a+b);
end;

function sum2(a,b: in std_logic_vector) return std_logic_vector is
begin
  return (a+b);
end;
end;
```

(b)
```
Library ieee;
Use ieee.std_logic_1164.ALL;
Use work.mypack.ALL;

Entity c1 is
  port( a,b:    in  integer range 0 to 127;
        c,d:    in  std_logic_vector(7 downto 0);
        q1,q2:  out integer range 0 to 127;
        q3,q4:  out std_logic_vector(7 downto 0));
end;

Architecture rtl of c1 is
begin
  q1<=average1(a,b);
  q2<=sum1(a,b);
  q3<=average2(c,d);
  q4<=sum2(c,d);
end;
```

9. See section 5.5.

10. Only alternative 2.

11. (a) See section 5.6.

(b) Yes, see section 5.6.

Chapter 6

1. See beginning of Chapter 6.

2.
```
Architecture rtl of top is
component c1
  port( a,b:  in    std_logic;
```

```
        q1:   out    std_logic);
end component;

component c2
   port( d1,d2:   in     std_logic;
         q:       out    std_logic);
end component;

signal i1,i2: std_logic;
For U1,U2: c1 Use entity work.c1(rtl);
For U3: c2 Use entity work.c2(rtl);
begin
      U1: c1 port map(a,b,i1);
      U2: c1 port map(c,d,i2);
      U3: c3 port map(i1,i2,q);
end;
```

3. (a) No.

   (b) Yes.

4. For simulation: For U1,U2: c1 Use entity work.c1(rtl); In the case of synthesis the last architecture to be entered in the tool is used.

5. Generic map.

6. See section 6.3.

7.

```
Architecture rtl of top is
  Component c1
     port( a,b:  in    std_logic;
             q:  out   std_logic);
  end component;
  begin
  For i in 0 to 4 generate
    U: c1 port map(a(i), b(i), q(i));
  end;
end;
```

Chapter 7

1. See in chapter 7.

2. The three methods are instantiation, register, and behavioural synthesis; see section 7.2.

3.

```
Library ieee;
Use ieee.std_logic_1164.ALL;

Entity c1 is
  port(  write,a0,a1:   in    std_logic;
         d_in:          in    std_logic_vector(7 downto 0);
         d_out:         out   std_logic_vector(7 downto 0));
```

```
Architecture rtl of RAM4 is
Component RAM4x1
  port(  d,a0,a1,we:     in     std_logic;
         q:              out    std_logic);
end component;
--for U1:RAM4x1 use Entity work.RAM4x1(rtl);   -- Simulation
                                               -- only.
begin
  for i in 0 to 7 generate
     RAM_block: RAM4x1 port map (d_in(i),a0,a1,write,d_out(i));
  end generate;
end;
```

Chapter 8

1. Input stimuli and verifying the outputs

2. See Table 8.1.

3. See section 8.4.

4. See section 8.5.

5. See in Chapter 8.

Chapter 9

1. See in Chapter 9.

2.
(a) No
(b) Moore and Mealy machines with clocked outputs and output=state machines.
(c) If the output signal drives three-state enable or is used as a clock.

3. Yes, see section 9.9.

4. See section 9.8.

5.
(a) No
(b) Output=state machine
(c) See section 9.7.
(d) Yes.
(e) Sequential: 3 and one-hot: 5 .

6. Mealy machine.

7.
(a) The output signals must not have the same output signal values in different states.
(b) Yes, see section 9.9.

8. The Moore machine is given a latch on the output, and on the Moore machine with clocked outputs the outputs are multiplexed back to the output flip-flop; see section 9.9.

9.
(a) See section 9.4
(b) See section 9.1
(c) See section 9.5.
(d) See section 9.2.
(e) See section 9.6.
(f)

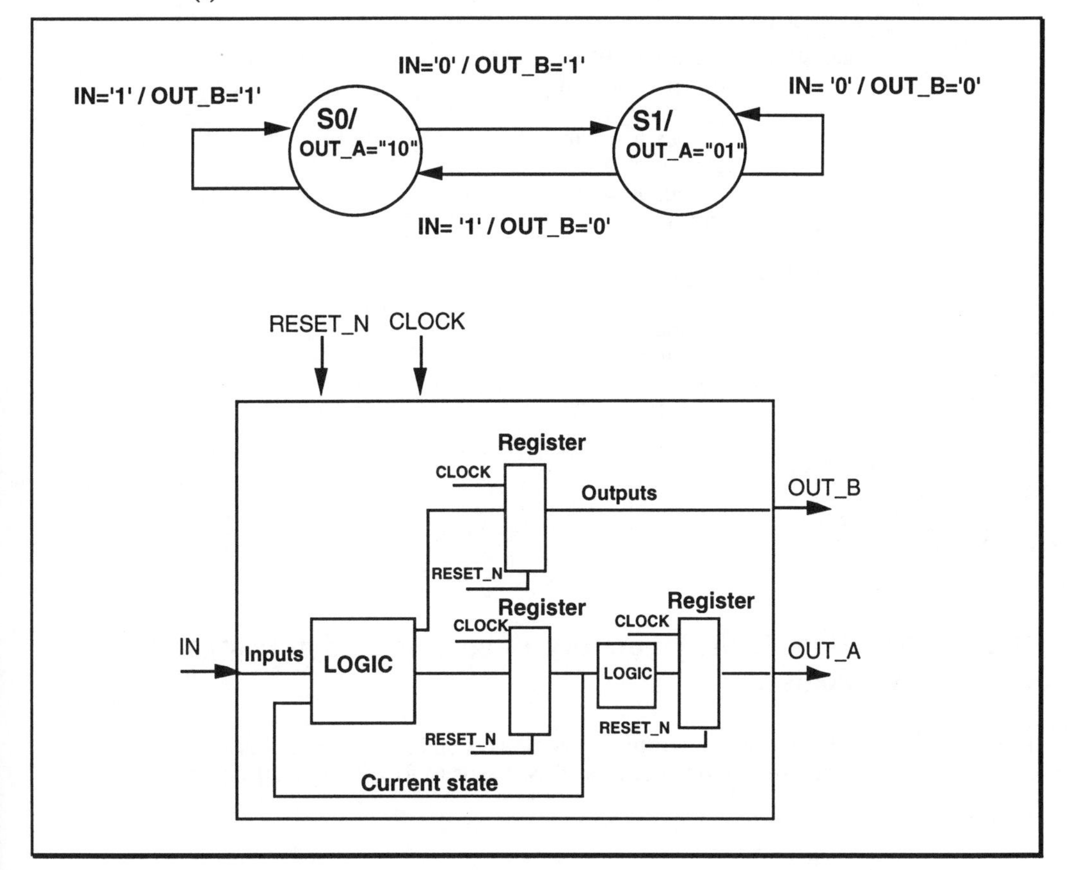


```
Entity  exercise_F  is
 port(  clock ,in1, reset:    in     std_logic
        out_a :               out    std_logic_vector(1  downto  0);
        out_b :               out    std_logic  )
end;
```

```
Architecture  rtl  of  exercise_F  is
type  state_type  is  (s0,  s1);
signal  state:state_type;
begin
   exerc_5_F_process:  process
   begin
     wait until  clock='1';

     if reset_n= '0' then
       state<=s0;
       out_a<= "10";
       out_b <= '1';

   else
       case  state  is
          when s0=>     out_a <= "10";
                        if   in1='0'then
                             state<=s1;
                             out_b <= '0';
                        else
                             out_b<='1';
                        end  if;

          when s1=>     out_a <= "01";
                        if  in1='1'then
                             state<=s0;
                             out_b <= '1';
                        else
                             out_b<='0';
                        end  if;
       end  case;
     end if;
   end process;
end;
```

19.2. Solutions to the laboratory assignments - Synopsys and Autologic 2

The solutions in this section are written for Synopsys and Autologic 2, but in the majority of cases they can also be used for other advanced synthesis tools without modification.

Lab 1

```
Library  ieee;
Use  ieee.std_logic_1164.ALL;

Entity  lab1  is
   port( a,b:  in      std_logic;
            c:  out     std_logic);
end;

Architecture  rtl  of  lab1  is
```

```
begin
   c<=a nor b;
end;
```

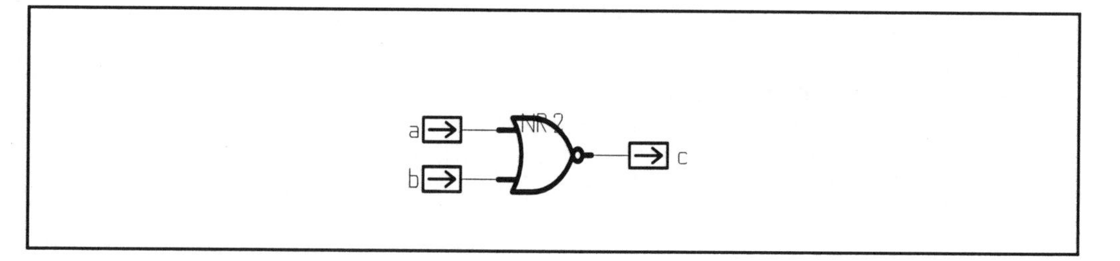

Lab 2

```
Library  ieee;
Use  ieee.std_logic_1164.ALL;

Entity  lab2  is
   generic  (Tdelay:time:=10  ns);
  port(  a,b,c,d:  in      std_logic;
         sel:      in      std_logic_vector(1  downto  0);
         q:        out     std_logic);
end;

Architecture  rtl  of  lab2  is
begin
 q<=   a  after  Tdelay  when  sel="00"  else
       b  after  Tdelay  when  sel="01"  else
       c  after  Tdelay  when  sel="10"  else
       d  after  Tdelay  when  sel="11"  else
       'X'  after  Tdelay;
end;
```

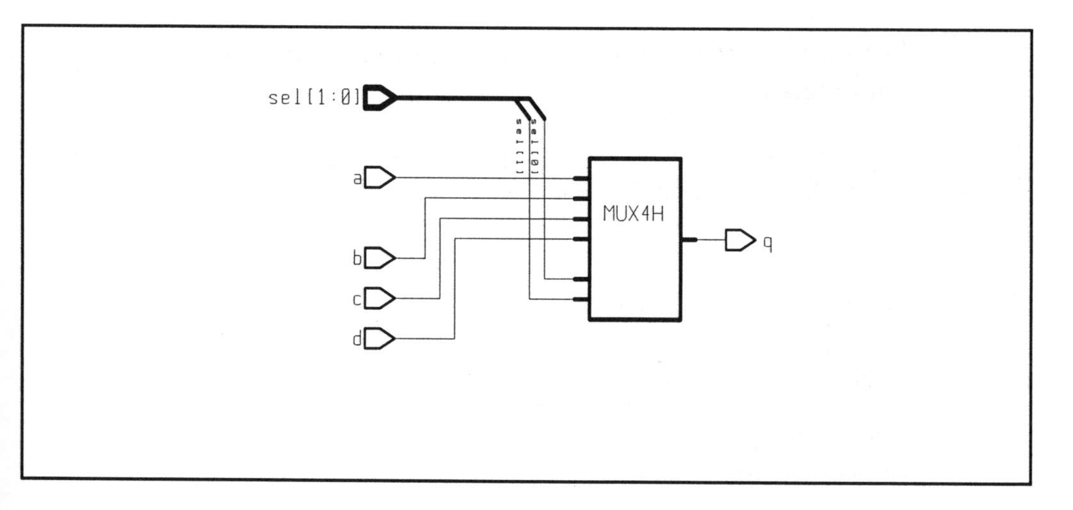

Lab 2 Supplementary study

```
Library ieee;
Use ieee.std_logic_1164.ALL;

Entity lab2 is
   generic (Tdelay:time:=10 ns);
  port(  a,b,c,d:   in    std_logic_vector(3 downto 0);
         sel:       in    std_logic_vector(1 downto 0);
         q:         out   std_logic_vector(3 downto 0));
end;

Architecture rtl of lab2b is
begin
 q<= a after Tdelay when sel="00" else
       b after Tdelay when sel="01" else
       c after Tdelay when sel="10" else
       d after Tdelay when sel="00" else
       (others=>'X') after Tdelay;
end;
```

Lab 3

```
Library ieee;
Use ieee.std_logic_1164.ALL;

Entity lab3 is
  port(  c,resetn,clk:  in        std_logic;
         a,b:           in        std_logic_vector(1 downto 0);
         q:             buffer    std_logic;
         data:          out       std_logic);
end;

Architecture rtl of lab3 is
signal int:std_logic;
begin
  int<='1' when a>b else '0';

   process(clk)
  begin
    if clk'event and clk='1' then
      if resetn='0' then
     q<='0';
    else
      q<=int;
     end if;
   end if;
  end process;

  data<=c and q;
end;
```

Input signal *q* is of type buffer because it is reread on the line data<=a and q. This means that it cannot be declared as out. Alternatively, it is possible to work with a copy of the signal and then assign *q* its value. If this method is used, *q* can be declared as out.

```
Architecture rtl of lab3 is
signal int,q_b:std_logic;
begin
 q<=q_b;

  int<='1' when a>b else '0';

   process(clk)
  begin
    if clk'event and clk='1' then
      if resetn='0' then
      q_b<='0';
    else
      q_b<=int;
    end if;
   end if;
  end process;

  data<=c and q_b;

end;
```

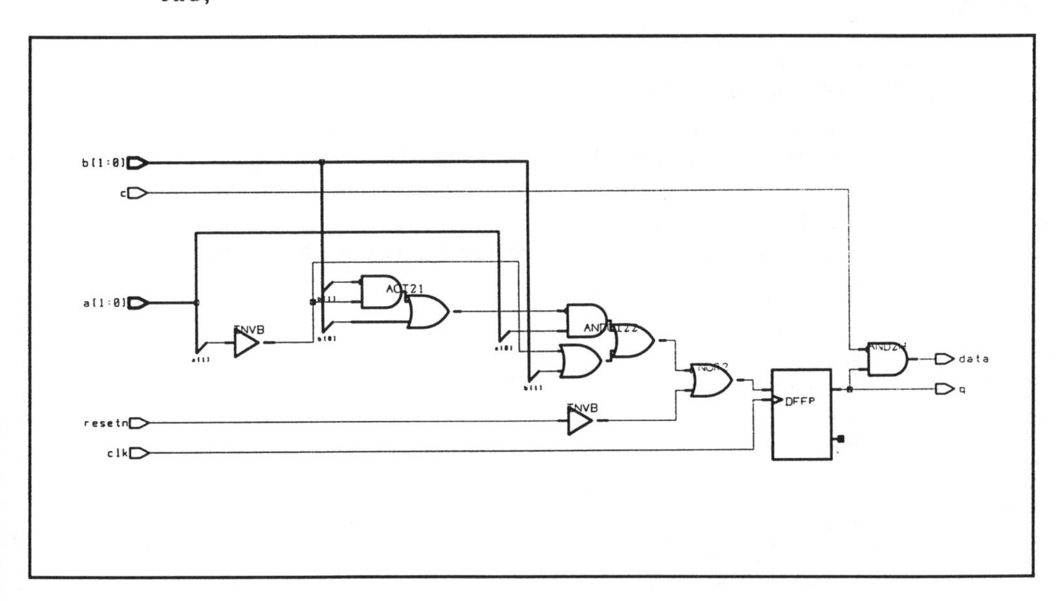

Lab 3 Supplementary study

```
Library ieee;
Use ieee.std_logic_1164.ALL;
```

```
Entity lab3b is
  port( c,resetn,clk: in   std_logic;
          a,b:        in   std_logic_vector(1 downto 0);
        data,q:       out  std_logic);
end;

Architecture rtl of lab3 is
signal int,q_b:std_logic;
begin
 q<=q_b;
 int<='1' when a>b else '0';

   process(clk,resetn)
  begin
    if resetn='0' then
    q_b<='0';
    elsif clk'event and clk='1' then
    q_b<=int;
   end if;
  end process;

 data<=a and q_b;
end;
```

Lab 4

```
Library ieee;
Use ieee.std_logic_1164.ALL;

Package mypack is
   function max(a,b:in std_logic_vector) return std_logic_vector;
end;

Package body mypack is
      function max(a,b:in std_logic_vector) return std_logic_vector is
    begin
     if a>b then
        return a;
    else
        return b;
    end if;
   end;
end;

Library ieee;
Use ieee.std_logic_1164.ALL;
Use work.mypack.ALL;

Entity lab4 is
  port( a,b: in  std_logic_vector(2 downto 0);
          c: out std_logic_vector(2 downto 0);
end;

Architecture rtl of lab4 is
begin
```

```
  c<=max(a,b);
end;
```

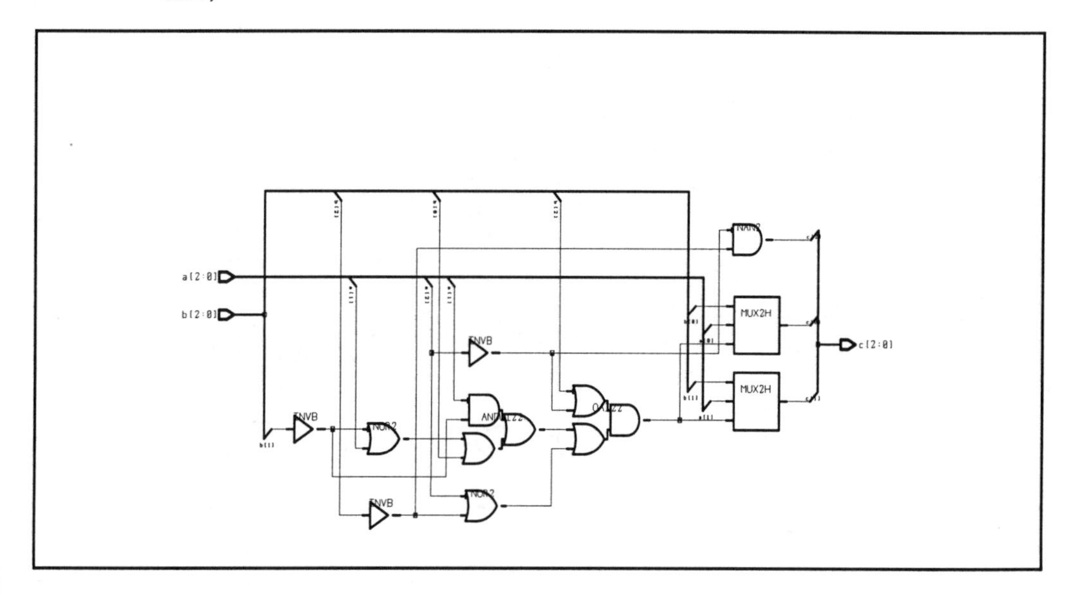

Lab 4 Supplementary study

```
Library  ieee;
Use  ieee.std_logic_1164.ALL;
Use  ieee.std_logic_unsigned.ALL;

Entity lab4b is
  port( clk,resetn:  in    std_logic;
        a,b:         in    std_logic_vector(1 downto 0);
        data:        out   std_logic_vector(2 downto 0));
end;

Architecture rtl of lab4b is
signal int:std_logic_vector(2 downto 0);
begin
  int<=('0' & a) + b;
   process(clk)
  begin
    if clk'event and clk='1' then
      if resetn='0' then
        data<=(others=>'0');
     else data<=int;
     end if;
   end if;
  end process;
end;
```

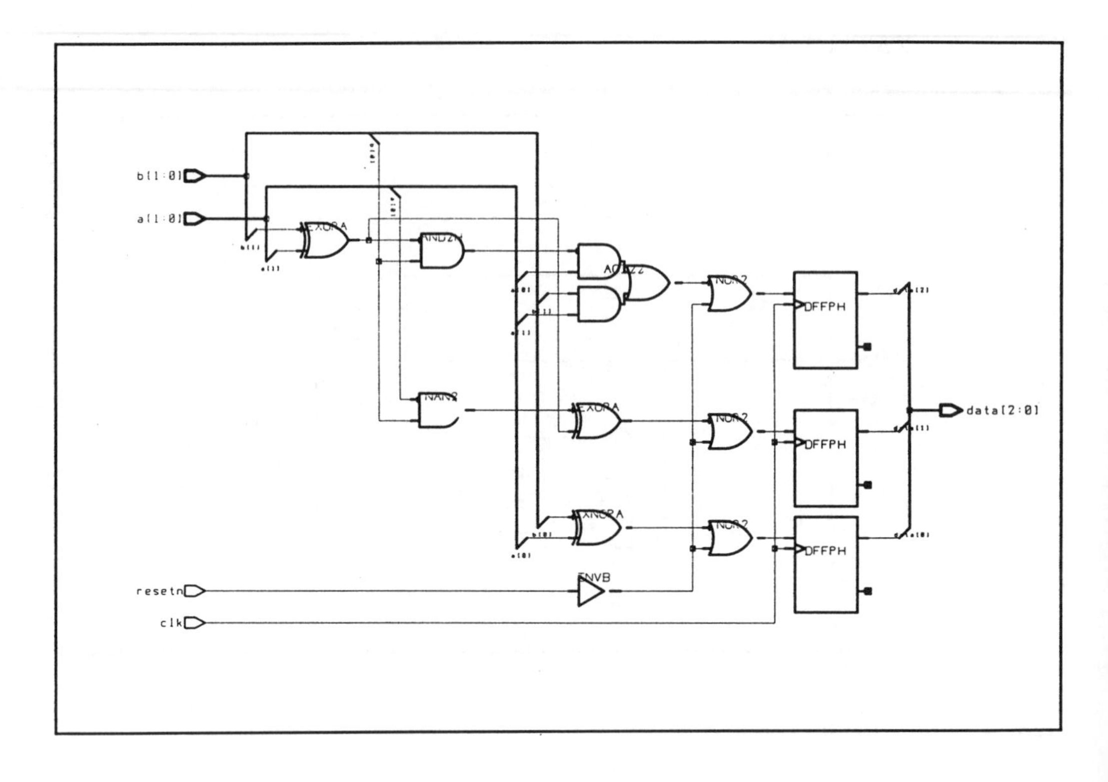

Lab 5

```
Library ieee;
Use ieee.std_logic_1164.ALL;
Entity k1 is
  port( a,b:  in    std_logic;
        c:    out   std_logic);
end;

Architecture rtl of k1 is
begin
  c<=a and b;
end;

Library ieee;
Use ieee.std_logic_1164.ALL;
Entity k2 is
  port( a,b:  in    std_logic;
        c:    out   std_logic);
end;

Architecture rtl of k2 is
begin
  c<=a or b;
end;
```

```
Library ieee;
Use ieee.std_logic_1164.ALL;
Entity k3 is
  port( a,b:  in   std_logic;
        c:    out  std_logic);
end;

Architecture rtl of k3 is
begin
 c<=a xor b;
end;

Library ieee;
Use ieee.std_logic_1164.ALL;

Entity lab5 is
  port( a,b,cin:   in   std_logic;
        cout,sum:  out  std_logic);
end;

Architecture rtl of lab5 is
component k1
   port( a,b:  in   std_logic;
         c:    out  std_logic);
end component;

component k2
   port(a,b: in  std_logic;
       c:  out   std_logic);
end component;

component k3
   port(a,b: in  std_logic;
       c:  out   std_logic);
end component;

for U1,U4:k1 Use entity work.k1(rtl);
for U2,U6:k2 Use entity work.k2(rtl);
for U3,U5:k3 Use entity work.k3(rtl);

signal n1,n2,n3,n4:std_logic;
begin
    U1:k1 port map(a,b,n1);
    U2:k2 port map(a,b,n2);
    U3:k3 port map(a,b,n3);
    U4:k1 port map(n2,cin,n4);
    U5:k3 port map(n3,cin,sum);
    U6:k2 port map(n1,n4,cout);
end;
```

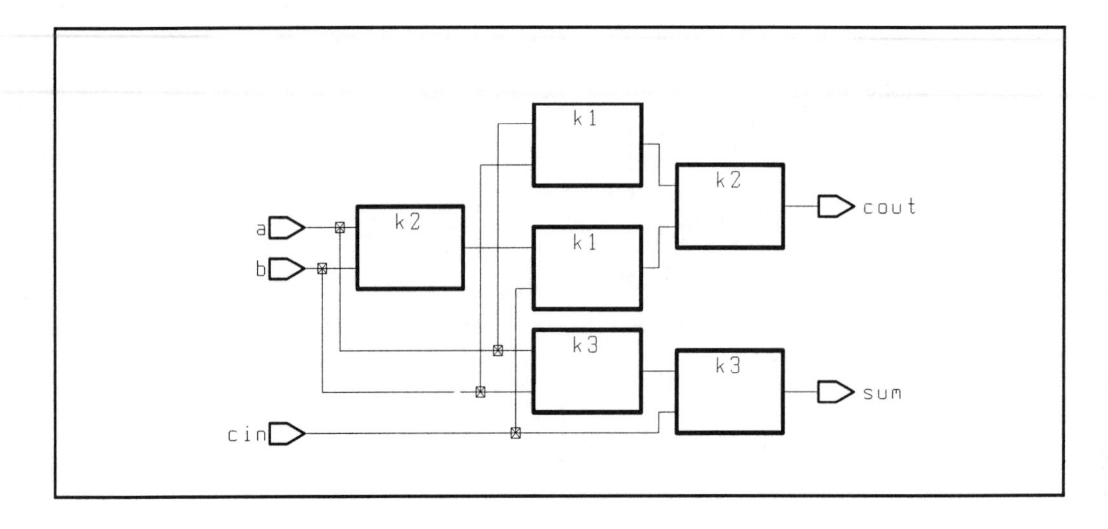

Lab 6

```
Library ieee;
Use ieee.std_logic_1164.ALL;

Entity lab6 is
  port( clk,resetn,in1,in2: in    std_logic;
        out1,out2:          out   std_logic);
end;

Architecture rtl of lab6 is
type state_type is (s0,s1,s2);
signal state:state_type;
begin
    process(clk)
   begin
      if clk'event and clk='1' then
         if resetn='0' then
         state<=s0;
      else
          case state is
           when s0 =>   if in1='1' then
                           state<=s1;
                        end if;
           when s1 =>   if in1='0' then
                           state<=s2;
                        end if;
           when s2 =>   if in1='1' and in2='1' then
                           state<=s0;
                        end if;
          end case;
       end if;
    end if;
   end process;
```

```
    process(state,in1,in2)
   begin
     case state is
       when s0 =>  if  in1='1'  then
                      out1<='0';
                      out2<='1';
                   else
                      out1<='0';
                      out2<='0';
                   end  if;
       when s1 =>  if  in1='0'  then
                      out1<='1';
                      out2<='1';
                   else
                      out1<='0';
                      out2<='0';
                   end  if;
       when s2 =>  if  in1='1'  and  in2='1'  then
                       out1<='1';
                      out2<='0';
                   else
                      out1<='0';
                      out2<='0';
                   end  if;
     end case;
   end process;
end;
```

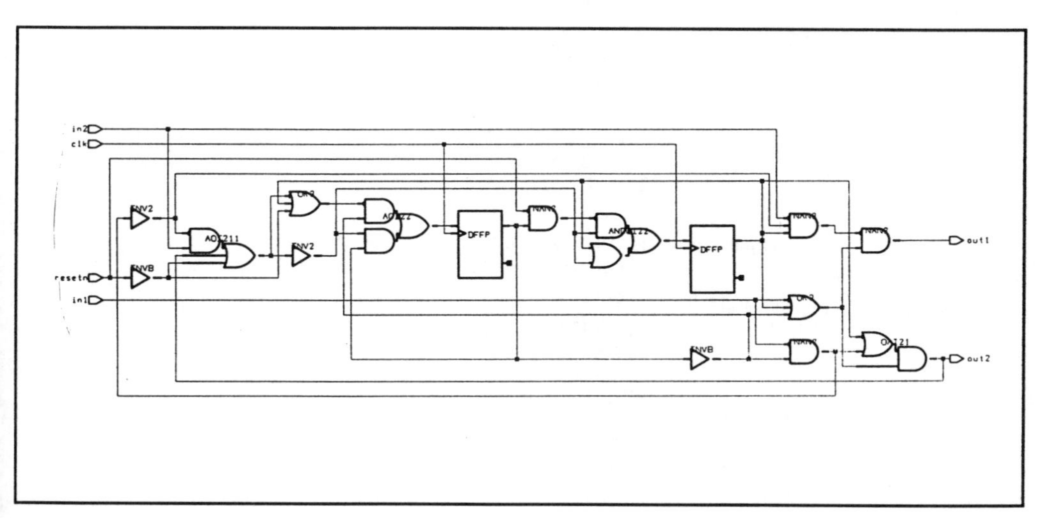

Lab 6 Supplementary study

```
Library  ieee;
Use  ieee.std_logic_1164.ALL;
```

```
Entity lab6b is
  port( clk,resetn,in1,in2:   in    std_logic;
        out1,out2:            out   std_logic);
end;

Architecture rtl of lab6b is
type state_type is (s0,s1,s2);
signal state:state_type;
begin
    process(clk)
   begin
     if clk'event and clk='1' then
       if resetn='0' then
      state<=s0;
      out1<='0';
      out2<='0';
     else
        case state is
         when s0 => if in1='1' then
                      state<=s1;
                      out1<='0';
                      out2<='1';
                    else
                      out1<='0';
                      out2<='0';
                    end if;

         when s1 => if in1='0' then
                      state<=s2;
                      out1<='1';
                      out2<='1';
                    else
                      out1<='0';
                      out2<='0';
                    end if;

         when s2 => if in1='1' and in2='1' then
                      state<=s0;
                      out1<='1';
                      out2<='0';
                    else
                      out1<='0';
                      out2<='0';
                    end if;
        end case;
      end if;
    end if;
   end process;
```

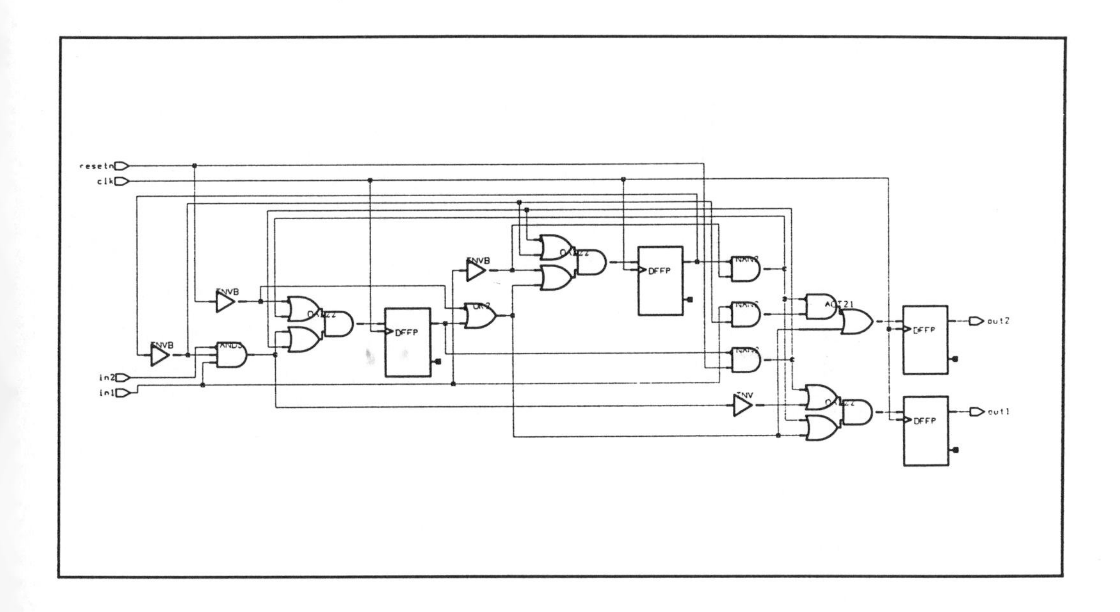

Lab 7

```
Library ieee;
Use ieee.std_logic_1164.ALL;
Use ieee.std_logic_unsigned.ALL;
Use work.mypack.ALL;

Entity lab7 is
  port( a,b,c,d:      in    std_logic_vector(7 downto 0);
        clk,resetn:   in    std_logic;
        overflow:     out   std_logic;
        abcd:         out   std_logic_vector(8 downto 0));
end;

Architecture rtl of lab7 is
signal ab_max,cd_max: std_logic_vector(7 downto 0);
signal count: std_logic_vector(2 downto 0);
signal overf,clear: std_logic;
signal abcd_b: std_logic_vector(8 downto 0);

constant s0: std_logic:='0';
constant s1: std_logic:='1';
signal state: std_logic;
begin
  abcd<=abcd_b;
  ab_max<=max(a,b);
  cd_max<=max(c,d);
  abcd_b<=('0' & ab_max) + cd_max;

    process(clk,resetn)
  begin
```

```
    if resetn='0' then
     count<='0';
    elsif clk'event and clk='1' then
      if abcd_b(0)='0' then      -- abcd_b even?.
         if count/=7 then
           count<=count+1;
        end if;
    else
         if count/=0 then
           count<=count-1;
        end if;
      end if;
   end if;
 end process;

  process(count)
 begin
   if count=0 then
     overf<='0';
     clear<='1';
    elsif count=7 then
     overf<='1';
     clear<='0';
   else
     overf<='0';
     clear<='0';
   end if;
 end process;

  process(clk,resetn)
 begin
   if resetn='0' then
    state<=s0;
    elsif clk'event and clk='1' then
     case state is
         when s0 => if overf='1' then
                      state<=s1;
                   end if;

         when others => if clear='1' then
                          state<=s0;
                        end if;
     end case;
   end if;
 end process;

 overflow<=state;
end;
```

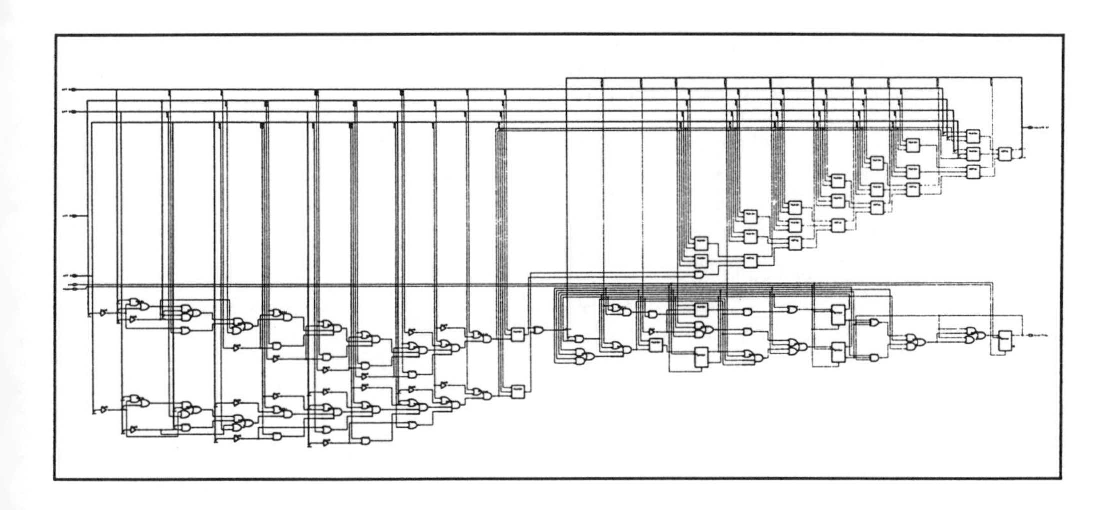

Lab 8

```
Library  ieee;
Use  ieee.std_logic_1164.ALL;

Entity  lab8  is
  port( clk,resetn,puls,dir:   in        std_logic;
        d_out:                 out       std_logic_vector(3  downto  0));
end;

Architecture  rtl  of  lab8  is
type  state_type  is  (s0,s1,s2,s3);
signal  state:state_type;
begin
  process(clk)  begin
    if clk'event  and  clk='1'  then
       if resetn='0' then
       state<=s0;
     else
          if puls='1' then
            case state is
             when s0 =>   if  dir='1'  then
                             state<=s1;
                          else
                             state<=s3;
                          end  if;

             when s1 =>   if  dir='1'  then
                             state<=s2;
                          else
                             state<=s0;
                          end  if;

             when s2 =>   if  dir='1'  then
                             state<=s3;
```

```
                         else
                            state<=s1;
                         end if;
          when s3 =>     if dir='1' then
                            state<=s0;
                         else
                            state<=s2;
                         end if;
         end case;

    end if;
       end if;
    end if;
  end process;

  process(state)
  begin
  case state is
    when s0 =>d_out<="0101";
    when s1 =>d_out<="0110";
    when s2 =>d_out<="1010";
    when s3 =>d_out<="1001";
  end case;
  end process;

end;
```

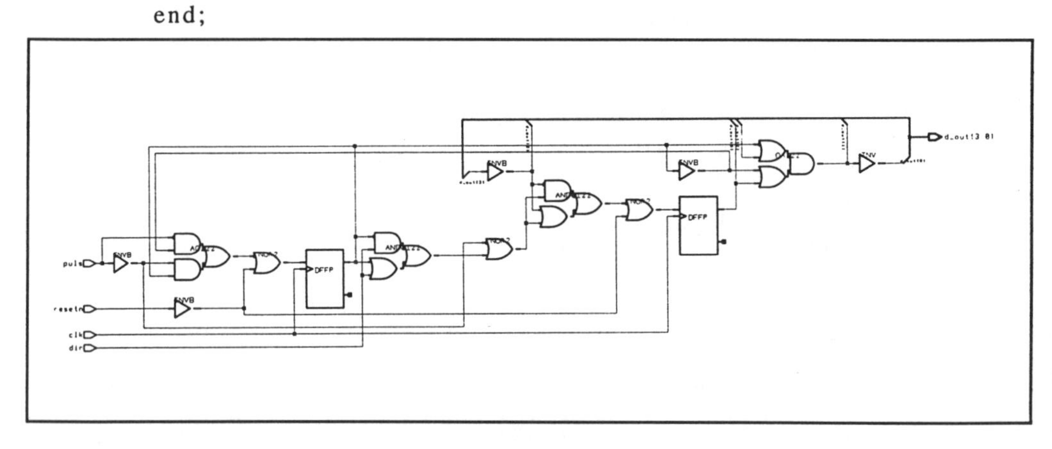

19.3. Solutions to the laboratory assignments - ViewLogic

The solutions in this section are written for ViewLogic's PC synthesis tool, but in the majority of cases they can also be used for other simple synthesis tools.

Lab 1

```
Entity lab1 is
  port(  a,b: in     vlbit;
```

```
        c:    out    vlbit);
end;

Architecture rtl of lab1 is
begin
  c<=a nor b;
end;
```

Lab 2

```
Entity lab2 is
   generic (Tdelay:time:=10 ns);
  port( a,b,c,d: in      vlbit;
        sel:     in      vlbit_vector(1 downto 0);
        q:       out     vlbit);
end;
Architecture rtl of lab2 is
begin
 q<=  a after Tdelay when sel(1)='0' and sel(0)='0' else
      b after Tdelay when sel(1)='0' and sel(0)='1' else
      c after Tdelay when sel(1)='1' and sel(0)='0' else
      d after Tdelay when sel(1)='1' and sel(0)='1' else
      'X' after Tdelay;
end;
```

Lab 2 Supplementary study

```
Entity lab2 is
   generic (Tdelay:time:=10 ns);
  port( a,b,c,d: in      vlbit_vector(3 downto 0);
        sel:     in      vlbit_vector(1 downto 0);
        q:       out     vlbit_vector(3 downto 0));
end;
Architecture rtl of lab2b is
begin
 q<=  a after Tdelay when sel(1)='0' and sel(0)='0' else
      b after Tdelay when sel(1)='0' and sel(0)='1' else
      c after Tdelay when sel(1)='1' and sel(0)='0' else
      d after Tdelay when sel(1)='1' and sel(0)='1' else
      "XXXX" after Tdelay;
end;
```

Lab 3

```
Entity lab3 is
  port( c,resetn,clk:  in    vlbit;
        a,b:           in    vlbit_vector(1 downto 0);
        q,data:        out   vlbit);
end;
```

```
Architecture rtl of lab3 is
signal intq_b: vlbit;
begin
 q<=q_b;
  int<='1' when a>b else '0';

   process
   begin
    wait until prising(clk);
     if resetn='0' then
       q_b<='0';
    else
       q_b<=int;
    end if;
   end process;

  data<=c and q_b;
end;
```

Lab 3 Supplementary study

```
Entity lab3 is
  port( c,resetn,clk:  in    vlbit;
        a,b:           in    vlbit_vector(1 downto 0);
        q,data:        out   vlbit);
end;

Architecture rtl of lab3 is
signal int,q_b: vlbit;
begin
 q<=q_b;
  int<='1' when a>b else '0';

   process
   begin
    wait until (prising(clk) or resetn='0');
     if resetn='0' then
       q_b<='0';
    else
       q_b<=int;
    end if;
   end process;

  data<=a and q_b;
end;
```

Lab 4

```
Package mypack is
   function max(a,b:in vlbit_vector) return vlbit_vector;
end;
```

```
Package body mypack is
      function max(a,b:in vlbit_vector) return vlbit_vector is
      variable int:vlbit_vector(2 downto 0);
    begin
     if a>b then
       int:=a;
     else
       int:=b;
     end if;
       return int;
    end;
end;

Use work.mypack.ALL;

Entity lab4 is
  port( a,b: in     vlbit_vector(2 downto 0);
        c:   out    vlbit_vector(2 downto 0);
end;

Architecture rtl of lab4 is
begin
  c<=max(a,b);
end;
```

Lab 4 Supplementary study

```
Library std;
Use std.standard.ALL;

Entity lab4b is
  port( clk,resetn:  in    vlbit;
        a,b:         in    vlbit_vector(1 downto 0);
        data:        out   vlbit_vector(2 downto 0));
end;

Architecture rtl of lab4b is
signal int:vlbit_vector(2 downto 0);
begin
  int<=addum(a,b);
  p1:process
  begin
    wait until prising(clk)
   if resetn='0' then
     data<="000";
   else
     data<=int;
   end if;
  end process;
end;
```

Lab 5

```
Entity k1 is
  port( a,b: in    vlbit;
          c: out   vlbit);
end;

Architecture rtl of k1 is
begin
 c<=a and b;
end;

Entity k2 is
  port( a,b: in    vlbit;
          c: out   vlbit);
end;

Architecture rtl of k2 is
begin
 c<=a or b;
end;

Entity k3 is
  port( a,b: in    vlbit;
          c: out   vlbit);
end;

Architecture rtl of k3 is
begin
 c<=a xor b;
end;

Entity lab5 is
   port(a,b,cin:     in  vlbit;
         cout,sum:   out vlbit);
end;

Architecture rtl of lab5 is
component k1
   port( a,b: in    vlbit;
           c: out   vlbit);
end component;

component k2
   port( a,b: in    vlbit;
           c: out   vlbit);
end component;

component k3
   port( a,b: in    vlbit;
           c: out   vlbit);
end component;

signal n1,n2,n3,n4:vlbit;
begin
    U1:k1 port map(a,b,n1);
```

```
    U2:k2 port map(a,b,n2);
    U3:k3 port map(a,b,n3);
    U4:k1 port map(n2,cin,n4);
    U5:k3 port map(n3,cin,sum);
    U6:k2 port map(n1,n4,cout);
end;
```

Lab 6

```
Entity lab6 is
  port( clk,resetn,in1,in2: in     vlbit;
        out1,out2:          out    vlbit);
end;

Architecture rtl of lab6 is
type state_type is array (1 downto 0) of vlbit;
constant s0: state_type:="00";
constant s1: state_type:="01";
constant s2: state_type:="10";
signal  state:state_type;
begin
   p1:process
   begin
     wait until prising(clk);
     if resetn='0' then
    state<=s0;
   else
    if state=s0 then
      if in1='1' then
        state<=s1;
      end if;

      elsif state=s1 then
        if in1='0 ' then
           state<=s2;
        end if;

      elsif state=s2 then
        if in1='1' and in2='1' then
           state<=s0;
        end if;

      else
        state<=s0;
      end if;
    end if;
   end process;

    process(state,in1,in2)
   begin
    if state=s0 then
      if in1='1' then
        out1<='0';
        out2<='1';
```

```
        else
          out1<='0';
          out2<='0';
        end if;

       elsif state=s1 then
        if in1='0' then
          out1<='1';
          out2<='1';
        else
          out1<='0';
          out2<='0';
        end if;

       elsif state=s2 then
        if in1='1' then
          out1<='1';
          out2<='0';
        else
          out1<='0';
          out2<='0';
        end if;

      else
        out1<='0';
        out2<='0';
      end if;
    end process;

end;
```

Lab 6 Supplementary study

```
Entity lab6b is
  port( clk,resetn,in1,in2: in     vlbit;
        out1,out2:          out    vlbit);
end;

Architecture rtl of lab6b is
type state_type is array (1 downto 0) of vlbit;
constant s0: state_type:="00";
constant s1: state_type:="01";
constant s2: state_type:="10";
signal state:state_type;
begin
   p1:process
   begin
     wait until prising(clk);
     if resetn='0' then
       state<=s0;
       out1<='0';
       out2<='0';
     else
       if state=s0 then
```

```
          if in1='1' then
             state<=s1;
             out1<='0';
             out2<='1';
          else
             out1<='0';
             out2<='0';
          end if;

        elsif state=s1 then
          if in1='0 ' then
             state<=s2;
             out1<='1';
             out2<='1';
          else
             out1<='0';
             out2<='0';
          end if;

        elsif state=s2 then
          if in1='1' and in2='1' then
             state<=s0;
             out1<='1';
             out2<='0';
          else
             out1<='0';
             out2<='0';
          end if;

        else
          state<=s0;
          out1<='0';
          out2<='0';
        end if;
      end if;
    end process;
end;
```

Lab 7

```
Use  work.mypack.ALL;

Entity lab7 is
   port( a,b,c,d:      in   vlbit_vector(7 downto 0);
         clk,resetn:   in   vlbit;
         overflow:     out  vlbit;
         abcd:         out  vlbit_vector(8 downto 0));
end;

Architecture rtl of lab7 is
signal ab_max,cd_max: vlbit_vector(7 downto 0);
signal count:vlbit_vector(3 downto 0);
signal overf,clear: vlbit;
signal abcd_b: vlbit_vector(8 downto 0);
```

```
constant  s0:vlbit:='0';
constant  s1:vlbit:='1';
signal  state:  vlbit;

begin
  abcd<=abcd_b;
  ab_max<=max(a,b);
  cd_max<=max(c,d);
  abcd_b<=addum(ab_max,cd_max);

  process
  begin
      wait until (resetn='0' or prising(clk));
     if resetn='0' then
      count<="0000";
   else
        if abcd_b(0)='0' then        -- even?
            if v1d2int(count)/=7 then
               count<=addum(count(2 downto 0),"001");
         end if;
     else
            if v1d2int(count)/=0 then
               count<=subum(count(2 downto 0),"001");
         end if;
      end if;
    end if;
 end process;

 process(count)
 variable  c_int:integer;
 begin
    c_int:v1d2int(count);

   if c_int=0 then
     overf<='0';
     clear<='1';

    elsif c_int=7 then
     overf<='1';
     clear<='0';

   else
     overf<='0';
     clear<='0';
   end if;
 end process;

 process
 begin
    wait until (resetn='0' and prising(clk));
   if resetn='0' then
    state<=s0;
  else
      if state=s0 then
         if overf='1' then
         state<=s1;
```

```
          end if;

        else
            if clear='1' then
            state<=s0;
           end if;
        end if;
      end if;
    end process;

    overflow<=state;
  end;
```

Lab 8

```
Entity lab8 is
  port( clk,resetn,puls,dir: in     vlbit;
        d_out:               out    vlbit_vector(3 downto 0));
end;

Architecture rtl of lab8 is
type state_type is array (1 downto 0) of vlbit;
constant s0:state_type:="00";
constant s1:state_type:="01";
constant s2:state_type:="10";
constant s3:state_type:="11";
signal state:state_type;
begin
  process
  begin
    wait until prising(clk);
    if resetn='0' then
      state<=s0;
    else
      if puls='1' then
        if state=s0 then
          if dir='1' then
            state<=s1;
          else
            state<=s3;
          end if;

        elsif state=s1 then
          if dir='1' then
            state<=s2;
          else
            state<=s0;
          end if;

        elsif state=s2 then
          if dir='1' then
            state<=s3;
          else
            state<=s1;
```

```
              end if;
           else
              if dir='1' then
                state<=s0;
              else
                state<=s2;
              end if;
           end if;
         end if;
       end if;
     end process;

     process(state)
     begin
       if state<=s0 then
           d_out<="0101";
       elsif state=s1 then
           d_out<="0110";
       elsif state=s2 then
           d_out<="1010";
       else
           d_out<="1001";
       end if;
     end process;
   end;
```

Appendix A
VHDL syntax

A.1. Library units

Architecture body

Library unit

Several architectures can be linked to an entity.

Architecture *architecture_name* of *entity_name*
architecture_declaration
begin
 architecture_statement
end *architecture_name*;

```
Architecture rtl of adder is
 constant 5v_const: std_logic:='1';
 type state_type is (s0,s1,s2,s3);
 signal state:state_type;
 signal a_int: std_logic_vector(4 downto 0);

 function add_f (a,b:in std_logic_vector) return std_logic_vector is
 begin
  return (('0' & a) + b);
 end;
begin
  a_int<=not a;
  q<=add_f(a_int,b);
end;
```

Configuration declaration

Library unit

Configuration *configuration_name* of *entity_name* is
end *configuration_name;*

```
Configuration config of ent is
```

```
 for rtl
       for all:comp
         use entity work.dff(rtl);
       end for;
 end for;
end config;
```

Entity declaration *Library unit*

Entity *Entity_name* is
entity_declaration
begin
 entity_statement
end *Entity_name*;

Only passive processes are permitted in the entity declaration.

```
Entity ent is
 port( clk,resetn,d_in    in  std_logic;
          d_out:              out      std_logic);
end;

Entity ent2 is
     generic(N:positive:=4);
     port(  a,b:   in       std_logic;
            q:     out  std_logic);
begin
     assert not (a='1' and b='0')
     report (ERROR a=1 and b=0 at the same time !)
     severity ERROR;
end;
```

Package body *Library unit*

```
Package body my_pack is
   function add_f (a,b: std_logic_vector) return std_logic_vector is
   begin
      return (('0' & a) + b);
   end;
end;
```

Package declaration *Library unit*

Package *package_name* is
end *package_name;*

```
Package my_pack is
  function add_f (a,b: std_logic_vector) return std_logic_vector;
  type my_array is array 4 downto 0 of std_logic_vector(3 downto 0);
```

```
  constant my_5v:std_logic_vector(7 downto 0):=(others=>'1');
end;
```

A.2. Declarations

Alias declaration *Declaration*

```
signal v:std_logic_vector(7 downto 0);
alias vect1 : std_logic_vector(8 downto 1) is v;
```

Attribute declaration *Declaration*

Attribute *attribute_name: type;*

```
Attribute version: version_number;
```

Component declaration *Declaration*

component *component_name*
 generic_clause
 port_clause
end component;

```
component my_comp
  generic (N:positive);
  port(a,b:in std_logic;
          q:out std_logic);
end component;
```

Constant declaration *Declaration*

```
constant my_5v:std_logic_vector(7 downto 0):=(others=>'1');
constant s0:std_logic_vector(1 downto 0):="01";
constant max_delay:time:=12 ns;
```

File declaration *Declaration*

file *file_name* : *file_type* is *mode file_logiskt_name*;

```
file prom: prom_file_type is in "prom_file.txt";
```

Signal declaration *Declaration*

It is not allowed to declare signals inside a process, subprogram or in the package body.

signal *signal_name*:type;

```
signal a,b: std_logic;
signal a_bus,b_bus: std_logic_vector(7 downto 0);
signal ab_bus: bit_vector(7 downto 0);
signal viewL_bus: vlbit_vector(7 downto 0);
```

Subprogram body *Declaration*

```
function add_f (a,b: std_logic_vector) return std_logic_vector is
begin
  return (('0' & a) + b;
end;

function reg (signal clk,resetn,d_in:in std_logic) return std_logic is
begin
  if clk'event and clk='1' then
    if resetn='0' then
      return '0';
    else
      return d_in;
    end if;
  end if;
end;

procedure add_p(a,b:in std_logic_vector; sum:out std_logic_vector) is
begin
  sum<=('0' & a) + b;
end;
```

Subprogram declaration *Declaration*

```
function add_f (a,b:in std_logic_vector) return std_logic_vector;
function reg (signal clk,resetn,d_in: std_logic) return std_logic;

procedure add_p(a,b:in std_logic_vector;
                    sum:out std_logic_vector);
```

Subtype declaration *Declaration*

```
subtype my_int is integer range 0 to 63;
subtype byte is std_logic_vector(7 downto 0);
```

Type declaration *Declaration*

```
type colour is (red,green,blue);
type state_type is (s0,s1,s2,s3);
```

```
type array_4 is array (3 downto 0) of std_logic_vector(4 downto 0);
type bit is ('0', '1');
type my_bit is ('X', 'Z', '0', '1');
```

Variable declaration *Declaration*

```
variable a:std_logic;
variable a_bus:std_logic_vector(7 downto 0):=(others=>'0');
variable my_int is integer range 0 to 255;
```

A.3. Sequential statements

The sequential VHDL statements can be used in the sequential part of VHDL, i.e. processes, functions and procedures. Note that some statements can be used in both the sequential and the concurrent part of VHDL. Such statements are included in both this section and the section for concurrent statements.

Assert statement *Sequential*

Assert *condition*
 report *expression*
 severity *severity_level*

```
Assert a='1'
 report "a/='1'"
 severity note;

Assert data/=0
 report "data=0"
 severity warning;

Assert boolean_sign
  report "boolean=false"
  severity error;

Assert state/=s4
 report "illegal state (s4)"
 severity failure;
```

Case statement *Sequential*

case *expression* is
 case_alternative
end case;

```
process(a,b)
variable a_int:integer range 0 to 12;
begin
```

```
  a_int<=a+b;
  case a_int is
     when 0         => en<="001";
     when 2 to 5  => en<="010";
     when 7 | 9     => en<="100";
     when others => en<="000";
   end case;
end process;
```

Exit statement *Sequential*

exit *loop_label* when *condition;*

```
loop
  exit when b=3;
  a<=b-c;
end loop;

l1:for i in 0 to 3 loop
  l2:for j in 0 to 5 loop
       a:=b+1;
       exit l2 when a>12;
  end loop l2;
end loop l1;

loop
  if a>=c-d then
    exit;
  else
    q:=c+1;
end loop;
```

If statement *Sequential*

```
process(clk,resetn)
begin
  if resetn='0' then
    data<=(others=>'0');
  elsif clk'event and clk='1' then
    data<=data_in;
  end if;
end process;

function my_f (a,b:in std_logic_vector) return std_logic_vector is
begin
   if a-b>12 then
     return "00";
   elsif a-b=12 then
     return "01";
   elsif a=12 then
```

```
    return "11";
  else
    if a<2 then
      return "00";
    else
      return "10";
    end if;
  end if;
end;
```

Loop statement *Sequential*

```
loop
  exit when b=3;
  a<=b-c;
end loop;

for i in 3 downto 0 loop
  q(i)<=a(i)-b;
end loop;

while a<12 loop
  a:=a+1;
  data(a)<=c(a+1);
end loop;
```

Next statement *Sequential*

The next statement interrupts the current loop and jumps straight back to the next loop iteration.

next *loop_label* when *condition*;

```
while a<12 loop
  next when a=4
  sum:=a-b;
  a:=a+cin(a);
end loop;
```

Null statement *Sequential*

```
process(a)
begin
  data<=(others=>'1');
  case a is
    when 2 | 4   => data<=d1;
    when 5 to 7  => data<=d2;
    when others => null;
  end case;
end process;
```

Return statement *Sequential*

```
function my_f (a,b:in std_logic_vector(2 downto 0) return std_logic_vector is
begin
  if a-b>12 then
    return "00";
  elsif a-b=12 then
    return "01";
  elsif a=12 then
    return "11";
  else
    if a<2 then
      return "00";
    else
      return "10";
    end if;
  end if;
end;
```

Signal assignment *Sequential*

```
process(a,b,c)
begin
  q1<=a and b after 10 ns;
  q2<=a & b(3 downto 0);
  q3<=c after Tdelay, b after Tdelay + 4 ns;
  q4<=my_func(a,b);
end process;
```

Variable assignment *Sequential*

```
process(a,b,c)
begin
  q1:=a and b after 10 ns;
  q2:=a & b(3 downto 0);
  q3:=c after Tdelay, b after Tdelay + 4 ns;
  q4:=my_func(a,b);
end process;
```

Wait statement *Sequential*

```
wait on a;

wait until clk='1';

Wait for 10 ns;

wait on a until b='1' for 10 ns;
```

Process(a,b) -- equal to have *wait on a,b;* last in the process.

A.4. Concurrent statements

Assert statement *Concurrent*

Severity can have four levels: note, warning, error and failure. On some VHDL simulators it is possible to determine which of the four levels should be regarded as so severe that the simulation must stop. The report statement is always output on the simulator when the assert statement is not true.

```
Architecture rtl of ex is
begin
  assert resetn/='0'
  report "The reset signal is activated"
  severity note;
...
```

Block statement *Concurrent*

block_label: block(*guard_expression*)
block_declaration part
begin

end block *block_label;*

```
Architecture rtl of ex is
begin
  b1:Block
  signal int:std_logic;
  begin
    int<='1' when a=3 else '0';
    q<=int and en;
  end block;

  q2<=q and b;
end;
```

Component instantiation *Concurrent*

```
U1: my_comp port map (a,b,q);

C1:add2 port map (a=>a,  b=>b, sum_int=>sum)

a1: generic map (N=>6)
      port map (a=>d1, b=>d2, q1=>dout1, q2=>open);
```

Generate statement *Concurrent*

```
l1: for i in 0 to 3 generate
   U1: comp1 port map (clk,resetn,din(i),q(i+1))
end generate;
```

Process statement *Concurrent*

```
process_label : process(sensitivity_list)
process_declaration
begin

end process process_label;

Process(a,b,c)
variable int:std_logic;
begin
  int:=a and b;
   q<=int nor c;
end process;

process(clk,resetn)
begin
  if resetn='0' then
     d_out<='0';
  elsif clk'event and clk='1' then
     d_out<=d_in;'
  end if;
end process;

p1:process
begin
  wait until clk='1';
   a<=not b;
end process;
```

Signal assignment *Concurrent*

```
q1<=a and b after 10 ns;
q2<=a & b(3 downto 0);
q3<=c after Tdelay, b after Tdelay + 4 ns;
q4<=my_func(a,b);
q5<=5 after 2 ns, 3 after 4 ns, 9 after 10 ns;
q6<=guarded d_out after 12 ns;
```

When else statement *Concurrent*

```
q<=a when "00" else
     b when "01" else
     c when "10" else
     d;
```

With select statement *Concurrent*

```
with sel select
 q<=a when "00",
     b when "01",
     c when "10",
     d when others;
```

Appendix B

VHDL-package

B.1. Std-package

```
-- This is Package STANDARD as defined in the VHDL 1987 Language
-- Reference Manual.

package standard is

-- predefined enumeration types:
type boolean is (FALSE, TRUE);
type bit is ('0', '1');

type character is (
    NUL, SOH, STX, ETX, EOT, ENQ, ACK, BEL,
    BS,  HT,  LF,  VT,  FF,  CR,  SO,  SI,
    DLE, DC1, DC2, DC3, DC4, NAK, SYN, ETB,
    CAN, EM,  SUB, ESC, FSP, GSP, RSP, USP,
    ' ', '!', '"', '#', '$', '%', '&', ''',
    '(', ')', '*', '+', ',', '-', '.', '/',
    '0', '1', '2', '3', '4', '5', '6', '7',
    '8', '9', ':', ';', '<', '=', '>', '?',
    '@', 'A', 'B', 'C', 'D', 'E', 'F', 'G',
    'H', 'I', 'J', 'K', 'L', 'M', 'N', 'O',
    'P', 'Q', 'R', 'S', 'T', 'U', 'V', 'W',
    'X', 'Y', 'Z', '[', '\', ']', '^', '_',
    '`', 'a', 'b', 'c', 'd', 'e', 'f', 'g',
    'h', 'i', 'j', 'k', 'l', 'm', 'n', 'o',
    'p', 'q', 'r', 's', 't', 'u', 'v', 'w',
    'x', 'y', 'z', '{', '|', '}', '~', DEL);
type severity_level is (NOTE, WARNING, ERROR, FAILURE);

-- predefined numeric types:
type integer is range implementation_defined
type real is range implementation_defined

--predefined type time
```

```
type time is range implementation_defined
        units
                fs;                         -- femtosecond
                ps   =   1000 fs;           -- picosecond
                ns   =   1000 ps            -- nanosecond
                us   =   1000 ns            -- microsecond
                ms   =   1000 us            -- millisecond
                sec  =   1000 ms            -- second
                min  =   60 sec             -- minute
                hr   =   60 min             -- hour
        end units;

-- function that returns the current simulation time:
function now return time;

-- predefined numeric subtypes:
subtype natural is integer range 0 to integer'high;
subtype positive is integer range 1 to integer'high;

-- predefined array types:
type string is array (positive range <>) of character;
type bit_vector is array (natural range <>) of bit;

end standard;
```

B.2. IEEE-package

B.2.1. Std_logic_1164

```
-- -------------------------------------------------------------------
--  Title       : std_logic_1164 multi-value logic system
--  Library     : This package shall be compiled into a library
--              : symbolically named IEEE.
--  Developers: IEEE model standards group (par 1164)
--  Purpose     : This packages defines a standard for designers
--              : to use in describing the interconnection data types
--              : used in vhdl modeling.
--
--  Limitation : The logic system defined in this package may
--              : be insufficient for modeling switched transistors,
--              : since such a requirement is out of the scope of this
--              : effort. Furthermore, mathematics, primitives,
--              : timing standards, etc. are considered orthogonal
--              : issues as it relates to this package and are therefore
--              : beyond the scope of this effort.
--
--  Note        : No declarations or definitions shall be included in,
--              : or excluded from this package. The "package declaration"
--              : defines the types, subtypes and declarations of
--              : std_logic_1164. The std_logic_1164 package body shall be
--              : considered the formal definition of the semantics of
--              : this package. Tool developers may choose to implement
--              : the package body in the most efficient manner available
```

```
--                  : to them.
--
-- -------------------------------------------------------------

PACKAGE std_logic_1164 IS
    -------------------------------------------------------------
    -- logic state system  (unresolved)
    -------------------------------------------------------------
    TYPE std_ulogic IS (   'U',    -- Uninitialized
                           'X',    -- Forcing  Unknown
                           '0',    -- Forcing 0
                           '1',    -- Forcing 1
                           'Z',    -- High Impedance
                           'W',    -- Weak Unknown
                           'L',    -- Weak 0
                           'H',    -- Weak 1
                           '-'     -- Don't care
                       );
    -------------------------------------------------------------
    -- unconstrained array of std_ulogic for use with the resolution function
    -------------------------------------------------------------
    TYPE std_ulogic_vector IS ARRAY ( NATURAL RANGE <> ) OF
    std_ulogic;

    -------------------------------------------------------------
    -- resolution function
    -------------------------------------------------------------
    FUNCTION resolved ( s : std_ulogic_vector ) RETURN std_ulogic;

    -------------------------------------------------------------
    -- *** industry standard logic type ***
    -------------------------------------------------------------
    SUBTYPE std_logic IS resolved std_ulogic;

    -------------------------------------------------------------
    -- unconstrained array of std_logic for use in declaring signal arrays
    -------------------------------------------------------------
    TYPE std_logic_vector IS ARRAY ( NATURAL RANGE <>) OF std_logic;

    -------------------------------------------------------------
    -- common subtypes
    -------------------------------------------------------------
    SUBTYPE X01     IS resolved std_ulogic RANGE 'X' TO '1';   -- ('X','0','1')
    SUBTYPE X01Z    IS resolved std_ulogic RANGE 'X' TO 'Z';
      -- ('X','0','1','Z')
    SUBTYPE UX01    IS resolved std_ulogic RANGE 'U' TO '1';
      -- ('U','X','0','1')
    SUBTYPE UX01Z   IS resolved std_ulogic RANGE 'U' TO 'Z';
        --   ('U','X','0','1','Z')

    -------------------------------------------------------------
    -- overloaded logical operators
    -------------------------------------------------------------

    FUNCTION "and"  ( l : std_ulogic; r : std_ulogic ) RETURN UX01;
```

```
    FUNCTION "nand" ( l : std_ulogic; r : std_ulogic ) RETURN UX01;
    FUNCTION "or"   ( l : std_ulogic; r : std_ulogic ) RETURN UX01;
    FUNCTION "nor"  ( l : std_ulogic; r : std_ulogic ) RETURN UX01;
    FUNCTION "xor"  ( l : std_ulogic; r : std_ulogic ) RETURN UX01;
--  function "xnor" ( l : std_ulogic; r : std_ulogic ) return ux01;
    function xnor   ( l : std_ulogic; r : std_ulogic ) return ux01;
    FUNCTION "not"  ( l : std_ulogic                 ) RETURN UX01;

    -------------------------------------------------------------------
    -- vectorized overloaded logical operators
    -------------------------------------------------------------------
    FUNCTION "and"  ( l, r : std_logic_vector  ) RETURN std_logic_vector;
    FUNCTION "and"  ( l, r : std_ulogic_vector ) RETURN std_ulogic_vector;

    FUNCTION "nand" ( l, r : std_logic_vector  ) RETURN std_logic_vector;
    FUNCTION "nand" ( l, r : std_ulogic_vector ) RETURN std_ulogic_vector;

    FUNCTION "or"   ( l, r : std_logic_vector  ) RETURN std_logic_vector;
    FUNCTION "or"   ( l, r : std_ulogic_vector ) RETURN std_ulogic_vector;

    FUNCTION "nor"  ( l, r : std_logic_vector  ) RETURN std_logic_vector;
    FUNCTION "nor"  ( l, r : std_ulogic_vector ) RETURN std_ulogic_vector;

    FUNCTION "xor"  ( l, r : std_logic_vector  ) RETURN std_logic_vector;
    FUNCTION "xor"  ( l, r : std_ulogic_vector ) RETURN std_ulogic_vector;

--  -----------------------------------------------------------------------
--  Note : The declaration and implementation of the "xnor" function is
--  specifically commented until at which time the VHDL language has been
--  officially adopted as containing such a function. At such a point,
--  the following comments may be removed along with this notice without
--  further "official" balloting of this std_logic_1164 package. It is
--  the intent of this effort to provide such a function once it becomes
--  available in the VHDL standard.
--  -----------------------------------------------------------------------
--  function "xnor" ( l, r : std_logic_vector  ) return std_logic_vector;
--  function "xnor" ( l, r : std_ulogic_vector ) return std_ulogic_vector;
    function xnor   ( l, r : std_logic_vector  ) return std_logic_vector;
    function xnor   ( l, r : std_ulogic_vector ) return std_ulogic_vector;

    FUNCTION "not"  ( l : std_logic_vector  ) RETURN std_logic_vector;
    FUNCTION "not"  ( l : std_ulogic_vector ) RETURN std_ulogic_vector;
    -------------------------------------------------------------------
    -- conversion functions
    -------------------------------------------------------------------
    FUNCTION To_bit       ( s : std_ulogic
                          ; xmap : BIT := '0'
                          ) RETURN BIT;

    FUNCTION To_bitvector ( s : std_logic_vector
                          ; xmap : BIT := '0'
                          ) RETURN BIT_VECTOR;

    FUNCTION To_bitvector ( s : std_ulogic_vector
                          ; xmap : BIT := '0'
                          ) RETURN BIT_VECTOR;

    FUNCTION To_StdULogic      ( b : BIT          ) RETURN std_ulogic;
```

```
    FUNCTION To_StdLogicVector  ( b : BIT_VECTOR        ) RETURN
        std_logic_vector;
    FUNCTION To_StdLogicVector  ( s : std_ulogic_vector ) RETURN std_logic_vector;
    FUNCTION To_StdULogicVector ( b : BIT_VECTOR        ) RETURN
        std_ulogic_vector;
    FUNCTION To_StdULogicVector ( s : std_logic_vector  ) RETURN
        std_ulogic_vector;
    -------------------------------------------------------------------
    -- strength strippers and type convertors
    -------------------------------------------------------------------

    FUNCTION To_X01  ( s : std_logic_vector  ) RETURN  std_logic_vector;
    FUNCTION To_X01  ( s : std_ulogic_vector ) RETURN std_ulogic_vector;
    FUNCTION To_X01  ( s : std_ulogic        ) RETURN  X01;
    FUNCTION To_X01  ( b : BIT_VECTOR        ) RETURN  std_logic_vector;
    FUNCTION To_X01  ( b : BIT_VECTOR        ) RETURN std_ulogic_vector;
    FUNCTION To_X01  ( b : BIT               ) RETURN  X01;
    FUNCTION To_X01Z ( s : std_logic_vector  ) RETURN  std_logic_vector;
    FUNCTION To_X01Z ( s : std_ulogic_vector ) RETURN std_ulogic_vector;
    FUNCTION To_X01Z ( s : std_ulogic        ) RETURN  X01Z;
    FUNCTION To_X01Z ( b : BIT_VECTOR        ) RETURN std_logic_vector;
    FUNCTION To_X01Z ( b : BIT_VECTOR        ) RETURN std_ulogic_vector;
    FUNCTION To_X01Z ( b : BIT               ) RETURN  X01Z;
    FUNCTION To_UX01  ( s : std_logic_vector  ) RETURN std_logic_vector;
    FUNCTION To_UX01  ( s : std_ulogic_vector ) RETURN std_ulogic_vector;
    FUNCTION To_UX01  ( s : std_ulogic        ) RETURN  UX01;
    FUNCTION To_UX01  ( b : BIT_VECTOR        ) RETURN std_logic_vector;
    FUNCTION To_UX01  ( b : BIT_VECTOR        ) RETURN std_ulogic_vector;
    FUNCTION To_UX01  ( b : BIT               ) RETURN  UX01;
    -------------------------------------------------------------------
    -- edge detection
    -------------------------------------------------------------------

    FUNCTION rising_edge  (SIGNAL s : std_ulogic) RETURN BOOLEAN;
    FUNCTION falling_edge (SIGNAL s : std_ulogic) RETURN BOOLEAN;
    -------------------------------------------------------------------
    -- object contains an unknown
    -------------------------------------------------------------------
    FUNCTION Is_X ( s : std_ulogic_vector ) RETURN  BOOLEAN;
    FUNCTION Is_X ( s : std_logic_vector  ) RETURN  BOOLEAN;
    FUNCTION Is_X ( s : std_ulogic        ) RETURN  BOOLEAN;
END std_logic_1164;
```

B.2.2. Std_logic_unsigned

```
--------------------------------------------------------------------------------------------
--
-- Copyright (c) 1990, 1991, 1992 by Synopsys, Inc.
-- All rights reserved.
-- This source file may be used and distributed without restriction
-- provided that this copyright statement is not removed from the file
-- and that any derivative work contains this copyright notice.
--
--      Package name: STD_LOGIC_UNSIGNED
--
--      Date:         09/11/92      KN
--                    10/08/92      AMT
--
--      Purpose:      A set of unsigned arithemtic, conversion,
--                    and comparision functions for STD_LOGIC_VECTOR.
--
--      Note:         comparision of same length discrete arrays is defined
--                    by the LRM. This package will "overload" those
--                    definitions
--
--------------------------------------------------------------------------------------------

library IEEE;
use IEEE.std_logic_1164.all;
use IEEE.std_logic_arith.all;

package STD_LOGIC_UNSIGNED is
   function "+"(L: STD_LOGIC_VECTOR; R: STD_LOGIC_VECTOR) return
      STD_LOGIC_VECTOR;
   function "+"(L: STD_LOGIC_VECTOR; R: INTEGER) return
      STD_LOGIC_VECTOR;
   function "+"(L: INTEGER; R: STD_LOGIC_VECTOR) return
      STD_LOGIC_VECTOR;
   function "+"(L: STD_LOGIC_VECTOR; R: STD_LOGIC) return
      STD_LOGIC_VECTOR;
   function "+"(L: STD_LOGIC; R: STD_LOGIC_VECTOR) return
      STD_LOGIC_VECTOR;
   function "-"(L: STD_LOGIC_VECTOR; R: STD_LOGIC_VECTOR) return
      STD_LOGIC_VECTOR;
   function "-"(L: STD_LOGIC_VECTOR; R: INTEGER) return
      STD_LOGIC_VECTOR;
   function "-"(L: INTEGER; R: STD_LOGIC_VECTOR) return
      STD_LOGIC_VECTOR;
   function "-"(L: STD_LOGIC_VECTOR; R: STD_LOGIC) return
      STD_LOGIC_VECTOR;
```

```
function "-"(L: STD_LOGIC; R: STD_LOGIC_VECTOR) return
     STD_LOGIC_VECTOR;
function "+"(L: STD_LOGIC_VECTOR) return STD_LOGIC_VECTOR;
function "*"(L: STD_LOGIC_VECTOR; R: STD_LOGIC_VECTOR) return
     STD_LOGIC_VECTOR;
function "<"(L: STD_LOGIC_VECTOR; R: STD_LOGIC_VECTOR) return
     BOOLEAN;
function "<"(L: STD_LOGIC_VECTOR; R: INTEGER) return BOOLEAN;
function "<"(L: INTEGER; R: STD_LOGIC_VECTOR) return BOOLEAN;
function "<="(L: STD_LOGIC_VECTOR; R: STD_LOGIC_VECTOR) return
     BOOLEAN;
function "<="(L: STD_LOGIC_VECTOR; R: INTEGER) return BOOLEAN;
function "<="(L: INTEGER; R: STD_LOGIC_VECTOR) return BOOLEAN;
function ">"(L: STD_LOGIC_VECTOR; R: STD_LOGIC_VECTOR) return
     BOOLEAN;
function ">"(L: STD_LOGIC_VECTOR; R: INTEGER) return BOOLEAN;
function ">"(L: INTEGER; R: STD_LOGIC_VECTOR) return BOOLEAN;
function ">="(L: STD_LOGIC_VECTOR; R: STD_LOGIC_VECTOR) return
     BOOLEAN;
function ">="(L: STD_LOGIC_VECTOR; R: INTEGER) return BOOLEAN;
function ">="(L: INTEGER; R: STD_LOGIC_VECTOR) return BOOLEAN;
function "="(L: STD_LOGIC_VECTOR; R: STD_LOGIC_VECTOR) return
     BOOLEAN;
function "="(L: STD_LOGIC_VECTOR; R: INTEGER) return BOOLEAN;
function "="(L: INTEGER; R: STD_LOGIC_VECTOR) return BOOLEAN;
function "/="(L: STD_LOGIC_VECTOR; R: STD_LOGIC_VECTOR) return
     BOOLEAN;
function "/="(L: STD_LOGIC_VECTOR; R: INTEGER) return BOOLEAN;
function "/="(L: INTEGER; R: STD_LOGIC_VECTOR) return BOOLEAN;
function SHL(ARG:STD_LOGIC_VECTOR;COUNT:
     STD_LOGIC_VECTOR)     return STD_LOGIC_VECTOR;
function SHR(ARG:STD_LOGIC_VECTOR;COUNT:
     STD_LOGIC_VECTOR)     return STD_LOGIC_VECTOR;
function CONV_INTEGER(ARG: STD_LOGIC_VECTOR) return
     INTEGER;
-- remove this since it is already in std_logic_arith
--  function CONV_STD_LOGIC_VECTOR(ARG: INTEGER; SIZE:   INTEGER)   -
- return STD_LOGIC_VECTOR;
end STD_LOGIC_UNSIGNED;
```

B.2.3. Std_logic_signed

```
--------------------------------------------------------------------------------------------------
-- Copyright (c) 1990, 1991, 1992 by Synopsys, Inc.
--                              All rights reserved.
-- This source file may be used and distributed without restriction
-- provided that this copyright statement is not removed from the file
-- and that any derivative work contains this copyright notice.
--                                                        --
--      Package name: STD_LOGIC_SIGNED
--      Date:           09/11/91 KN
--                      10/08/92 AMT change std_ulogic to signed std_logic
--                      10/28/92 AMT added signed functions, -, ABS

--      Purpose:        A set of signed arithemtic, conversion,
--                      and comparision functions for STD_LOGIC_VECTOR.
--                                                        --
--      Note:           Comparision of same length std_logic_vector is defined --
                        in the LRM. The interpretation is for unsigned vectors
--                      This package will "overload" that definition.
--------------------------------------------------------------------------------------------------

library IEEE;
use IEEE.std_logic_1164.all;
use IEEE.std_logic_arith.all;

package STD_LOGIC_SIGNED is
    function "+"(L: STD_LOGIC_VECTOR; R: STD_LOGIC_VECTOR) return
        STD_LOGIC_VECTOR;

    function "+"(L: STD_LOGIC_VECTOR; R: INTEGER) return
        STD_LOGIC_VECTOR;

    function "+"(L: INTEGER; R: STD_LOGIC_VECTOR) return
        STD_LOGIC_VECTOR;

    function "+"(L: STD_LOGIC_VECTOR; R: STD_LOGIC) return
        STD_LOGIC_VECTOR;

    function "+"(L: STD_LOGIC; R: STD_LOGIC_VECTOR) return
        STD_LOGIC_VECTOR;

    function "-"(L: STD_LOGIC_VECTOR; R: STD_LOGIC_VECTOR) return
        STD_LOGIC_VECTOR;

    function "-"(L: STD_LOGIC_VECTOR; R: INTEGER) return
        STD_LOGIC_VECTOR;

    function "-"(L: INTEGER; R: STD_LOGIC_VECTOR) return
        STD_LOGIC_VECTOR;

    function "-"(L: STD_LOGIC_VECTOR; R: STD_LOGIC) return
        STD_LOGIC_VECTOR;

    function "-"(L: STD_LOGIC; R: STD_LOGIC_VECTOR) return
        STD_LOGIC_VECTOR;

    function "+"(L: STD_LOGIC_VECTOR) return STD_LOGIC_VECTOR;
```

```
    function "-"(L: STD_LOGIC_VECTOR) return STD_LOGIC_VECTOR;
    function "ABS"(L: STD_LOGIC_VECTOR) return STD_LOGIC_VECTOR;
    function "*"(L: STD_LOGIC_VECTOR; R: STD_LOGIC_VECTOR) return
        STD_LOGIC_VECTOR;
    function "<"(L: STD_LOGIC_VECTOR; R: STD_LOGIC_VECTOR) return
        BOOLEAN;
    function "<"(L: STD_LOGIC_VECTOR; R: INTEGER) return BOOLEAN;
    function "<"(L: INTEGER; R: STD_LOGIC_VECTOR) return BOOLEAN;
    function "<="(L: STD_LOGIC_VECTOR; R: STD_LOGIC_VECTOR) return
        BOOLEAN;
    function "<="(L: STD_LOGIC_VECTOR; R: INTEGER) return BOOLEAN;
    function "<="(L: INTEGER; R: STD_LOGIC_VECTOR) return BOOLEAN;
    function ">"(L: STD_LOGIC_VECTOR; R: STD_LOGIC_VECTOR) return
        BOOLEAN;
    function ">"(L: STD_LOGIC_VECTOR; R: INTEGER) return BOOLEAN;
    function ">"(L: INTEGER; R: STD_LOGIC_VECTOR) return BOOLEAN;
    function ">="(L: STD_LOGIC_VECTOR; R: STD_LOGIC_VECTOR) return
        BOOLEAN;
    function ">="(L: STD_LOGIC_VECTOR; R: INTEGER) return BOOLEAN;
    function ">="(L: INTEGER; R: STD_LOGIC_VECTOR) return BOOLEAN;
    function "="(L: STD_LOGIC_VECTOR; R: STD_LOGIC_VECTOR) return
        BOOLEAN;
    function "="(L: STD_LOGIC_VECTOR; R: INTEGER) return BOOLEAN;
    function "="(L: INTEGER; R: STD_LOGIC_VECTOR) return BOOLEAN;
    function "/="(L: STD_LOGIC_VECTOR; R: STD_LOGIC_VECTOR)
        return BOOLEAN;
    function "/="(L: STD_LOGIC_VECTOR; R: INTEGER) return BOOLEAN;
    function "/="(L: INTEGER; R: STD_LOGIC_VECTOR) return BOOLEAN;
    function SHL(ARG:STD_LOGIC_VECTOR;COUNT:
        STD_LOGIC_VECTOR) return STD_LOGIC_VECTOR;
    function SHR(ARG:STD_LOGIC_VECTOR;COUNT:
        STD_LOGIC_VECTOR)    return STD_LOGIC_VECTOR;
    function CONV_INTEGER(ARG: STD_LOGIC_VECTOR) return INTEGER;
-- remove this since it is already in std_logic_arith
--    function CONV_STD_LOGIC_VECTOR(ARG: INTEGER; SIZE: INTEGER)
--        return STD_LOGIC_VECTOR;
end STD_LOGIC_SIGNED;
```

Appendix C

Keywords in VHDL-87

abs
access
after
alias
all
and
architecture
array
assert
attribute
begin
block
body
buffer
bus
case
component
configuration
constant
disconnect
downto
else
elsif
end
entity
exit
file
for
function
generate
generic
guarded
if
in
inout
is
label
library
linkage
loop
map
mod
nand
new
next
nor
not
null
of
on
open
or
others
out
package
port
procedure
process
range
record
register
rem
report
return
select
severity
signal
subtype
then
to
transport
type
units
until
use
variable
wait
when
while
with
xor

Appendix D

Additional keywords in VHDL-93

group
impure
inertial
litteral
postponed
pure
reject
rol
ror
shared
sla
sll
sra
srl
unaffected
xnor

Index